Incommensurate Crystals, Liquid Crystals, and Quasi-Crystals

NATO ASI Series

Advanced Science Institutes Series

A series presenting the results of activities sponsored by the NATO Science Committee, which aims at the dissemination of advanced scientific and technological knowledge, with a view to strengthening links between scientific communities.

The series is published by an international board of publishers in conjunction with the NATO Scientific Affairs Division

A	**Life Sciences**	Plenum Publishing Corporation
B	**Physics**	New York and London
C	**Mathematical and Physical Sciences**	D. Reidel Publishing Company Dordrecht, Boston, and Lancaster
D	**Behavioral and Social Sciences**	Martinus Nijhoff Publishers
E	**Engineering and Materials Sciences**	The Hague, Boston, Dordrecht, and Lancaster
F	**Computer and Systems Sciences**	Springer-Verlag
G	**Ecological Sciences**	Berlin, Heidelberg, New York, London,
H	**Cell Biology**	Paris, and Tokyo

Recent Volumes in this Series

Series B: Physics

Incommensurate Crystals, Liquid Crystals, and Quasi-Crystals

Edited by

J. F. Scott and
N. A. Clark

University of Colorado
Boulder, Colorado

Plenum Press
New York and London
Published in cooperation with NATO Scientific Affairs Division

CHEMISTRY

Proceedings of a NATO Advanced Research Workshop on
Incommensurate Crystals, Liquid Crystals, and Quasi-Crystals
held July 7–11, 1986,
in Boulder, Colorado

Library of Congress Cataloging in Publication Data

NATO Advanced Research Workshop on Incommensurate Crystals, Liquid
Crystals, and Quasi-Crystals (1986: Boulder, Colo.)
   Incommensurate crystals, liquid crystals, and quasi-crystals / edited by
J.F. Scott and N. A. Clark.
      p.   cm.—(NATO ASI series. Series B, Physics; vol. 166)
   "Proceedings of a NATO Advanced Research Workshop on Incommen-
surate Crystals, Liquid Crystals, and Quasi-Crystals, held July 7–11, 1986, in
Boulder, Colorado"—Copr. p.
   Bibliography: p.
   Includes index.
   ISBN 0-306-42760-5
   1. Crystals—Defects—Congresses. 2. Liquid crystals—Congresses. I.
Scott, J. F. (James Floyd), 1942–   . II. Clark, N. A. (Noel Anthony),
1940–   . III. Title. IV. Series. NATO ASI series. Series B, Physics; v. 166.
QD921.N378   1986
548—dc19   87-24391
                                                                        CIP

© 1987 Plenum Press, New York
A Division of Plenum Publishing Corporation
233 Spring Street, New York, N.Y. 10013

Printed in the United States of America

PREFACE

In this NATO-sponsored Advanced Research Workshop we succeeded in bringing together approximately forty scientists working in the three main areas of structurally incommensurate materials: incommensurate crystals (primarily ferroelectric insulators), incommensurate liquid crystals, and metallic quasi-crystals. Although these three classes of materials are quite distinct, the commonality of the physics of the origin and description of these incommensurate structures is striking and evident in these proceedings. A measure of the success of this conference was the degree to which interaction among the three subgroups occurred; this was facilitated by approximately equal amounts of theory and experiment in the papers presented.

We thank the University of Colorado for providing pleasant housing and conference facilities at a modest cost, and we are especially grateful to Ann Underwood, who retyped all the manuscripts into camera-ready form.

J. F. Scott

N. A. Clark

Boulder, Colorado

CONTENTS

PART II: INCOMMENSURATE LIQUID CRYSTALS

A. Theory

B. Experiment

PART III: INCOMMENSURATE QUASI-CRYSTALS

PART IV: TWO-DIMENSIONAL AND LAYERED SYSTEMS

A PHENOMENOLOGICAL THEORY OF THE TRANSITION SEQUENCE INCLUDING AN

INCOMMENSURATE (COMMENSURATE) PHASE SANDWICHED BY REENTRANT COMMENSURATE

(INCOMMENSURATE) PHASE

Yoshihiro Ishibashi

Synthetic Crystal Research Laboratory
Faculty of Engineering
Nagoya University, Chikusa-ku
Nagoya 464, Japan

Abstract   A two-sublattice model is presented for interpreting the
appearance of the phase sequence Normal-Commensurate-Incommensurate-
Commensurate phases such as seen in $(C_3H_7NH_3)_2MnCl_4$.

Temperature dependences of transition parameters, a strain
component and soft mode frequencies are derived phenomenologically.
The present model can be easily converted to the one reproducing the
phase sequence Normal-Incommensurate-Commensurate-Incommensurate
phases.

I.   INTRODUCTION

One of the most characteristic features common to incommensurate (IC)
phases seen in various dielectric materials is that an IC phase usually
appears as a phase sandwiched by a high symmetry normal (N) phase and a low
symmetry commensurate (C) phase, the stability of the IC phase thus being
limited to some range of temperature, pressure and other controllable
external parameters [1,2]. To obtain an overall understanding of the transition sequence of N-IC-C phases the continuum description is helpful. In
such a description the thermodynamic potential functional is written in
terms of the transition parameters which transform upon symmetry operations
as the bases of an irreducible representation, which induces the N-C transition. To describe the IC phase it is necessary to include suitable terms
consisting of spatial derivative of the parameters.

It is possible to classify the IC phases into two groups according to the form of the thermodynamical potential functional. One group includes the Lifshitz invariant, as is represented by $(NH_4)_2BeF_4$, while the other group does not, as is represented by $NaNO_2$.

Behaviors of physical quantities in the N-IC-C transition sequence can be well described by such thermodynamic potential functionals. It is known that some physical properties of an IC phase are strongly dependent upon physical properties of the C phase which follows. If the C phase is ferro-electric, for example, the dielectric susceptibility shows a tendency to divergence just above the IC-C transition temperature, while if the C phase is ferroelastic, an elastic compliance will diverge there.

As is readily understood from the above, it is most important to write down the thermodynamic potential functional, based upon the symmetry con-sideration of the C phase. Recently, however, an unusual transition sequence, including an IC phase, has been reported. For example, in propyl-ammonium manganese chloride $(C_3H_7NH_3)_2MnCl_4$ (abbreviated to PAMC) an IC phase is sandwiched by the same C phase, as shown in Table 1 [3], i.e., the C phase has a reentrant nature, and the transition from the IC phase to the low temperature C phase is not the lock-in type. In other words, there is no locked-in C phase, and therefore there is no clue required to construct a thermodynamic potential functional.

A phenomenological theory for the transition of PAMC has already been published by Muralt et al. [3]. But it only treats the transition sequence starting at the high temperature C phase. Here, we would like to present a different interpretation of the transition sequence N-C-IC-C in the form which is more general (not limited only to the case of PAMC).

2. THEORY

A basic idea of the present approach lies in adopting sublattices in the N phase [4]. These sublattices must comply with the way N-C transition takes place. If the N-C transition takes place at the Γ point of the Brillouin zone, sublattices must include atoms or dipoles located in a single unit cell, while if it occurs at, say, the zone boundary, sublat-tices have to include atoms or dipoles located in two unit cells of the N phase. Let us take two parameters q_1 and q_2, which are regarded to repre-sent the displacement of two adopted sublattices.

Next we have to examine the symmetry property of these sublattices. Based upon it, we would be able to write down a thermodynamic potential functional like [4]

Table 1. The phase sequence of $(C_3H_7NH_3)_2MnCl_4$.

C	IC	C	IC	C	N
ζ	ε	δ	γ	β	α

C	IC	C	IC	C	N
$P112_1/n$		Cmca		Cmca	I4/mmm
Z=6		Z=2		Z=2	Z=1
$q=(a^*+b^*)/3$	$q=(1/3+\delta)a^*$		$q=b^*+(1/6+\delta)c^*$		

114	163	343		393	440 T(K)

$$f = \frac{\alpha}{2}(q_1^2 + q_2^2) + \beta q_1 q_2$$

$$+ \varepsilon_1(q_1 - q_2)^4 + \varepsilon_2(q_1 + q_2)^4 + \varepsilon_3 q_1^2 q_2^2 + f_6$$

$$\tag{1}$$

$$-\delta(q_1\frac{\partial q_2}{\partial x} - q_2\frac{\partial q_1}{\partial x}) + \frac{\kappa}{2}[(\frac{\partial q_1}{\partial x})^2 + (\frac{\partial q_2}{\partial x})^2],$$

$$(\beta,\ \delta,\ \kappa > 0)$$

where f_6 consists of the sixth order terms in q_1 and q_2, and only α is assumed to be temperature dependent as

$$\alpha = a(T - T_0), \tag{2}$$

and x represents the direction of modulation given rise to in the IC phase. The transition sequence N-C-IC-C and behaviors of physical quantities should be reproduced from the above functional. In what follows, let us analyze it.

By inspection of the pseudo-Lifshitz invariant (to be discussed later), $q_1\frac{\partial q_2}{\partial x} - q_2\frac{\partial q_1}{\partial x}$ and the cross-term $q_1 q_2$, it is seen that, if $\beta > 0$, q_1 and q_2 are expressed as

$$q_1 = q + \rho \cdot \cos kx,$$

$$\tag{3}$$

$$q_2 = -q + \rho \cdot \sin kx,$$

3

where q represents the C component and ρ the IC component of displacements. The wavenumber k has to be chosen so that it minimizes f, i.e.,

$$k = \frac{\delta}{\kappa}.$$

Putting this into (1), it is reduced to the form

$$f = (\alpha - \beta)q^2 + \frac{\gamma_1}{2}q^4 + \gamma_3 q^2\rho^2$$

$$+ \frac{\alpha - \alpha_I}{2}\rho^2 + \frac{\gamma_2}{4}\rho^4 + \frac{\eta}{2}q^4\rho^2 \tag{4}$$

where

$$\alpha_I = \frac{\delta^2}{\kappa}, \tag{5}$$

and in (4) the fourth order terms have been rearranged, and only one sixth order term $\rho^2 q4$ is taken into account.

3. DISCUSSION

If $\beta > \alpha_I$, firstly the N-C transition takes place, and q in the C phase ($\rho = 0$) is given as

$$q^2 = - \frac{\alpha - \beta}{\gamma_1}. \tag{6}$$

When q^2 grows, this C phase loses its stability when

$$\left.\frac{d^2 f}{d\rho^2}\right|_{\rho=0} = \left.\frac{\partial^2 f}{\partial\rho^2}\right|_{\rho=0} \tag{7}$$

$$= \alpha - \alpha_I + 2\gamma_3 q^2 + \eta q^4$$

is satisfied. Whether the C-IC transition is of the second order or the first order is governed by the sign of the coefficient of the fourth order term in the power series expansion of f in terms of ρ. It is given, after a simple mathematics respecting the functional relation between q and ρ, as

$$\left.\frac{d^4 f}{d\rho^4}\right|_{\rho=0} = -12\frac{(\gamma_3 + \eta q^2)^2}{\gamma_1} + 6\gamma_2. \tag{8}$$

4

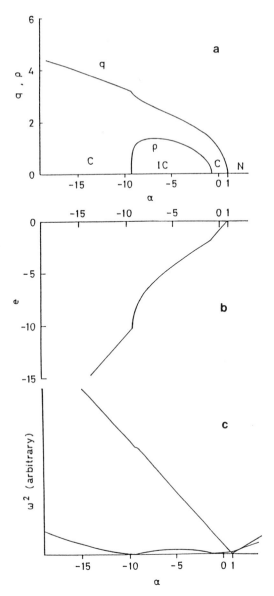

Fig. 1(a). Temperature dependences of the transition parameters q and ρ.
    (b).   Temperature dependence of a tensile strain component.
    (c).   Temperature dependences of soft mode frequencies.
            Adopted parameter values: $\alpha_I = 0.1$, $\beta = 1$, $\gamma_1 = 1$, $\gamma_2 = 1$,
            $\gamma_3 = 0.2$ and $\eta = 0.05$. In this case the IC-C transition at
            low temperature is weakly of the first order.

As is seen from the sixth order term $\rho^2 q^4$ the potential becomes large
if both ρ^2 and q^4 grow with decreasing temperature, implying that the IC
phase (actually it is the C + IC phase) becomes unstable and the C phase
will be re-entered.

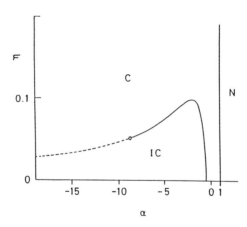

Fig. 2. The α-η phase diagram. The solid and the dotted lines indicate the second and the first order transitions, respectively. An open circle indicates the tricritical point. Adopted paramater values are the same, except for η, and given in the caption of Fig. 1

The temperature dependences of q, ρ , and a tensile strain component e given as

$$e = \xi_1 q^2 + \xi_2 \rho^2 \tag{9}$$

are shown in Figs. 1(a) and (b) for given values of coefficients. Note a reasonable resemblance of these temperature dependences with the ones actually observed in PAMC. It indicates that only reinterpretation of q_1 and q_2 is required to apply the present model to PAMC. In Fig. 1(c) the temperature dependences of two soft modes are shown, one related to the C lattice wave and the other to the IC lattice wave.

The stability range of the IC phase is mostly governed by the sixth order term. An α-η phase diagram is depicted in Fig. 2, from which it is seen that the η-term is substantial in giving rise to the N-C-IC-C sequence. (If $\eta = 0$, no re-entrance to the C phase takes place.)

In an IC phase the modulation wavenumber k usually depends upon temperature. However, in the present case where the stability range of the IC phase is limited to somewhat narrow temperature range, the change in k will be safely ignored except for small η where the IC phase is stable down to low temperature range.

It is readily seen that in the thermodynamic potential function (4) q and ρ appear on equal footing except for the $q^4 \rho^2$ term. If we replace the $q^4 \rho^2$ term with the $q^2 \rho^4$ term, therefore, the transition sequence N-IC-C-IC, instead of N-C-IC-C, will be easily reproduced. Such a phase sequence is shown in Fig. 3, where the temperature dependences of q and ρ

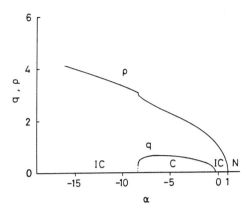

Fig. 3. Temperature dependences of the transition parameters q and .
Adopted parameters values: $\alpha_I = 1$, $\beta = 0.1$, $\gamma_1 = 5$, $\gamma_2 = 1$,
$\gamma_3 = 0.2$ and $\eta = 0.15$. In this case the C-IC transition at
low temperature is of the first order.

are shown. It is also easy to depict figures corresponding to Figs. 1(a)-
(c), though none of them is actually presented here.

In this case, one should take into account the spatial higher harmonics in q_1 and q_2 and the temperature change of the modulation wave-number
k, since the IC phase is stable down to lower temperatures where ρ grows so
much that the contribution from higher order terms becomes important.

Before concluding, let us examine the nature of the $q_1 \frac{\partial q_2}{\partial x} - q_2 \frac{\partial q_1}{\partial x}$
term. The parameters r_1 and r_2 which transform as the bases of irreducible
representations are considered to be made of q_1 and q_2 as

$$r_1 = \frac{1}{\sqrt{2}}(q_1 + q_2),$$

$$r_2 = \frac{1}{\sqrt{2}}(q_1 - q_2).$$

(10)

Then,

$$q_1 \frac{\partial q_2}{\partial x} - q_2 \frac{\partial q_1}{\partial x} = r_2 \frac{\partial r_1}{\partial x} - r_1 \frac{\partial r_2}{\partial x} .$$

(11)

This is nothing but a pseudo-Lifshitz invariant if r_1 and r_2 belong to different irreducible representations [5].

Finally it should be noted that if q and ρ appear simultaneously at
the transition from the N phase where $q = \rho = 0$, such a transition must be
of the first order.

4. REFERENCES

1. Y. Ishibashi. Ferroelectrics $\underline{24}$, 119 (1980).
2. Y. Ishibashi. Ferroelectrics $\underline{35}$, 111 (1981).
3. P. Muralt, R. Kind and R. Blinc. Phys. Rev. Letters $\underline{49}$, 1019 (1982).
4. Y. Ishibashi. J. Phys. Soc. Jpn. $\underline{55}$, 4309 (1986).
5. A. P. Levanyuk and D. G. Sannikov. Sov. Phys. Solid States $\underline{18}$, 1122 (1976).

DAUPHINÉ-TWIN DOMAIN CONFIGURATIONS IN QUARTZ AND ALUMINUM PHOSPHATE

M. B. Walker

Department of Physics and Scarborough College
University of Toronto
Toronto, Ontario M5S 1A7
Canada

Abstract   The properties of Dauphiné-twin domain configurations in
quartz and aluminum phosphate are reviewed. Also, the superspace
symmetry of the incommensurate phases of quartz and aluminum phosphate
is determined, and some consequences of the superspace symmetry are
noted.

My talk will focus on the interpretation of the Dauphiné twin domain
configurations which have been observed in quartz and aluminum phosphate by
Van Tendeloo, Van Landuyt and Amelinckx [1,2]. Although now a decade old,
their observations gave a new impetus, which is still continuing, to
studies of quartz. Since a number of experts in this area will be following me in the program, I will use my time to introduce the subject by
reviewing some elementary but basic properties [3,4] of the walls between
Dauphiné twins. The domain-wall approach not only gives a good qualitative
account of the periodic triangular structure (i.e., the incommensurate
phase) observed in a small temperature interval between the α- and β-phases
at approximately T = 845K, but provides a framework for the description of
the irregular domain configurations occurring during the phase transition
between the incommensurate to α-phase, and even describes well the coarse
domain structures which are observed [1,2] at temperatures well below T =
845K.

Although I will explicitly discuss only quartz, the discussion will be
applicable to aluminum phosphate as well since the properties treated are
consequences of symmetry, and quartz and aluminum phosphate have the same
symmetry.

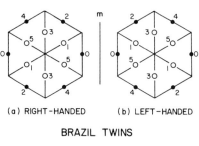

(a) RIGHT-HANDED   (b) LEFT-HANDED

BRAZIL TWINS

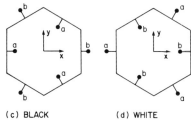

(c) BLACK   (d) WHITE

DAUPHINÉ TWINS

Fig. 1. Projections onto the basal plane of the positions of the ions in the Brillouin--Wigner unit cells of various quartz structures. The β-phase structure of quartz, stable above T = 845K, is shown in (a) and (b); the left-handed structure is obtained from its right-handed Brazil twin by reflecting in the mirror plane labeled m. The α-phase of quartz, shown in Figs. (c) or (d) is obtained by starting in the β-phase with quartz of a given handedness, say right-handed, and lowering the temperature below 845K (the oxygen-ion positions are omitted in (c) and (d)); the two degenerate ground states labeled black and white are called the Dauphiné twins. The high-symmetry directions x and y are also defined by this figure.

The β-phase of quartz is stable at temperatures greater than 845K. The right-handed and left-handed β-phase structures shown in Figs. 1(a) and (b) are called the Brazil twins and are mirror images of each other. In the right-handed structure, the silicon ions on the edges of the unit cell lie on a right-handed helix. Detailed descriptions of repeated Brazil twinning in amethyst quartz, accompanied by a beautiful plate of figures which have been hand-colored in water colors, have been given by Brewster [5]. In the remainder of this article I assume that one is dealing with a single crystal of quartz of a definite handedness.

If, starting in the β-phase, the temperature is lowered below 845K, a phase transition to the α-phase, in which the silicon ions labeled a are displaced either towards the center of the unit cell (Fig. 1(c)) or away from it (Fig. 1(d)) occurs. The two possible ground states of the α-phase are called the Dauphiné twins. I have labeled the two twins black and white because one appears black and the other white in electron microscope

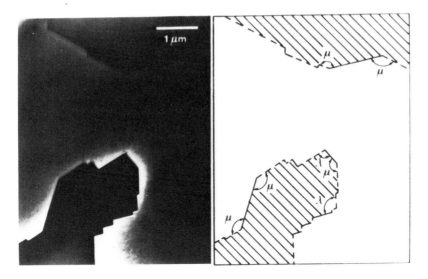

Fig. 2. Dark field image of coarse Dauphine twins in a foil of quartz cut
perpendicular to the c-axis. All of the labeled angles were
measured with the results $\lambda = 120° - 2\theta_0$ and $\mu = 120° + 2\theta_0$
where $2\theta_0 = 19°$. (After Van Landuyt et al. [4]).

images (see, for example, Fig. 2). The fact that the walls between the
two types of Dauphiné twins have certain well-defined orientations (see
Fig. 2) is one of the facts that must be explained by a successful model
of Dauphiné-twin domain structures.

The phase transition at 845K between the α- and β-phases has been
intensively studied at various periods since its discovery by Le Chatelier
[6] in 1889. Scott's [7] review article in 1974 gives a picture of some of
the properties of this phase transition as they were known before the electron microscopy studies of Van Tendeloo et al. [1,2]. A particularly
interesting aspect of this early work was the determination, by Shapiro and
Cummins [8], that the intense light scattering observed in the neighborhood
of the phase transition was not scattering from time-dependent critical
fluctuations, but was in fact elastic scattering. This observation, combined with a knowledge of Young's diffraction experiments [9], led to the
conclusion [8] that extensive twinning of the Dauphiné type occurs in the
neighborhood of the α-β phase transition.

Nearly a decade later this extensive twinning was indeed observed by
Van Tendeloo et al. [1,2] (see Fig. 3). Their observations of a periodic
triangular domain structure [1] indicating the existence of a new phase
occurring in between the α- and β-phases was particularly striking; in

addition, a wide variety of irregular domain configurations were observed on the low temperature edge of the regular triangular patterns (see Fig. 3).

One might expect to be able to diffract neutrons or X-rays from this periodic triangular domain structure in the same way as one can diffract neutrons or X-rays from a periodic crystal lattice, and this has been done by Dolino et al. [10,11]. In the neutron measurements, the satellite wave vectors were found to be along reciprocal-lattice-vector directions, which is consistent with the original electron microscopy observation [1,2] that the domain walls are approximately normal to these directions.

A phenomenological Landau-type of free energy aimed at providing a basis for studying the incommensurate phase of quartz has been given by Aslanyan and Levanyuk [12]. Two complementary methods of analysis have been applied to the problem: 1) Aslanyan et al. [13] have studied the periodic triangular domain structure by approximating the order parameter by the first few terms of a Fourier series. 2) Walker [3] has focused on using the model to obtain the properties of an isolated domain wall. The second approach is the appropriate one if one is interested in obtaining an interpretation of the wide variety of Dauphiné-twin domain configurations observed in electron microscopy studies, and I will thus quote a few results of this approach.

It should be recalled that the original electron microscopy experiments [12] and the neutron diffraction experiments [10,11] had shown that the domain walls were parallel to a high symmetry direction in quartz. However, an important result of the theory is that such an orientation cannot be a stable orientation for domain walls [3]. This is a consequence of symmetry and can be understood by testing the stability of the high symmetry orientation of domain walls as in Fig. 4, which shows three different orientations of domain walls. In Fig. 4(a), the domain wall is parallel to the x-axis of Fig. 1. In Fig. 4(b) the domain wall has been rotated (about the c-axis) by the angle $+\theta_1$; the silicon ion displacements shown here have a component toward the solid line defining the wall center, and this wall is therefore called a heavy wall. It can be seen in Fig. 4(c) that a rotation of the wall by the angle $-\theta_1$ results in a light wall; in this case the silicon ions have components of their displacements away from the center of the wall.

Two possibilities for the free energy per unit area are shown in Fig. 5. In (a) the free energy is the same for orientations $+\theta_1$ and $-\theta_1$, but in (b), this is not so. Since it has just been shown for domain walls in quartz that rotations by $+\theta_1$ and $-\theta_1$ produce physically inequivalent

Fig. 3. Dark field image of a foil of quartz taken looking along the
c-axis. The temperature difference between the top (cooler) and
bottom (warmer) parts of the figure is estimated to be about 5°.
Regions having a regular equilateral triangular domain structure
are observable, as are more irregular domain patterns at the top
of the picture. The quantity $2\theta_0$ (i.e., twice the domain wall
rotation angle) is shown to vary from 10° to 20° as one lowers
the temperature in going from the bottom to the top of the picture. (After Van Landuyt et al. [4].)

states (corresponding to light walls and heavy walls, respectively), the
curve of Fig. 5(a) will not in general be applicable, and the appropriate
curve for quartz is that of Fig. 5(b). The equilibrium orientation of a
domain wall (found by minimizing $F(\theta)$) is therefore not exactly parallel
to the x-axis [3,4]. According to an analysis [3] of a phenomenological
model, the deviation of the wall orientation from the x-axis direction is

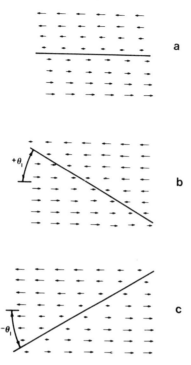

Fig. 4. Domain walls of three different orientations. Each wall separates
a black twin on the bottom from a white twin on the top. The
arrows represent displacements of silicon ions from their average
phase positions. For simplicity, only the displacements of
those silicon ions which are displaced parallel to the x-axis of
Fig. 1(c) and (d) are shown, while displacements of those silicon
ions displaced at ± 120° to the x-axis are omitted. It should be
emphasized that the solid line indicates the center of a domain
wall, and that the transition from black to white is gradual.
(After Van Landuyt et al. [4]).

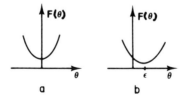

Fig. 5. Two possibilities for the free energy per unit area of a domain
wall as a function of its orientation. (After Van Landuyt et
al. [4]).

14

given approximately by the formula

$$\theta_o = C(\eta_o/w)$$

where C is a constant, η_o is the order parameter and w is the wall thickness. This formula shows that θ_o is expected to increase as the temperature is lowered from the β-phase to incommensurate-phase transition temperature since η_o increases and the domain walls are expected to sharpen up.

The first rough estimate of the domain wall rotation angle θ_o based on electron microscopy measurements was given by Walker [3], and more detailed measurements were given by Van Landuyt et al. [4] as shown in Fig. 3. The high resolution Laue measurements of Gouhara and Kato [14] are at present the best measurements of θ_o as a function of temperature; the unusual curvature of $\theta_o(T)$ found in these measurements [14] has yet to be explained by theory.

Given the equilibrium orientation of a domain wall, it follows from the symmetry of quartz that there are exactly six distinct domain walls which have the same energy; these are shown in Fig. 6(a) and (b). The next step is to study the different ways in which domain walls can be joined together at vertices. It turns out that only two, four, or six walls can be joined together at a vertex. Figure 6(c) to (v) shows a number of possible intersections. Note in particular that two walls can intersect only at angles of $2\theta_o$, 60°, 120° + $2\theta_o$ and 120° − $2\theta_o$.

In the coarse domain structure of quartz which is observed at room temperature, only two domain wall intersection angles, namely λ = 120° − $2\theta_o$ and μ = 120° + $2\theta_o$ (with $2\theta_o \approx 19°$), are observed (see Fig. 2). In this temperature region the surface energy of a domain wall is positive (this is evident since the system tries to reduce the number of domain walls as much as possible--the only remaining domain walls are trapped by pinning centers or crystal imperfections of some sort); the total surface energy is reduced by configurations which are as round as possible (as in the liquid drop problem) and hence the domain wall intersection angles $2\theta_o$ and 60° are not observed.

As the temperature is increased towards the α-phase to incommensurate-phase transition temperature, the domain wall energy decreases until it becomes negative, at which point domain walls start to form in order to lower the energy of the system. The triangular domain structure characteristic of the incommensurate phase is one way of packing a high density of domain walls into the system. The triangular domain structure is observable in Fig. 3. In fact, two distinct triangular structures rotated relative to each other by 180° ± $2\theta_o$ are observed, giving rise to what has been called a macrodomain structure [3,4].

15

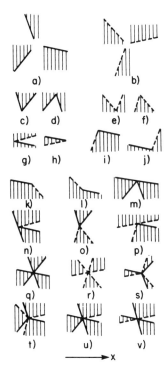

Fig. 6. The six distinct domain walls (all of which have the same energy per unit area) which occur in quartz are shown in (a) and (b); the shaded and unshaded regions which occur on either side of a domain wall correspond to the black and white Dauphiné twins of Fig. 1. Also, the x-direction at the bottom of the figure corresponds to the x-direction in Fig. 1. A number of possible wall vertices are shown in (c) to (v). The domain wall intersection angles are 60° in (c) to (f), $2\theta_0$ in (g) and (h), $120° - 2\theta_0$ in (i) and (j), and $120° + 2\theta_0$ in (k) and (l).

At the low temperature edge (top) of Fig. 2, a number of domains of unusual shapes can be seen. These can all be interpreted in terms of the vertices of Fig. 6. For a further discussion of the interpretation of domain wall patterns, as well as for a description of the interesting differences in domain configurations which occur on heating and on evolving across the α-phase to incommensurate-phase transition temperature, see Reference 3.

It has been shown by de Wolff, Janner and Janssen that the appropriate description of the symmetry of incommensurate structures is in terms of superspace groups (e.g., see Ref. 15). I therefore close with a description of the superspace symmetry of quartz [16], which has not yet been reported in the literature. Basis vectors describing the five-dimensional superspace Bravais lattice and reciprocal lattice are

$$A1 = (\vec{a}_1, -\vec{a}_1) \qquad A_1{}^* = (\vec{a}_1{}^*, 0)$$
$$A_2 = (\vec{a}_2, -\vec{a}_2) \qquad A_2{}^* = (\vec{a}_2{}^*, 0)$$
$$A_3 = (a_3, 0) \qquad A_3{}^* = (\vec{a}_3{}^*, 0)$$
$$A_4 = (0, \vec{b}_1) \qquad A_4{}^* = (\vec{b}_1{}^*, \vec{b}_1{}^*)$$
$$A_5 - (0, \vec{b}_2) \qquad A_5{}^* = (\vec{b}_2{}^*, \vec{b}_2{}^*)$$

Here, $\vec{a}_1$, $\vec{a}_2$ and $\vec{a}_3$ are basis vectors describing the Bravais lattice of the α and β phases; $\vec{a}_1$ can be taken to lie along the x-direction of Fig. 1 and $\vec{a}_2$ is $\vec{a}_1$ rotated by $120°$ about the c-axis; $\vec{a}_3$ is parallel to the c-axis. The vectors $\vec{b}_1$ and $\vec{b}_2$ describe the two-dimensional Bravais lattice charac-terizing the modulation as shown in Fig. 7. The vectors a_i^* and b_j^* are sets reciprocal to $\vec{a}_i$ and $\vec{b}_j$, respectively. The superspace group of quartz is generated by a single generator, namely,

$$g = (\{C_6 | \tfrac{1}{3} \vec{a}_3\}, [C_6 | 0])$$

where the first operator $\{C_6 | \tfrac{1}{3} \vec{a}_3\}$ acts on the external space [15] and the second operator $[C_6 | 0]$ acts on the internal space [15]. The occurrence of a six-fold axis of symmetry is perhaps surprising since the pattern shown in Fig. 7 has only three-fold axes of symmetry. Note, however, that the pattern has a six-fold axis of color symmetry in which a six-fold rotation is accompanied by the interchange of black and white; by going back to Fig. 1(c) and (d) one can see that a six-fold rotation does indeed inter-change black and white so that the six-fold symmetry axis is in fact easily understood.

The fact that the point group of the incommensurate phase is C_6 has the immediate consequence that the incommensurate phase should be ferro-electric, with a spontaneous polarization allowed along the c-axis. The ferroelectricity of the incommensurate phase was anticipated from Landau theory considerations as described in a review by Dolino [17], and the ferroelectric properties of the domain walls have been described by Walker and Gooding [18].

The superspace group also determines the systematic absences in the X-ray and neutron diffraction pattern. For example, as a result of the re-tention of the six-fold screw axis in the incommensurate phase of quartz, the Bragg scattering at $(0, 0, \ell)$ is absent unless $\ell = 3n$. On the other hand, the apparent absence of the satellites along $\vec{a}^*$ in the neighborhood of (300) noted by Dolino et al. [11] (see their Fig. 6) is not a conse-quence of symmetry; the intensity of this satellite is therefore not zero, although it results from a higher-order effect which makes it too small to be observable in that particular experiment.

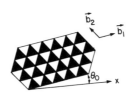

Fig. 7. A portion of an infinite periodic triangular domain structure shown in a plane perpendicular to the c-axis of quartz; the x-direction corresponds to that shown in Fig. 1. The periodic structure shown is invariant with respect to translations $n_1\vec{b}_1 + n_2\vec{b}_2$ where n_1 and n_2 are integers.

ACKNOWLEDGMENTS

I thank G. Van Tendeloo, J. Van Landuyt and S. Amelinckx for providing the electron microscopy photographs appearing in this article, and for a stimulating collaboration on the problems discussed here. This research was partially supported by the Natural Sciences and Engineering Research Council of Canada.

REFERENCES

1. G. Van Tendeloo, J. Van Landuyt and S. Amelinckx, Phys. Stat. Sol. A30, K11 (1975).
2. G. Van Tendeloo, J. Van Landuyt and S. Amelinckx, Phys. Stat. Sol. A33, 723 (1976).
3. M. B. Walker, Phys. Rev. B28, 6407 (1983).
4. J. Van Landuyt, G. Van Tendeloo, S. Amelinckx and M. B. Walker, Phys. Rev. B31, 2986 (1985).
5. D. Brewster, Transactions of Royal Society of Edinburgh 9, 139 (1823).
6. H. Le Chatelier, Comptes Rendus, Academie des Sciences (Paris), 108, 1046 (1889).
7. J. F. Scott, Rev. Mod. Phys. 46, 83 (1974).
8. S. M. Shapiro and H. Z. Cummins, Phys. Rev. Letts. 21, 1578 (1968).
9. R. A. Young, unpublished, quoted in Reference 8.
10. G. Dolino, J. P. Bachheimer and C. M. E. Zeyen, Solid State Commun. 45, 295 (1983).
11. G. Dolino, J. P. Bachheimer, B. Berge and C. M. E. Zeyen, J. Physique 45, 361 (1984).
12. T. A. Aslanyan and A. P. Levanyuk, Solid State Commun. 31, 547 (1979).
13. T. A. Aslanyan, A. P. Levanyuk, M. Vallade and J. Lajzerowicz, J. Phys. C: Solid State Phys. 16, 6705 (1983).
14. K. Gouhara and N. Kato, J. Phys. Soc. Japan 53, 2177 (1984).
15. A. Janner and T. Janssen, Acta Cryst. A36, 399 (1980).
16. The work on superspace symmetry was carried out in collaboration with R. H. Gooding.
17. G. Dolino, in Incommensurate Phases in Dielectrics, Vol. II of Modern Problems in Condensed Matter Sciences (R. Blinc and A. P. Levanyuk, eds.) (North-Holland, Amsterdam, 1985).
18. M. B. Walker and R. H. Gooding, Phys. Rev. B32, 7408 (1985).

ELASTIC AND INELASTIC SCATTERING FROM QUASI-PERIODIC STRUCTURES

T. Janssen

Institute for Theoretical Physics, University of Nijmegen
6525 ED Nijmegen, Holland

R. Currat
Institut Max von Laue—Paul Langevin
Avenue des Martyrs
38042 Grenoble Cedex, France

Abstract   Elastic and inelastic scattering from incommensurate crystal phases, nowadays sometimes called quasi-periodic structures, is discussed both from a theoretical and an experimental point of view. The excitations of these structures are treated in the harmonic approximation in the framework of higher-dimensional space groups. Expressions are given for the static structure factor, the differential scattering cross section and for the Debye-Waller factors. The results are exemplified on a simple model, the DIFFFOUR model, and illustrated with experiments with neutrons on incommensurate phases of $ThBr_4$.

1.   INTRODUCTION

A convenient way to characterize incommensurate crystal phases is via their Fourier transform. A structure is an incommensurate crystal phase if its Fourier transform is a sharply peaked function where the peaks belong to a quasi-lattice. The latter is the set of all linear integral combinations of a finite number of vectors. So in general the wave vectors of the Fourier transform are of the form

$$\vec{Q} = \sum_{i=1}^{n} h_i \vec{a}_i^{*}; \quad (h_i \text{ integer}).$$
(1)

Ideally the peaks are delta-functions. The rank of the quasi-lattice is the number of rationally independent vectors. If this rank is equal to the number of linearly independent vectors over the real numbers, the quasi-lattice is an ordinary lattice and the structure is periodic. Otherwise the term quasi-periodic or incommensurate is used. These are synonymous.

A class of incommensurate crystals is formed by the incommensurate modulated ones. Here one can distinguish a lattice spanned by the vectors $\vec{a}_1^*, \vec{a}_2^*, \vec{a}_3^*$ to which the main reflections belong, whereas for vectors with one of the $h_4, \ldots, h_n$ unequal to zero one has satellites. In this case the number of rationally independent vectors corresponding to the modulation is $d = n-3$, which is called the dimension of the modulation.

Another class of IC (incommensurate crystal) phases is that of quasi-crystals, as have become of interest in the recent years. The diffraction of the icosahedral quasi-crystals has icosahedral point group symmetry and, therefore, it cannot contain a lattice of main reflections.

Experimentally it is, of course, hard to prove that a structure is really incommensurate. For a real crystal the lines have always a certain width due to the finite size of the specimen or to imperfections in it. Inside the width of the peak there is always a vector with rational components. So a commensurate superstructure approximation is always possible. This is even the case when the diffraction vectors depend on thermodynamical variables, like temperature and pressure. Then one cannot exclude the possibility that the wave vectors, although apparently changing in a smooth way, jump from one commensurate position to the other. However, it is useful, even in commensurate cases, to use the description of incommensurate or quasi-periodic structures, because many properties do not depend on the exact values of the wave vectors.

One other condition that should be satisfied in order to describe the structure as quasiperiodic in a meaningful way is that the peaks of the Fourier transform do not overlap substantially. In principle the points of a quasi-lattice with rank larger than the dimension of the space form a dense set. In practice the values of $h_1, \ldots, h_n$ for which the peak height is not negligible give a discrete set of vectors.

A quasi-lattice in 3 dimensions is the projection of an n-dimensional ordinary lattice in n dimensions on a 3-dimensional hyperplane. This means that the density function of the IC phase is the restriction of a periodic function in n dimensions to a 3-dimensional hyperplane parallel to the first hyperplane and going through the origin. This function in n dimensions, therefore, has space group symmetry and consequently relations between and restrictions on the Fourier components exist. Since the latter are also the Fourier components of the 3-dimensional quasi-periodic structure, the same relations and conditions are present in 3 dimensions. This is the rationale behind the description of IC phases in higher dimensional space. This was done already for IC modulated phases and has been extended recently to the class of quasi-crystals [1-8].

Also physical properties may be studied in the higher-dimensional formulation. In particular, scattering of neutrons and X-rays from IC phases should reflect the underlying symmetry. For the elastic scattering this is a geometric problem, for inelastic scattering one has to study the excitations of the IC phases, which differ qualitatively from those in periodic structures.

2. STATIC STRUCTURE FACTOR

The Fourier transform of a quasi-periodic function is the projection of the Fourier transform of a periodic function in a space of higher dimension. This is in particular true for the geometric structure factor. When the vector $\vec{Q}$ of the quasi-lattice is the projection of the n-dimensional vector $(\vec{Q}, \vec{Q}_I)$ the structure factor $F(\vec{Q})$ is given by

$$F(\vec{Q}) = \sum_j \frac{1}{\Omega_j} \int_{\Omega_j} \exp[i\vec{Q}.\vec{r}_j \ (\vec{r}_I) + i\vec{Q}_I.\vec{r}_I] \ d\vec{r}_I. \tag{2}$$

Here the summation is over the particles in the unit cell. It is assumed that the quasi-periodic (i.e., IC) structure consists of discrete points which result as the intersection of a 3-dimensional hyperplane with a set of discrete d-dimensional hyperplanes. For an IC modulated structure with a one-dimensional modulation the positions $\vec{n}+\vec{r}_j+\vec{f}_j(\vec{q}.\vec{n})$ are the intersections of the lines $(\vec{n}+\vec{r}_j+\vec{f}_j[\vec{q}.\vec{n}+t],t)$ with the hyperplane t=0. The integration in Eq. (2) is over the discrete elements in the n-dimensional unit cell. For the IC modulated crystal this means an integration from 0 to 2π and $\Omega_j=2\pi$. The structure factor for an IC modulated crystal with one-dimensional modulation with wave vector $\vec{q}$ then becomes

$$F(\vec{Q}) = \frac{1}{2\pi} \sum_j \int_0^{2\pi} \exp[i\vec{Q}.\vec{r}_j + i\vec{Q}.\vec{f}_j(\tau)+im\tau]d\tau; \quad (\vec{Q}= \sum_{i=1}^3 h_i\vec{a}_i^* + m\vec{q}). \tag{3}$$

21

This is the formula already used by de Wolff [0] and by Yamamoto [10]. For a sinusoidal modulation this leads to the well known Bessel function dependence of the satellites: if $\vec{f}_j(\tau)=\vec{A}_j \sin(\tau+\phi j)$ then

$$F(\vec{Q}) = \sum_j e^{i\vec{Q}\cdot\vec{r}_j} J_{-m}(\vec{Q}\cdot\vec{A}_j)\exp(-im\phi_j).$$ (4)

In another limit the modulation is no longer sinusoidal but the structure is nearly commensurate in domains. The commensurate domains are separated by discommensurations. An extreme case is present if the width of the discommensurations goes to zero and only the phase is modulated. This situation is given by a modulation function $f(\tau)$ that is constant on p intervals of equal length $2\pi/p$. On the j-th interval the function has the value $\vec{A} \sin(\phi_0+2\pi j/p)$. For the case p=6, for example, and longitudinal modulation of value d,2d,d,-d,-2d,-d the structure factor for $\vec{Q}=\sum h_i\vec{a}_i^*+m\vec{q}$ becomes

$$F(\vec{Q}) = \frac{2\sin(\pi m/6)}{\pi m} [\cos(Qd+\pi m/6)+\cos(Qd+5\pi m/6)+\cos(2Qd+\pi m/2)].$$ (5)

For Qd≪1 the main reflections (m=0) have $F(\vec{Q})$ of absolute value approximately one, the satellites m=6n vanish, satellites with m=1,5,7 or 11 (mod 12) are linear in Qd and the other quadratic in Qd. Therefore, the relatively strong satellites are those of order ±1 or ±5 (mod 12). The difference between the basic satellite and the satellite $100\bar{5}$ is $2\pi(1-q/q_c)=2\pi/b$, where b is the distance between the discommensurations and q_c the wave vector of the commensurate phase. Therefore, close to, or even on the shoulders of, the basic satellite appear higher order satellites at distances that are the inverse of the discommensuration lattice constant. Their intensity becomes important because they are at positions where the peaks of the incommensurate modulation nearly coincide with the peaks of the discommensuration lattice. This reminds of the mechanism studied by Hendricks and Teller [11] for random sequences of intervals.

For quasi-crystals one may apply the same procedure for the calculation of the structure factor. In first approximation, quasi-crystals may be obtained as intersection of a periodic structure of discrete elements of dimension d, and parallel to the additional space, and 3-dimensional space [6-8]. Then the structure factor becomes

$$F(\vec{Q}) = \sum_{j=1} e^{i\vec{Q}\cdot\vec{r}_j} \frac{1}{\Omega_j} \int_{\Omega_j} e^{i\vec{Q}_I\cdot\vec{\tau}} \, d\vec{\tau} = \sum_j e^{i\vec{Q}\cdot\vec{r}_j} a_j(\vec{Q}_I), \qquad (6)$$

where $a_j(\vec{Q}_I)$ may be seen as an atomic scattering factor, depending on $\vec{Q}_I$ which is also known in the literature as "Q perpendicular." It is the integration over the discrete element parallel to the additional space with integrand $\exp(\vec{Q}_I\cdot\vec{\tau})$. For the one-dimensional Fibonacci chain it is

$$a(n\vec{a}_{1I}^* + m\vec{a}_{2I}^*) = \frac{\sin(\pi y)}{\pi y}, \quad \text{with } y=(n-m+n\tau)/(2-\tau), \quad \tau = \frac{\sqrt{5}-1}{2}. \qquad (7)$$

For a more general quasi-periodic structure this factorization does not occur.

3. EXCITATIONS

In order to calculate the dynamic structure factor one has to know the excitations. Because there is no lattice translation symmetry the usual description in terms of dispersion curves in a Brillouin zone breaks down. One can treat the excitations in the context of the phenomenological theory of second order phase transitions. This has been done, for example, by Axe et al. [12]. One can also generalize the 3-dimensional description [13] and use the symmetry in more dimensions [14]. The vibrations of an IC modulated crystal are described by time dependent displacements $\vec{u}(\vec{n}j\tau)$, which are in the form of Bloch functions because of the periodicity of the 4-dimensional problem. Hence,

$$\vec{u}(\vec{n}j\tau) = e^{i\vec{k}\cdot\vec{n}} \vec{U}_j (\tau+\vec{q}\cdot\vec{n}). \qquad (8)$$

The reason why the mode is characterized by the 3-vector $\vec{k}$ rather than a 4-vector is that the fourth component of such a vector drops out of the equations of motion. Because the function $\vec{U}(\tau)$ is periodic it may be decomposed in a Fourier series. For a mode characterized by $\vec{k}$ and branch label ν one may write

$$\vec{U}_j^{\vec{k}\nu} = \sum_\ell \vec{A}_{\ell j}^{\vec{k}\nu} e^{i\ell\tau}. \qquad (9)$$

For an unmodulated crystal the displacement is independent of τ. Then $\vec{A}^{\vec{k}\nu}_{\ell j}$ is different from zero for $\ell=0$ only and for $\ell=0$ it is exactly the polarization vector $\vec{e}(\vec{k}\nu;j)$. It should be noted that one obtains the same value of the displacement at $\tau=0$ when $\vec{k}$ is replaced by $\vec{k}+\vec{q}$ and ℓ by $\ell-1$. Therefore, the excitation is equally well characterized by a vector in the Brillouin zone of the modulated structure. This Brillouin zone is at most of dimension two when the modulation is incommensurate. When the modulation is commensurate, the wave vector may be chosen in the Brillouin zone of the superstructure.

For small amplitude of the modulation $\vec{A}^{\vec{k}\nu}_{\ell j}$ is maximal for one value of $\vec{k}+\ell\vec{q}$ but there is admixture of $\vec{k}+(\ell+1)\vec{q}$. In the original zone this mode is then characterized by $k+\ell q$ (modulo reciprocal lattice). Then the frequency is again a function $\omega_{\vec{k}\nu}$ in the (original) Brillouin zone. In contrast to a periodic system the frequency function is not continuous inside the zone, but has gaps. For small amplitude, the gaps of substantial value are small in number, however. The gaps occur at multiples of $\tfrac{1}{2}\vec{q}$. Further away from the normal-incommensurate phase transition the amplitude increases, discommensurations occur and $\vec{A}_{\ell j}$ has contributions from more and more values of ℓ [15]. Also the number of gaps increases.

One may study the behavior of the excitation spectra and eigenvectors on a simple model, that is, however, sophisticated enough to show the consequences of the modulation. This model, the discrete frustrated ϕ^4 (DIFFFOUR)model has been studied by Janssen and Tjon [16]. It is essentially a one-dimensional chain of particles connected by short range interactions. When the displacement of the n-th particle from an equidistant array is u_n the potential energy V may be written in terms of the difference variable $x_n=u_n-u_{n-1}$ as

$$V = \sum_n [\frac{\alpha+2\beta+3\delta}{2}\, x_n^2 + \frac{1}{4}\, x_n^4 + (\beta+2\delta)x_n x_{n-1} + \delta x_n x_{n-2}] \tag{10}$$

The equilibrium positions for which V is minimal may be modulated if α/δ and β/δ are of comparable value and in competition. For a restricted range of β/δ one finds the P-phase (unmodulated) for large α/δ, a commensurate or incommensurate modulated structure for intermediate values and a ferro- or antiferrodistortive structure for large negative values. The excitation spectrum is obtained as the spectrum of the matrix

$$\partial^2 V/\partial u_n \partial u_m$$

evaluated at the equilibrium positions. We shall use this model to illus-

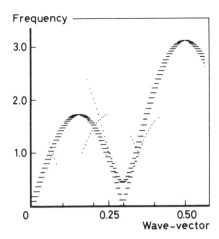

Fig. 1. Dispersion of the translationally invariant DIFFFOUR model for
$\alpha=1.2$, $\beta=-1$, $\delta=1.726$. For these parameters the groundstate is
sinusoidally modulated with wave vector $q=0.2963 \times 2\pi$. Modes k
are indicated at the position $\omega=\omega_{k\nu}$ and wave vector $k+mq$ by a
dash of length proportional to the absolute value of $A_m^{k\nu}$.
Dashes below a certain length have been omitted.

trate the effects of the excitations on the diffraction in an IC modulated
phase. The formalism has to be adapted slightly to be used also for quasi-
crystals, but essentially one obtains in that case results that are very
similar to those in the IC phase when in the discommensuration region.

For two cases the dispersion $A_\ell^{k\nu}$ has been calculated for the DIFFFOUR
model. They are presented in Figures 1 and 2. Figure 1 corresponds to a
situation just below the normal-incommensurate phase transition. Indicated
are values of $A_\ell^{k\nu}$ that are larger than a certain threshold value: they are
indicated by dashes with the absolute value of $A_\ell^{k\nu}$ as length at the position $\omega=\omega_{k\nu}$; wavevector $= k+\ell q$. Figure 2 does the same for a case close to
the lock-in transition. Because the model has translational symmetry there
is an acoustic mode. The original soft mode dispersion curve is still
recognizable, but now for every wave vector there is more than just one
frequency. Because in the sinusoidal region the frequencies can be
obtained from a perturbation of the unmodulated structure the phason branch
is strong left from q and the amplitudon branch is strong right from q.

A related model is the DIFFFOUR model where the squares of the fre-
quencies are the eigenvalues of $\partial^2/\partial x_n \partial x_m$, i.e., derivatives with respect
to the difference variables. Since this model does not have translational
symmetry, there is no acoustic mode. It may be considered as a model for
an IC phase resulting from an instability in an optic branch (see for
example biphenyl [17]). Figure 3 corresponds to the sinusoidal region,
Fig. 4 to the discommensuration region. In the sinusoidal region the

original dispersion relation is clearly recognizable. The soft mode branch has split up into a phason branch which goes to zero and an amplitudon branch. This can be seen from the function $U(\tau)$ which is essentially the eigenvector of the excitation. In the discommensuration region the original modes mix up in a complicated way, a number of gaps appear and there are very narrow bands, for example the low frequency band in Fig. 4, which corresponds to long wave length vibrations of the discommensuration lattice.

4. DYNAMIC STRUCTURE FACTOR

The coherent inelastic neutron scattering is governed by the differential cross section. A derivation for the expression for an ordinary crystal can be found in reference [18]. The derivation can be generalized for IC modulated phases using the expression

$$\vec{u}(\vec{n}j\tau) = \frac{1}{\sqrt{Nm_j}} \sum_{\vec{k}\nu} (\frac{h}{2})^{\frac{1}{2}} e^{i\vec{k}.\vec{n}}[\vec{U}_j^{\vec{k}\nu} (\tau+\vec{q}.\vec{n}) (a_{\vec{k}\nu} + a_{-\vec{k}\nu}^+)]. \tag{11}$$

Then the differential cross section becomes

$$\frac{d^2\sigma}{d\Omega dE}\Big|_{coh}^{inel} = \frac{k_{out}}{k_{in}} \frac{(2\pi)^3}{v_o} \sum_{\vec{G}\vec{k}\nu} \delta(\vec{Q}+\vec{k}-\vec{G}) |H_{\vec{k}\nu}(\vec{Q})|^2 S_\nu(\vec{k},\omega), \tag{12}$$

where $\vec{G}$ runs over the quasi-lattice (1) and where the matrix elements are given by

$$S_\nu(\vec{k},\omega) = \frac{1}{2\omega_{\vec{k}\nu}} [n_{\vec{k}\nu} \delta(\omega+\omega_{\vec{k}\nu}) + (n_{\vec{k}\nu}+1)\delta(\omega-\omega_{\vec{k}\nu})]; \tag{13}$$

and

$$H_{\vec{k}\nu}(\vec{Q}) = \sum_{j=1}^{s} \frac{\bar{b}_j}{m_j^{\frac{1}{2}}} e^{-W_j(\vec{Q})} e^{i\vec{Q}.\vec{r}_j} \sum_n e^{i\vec{Q}.(\vec{n}+\vec{f}_j(\vec{q}.\vec{n}))} \sum_m \vec{Q}.\vec{A}_{mj}^{\vec{k}\nu} e^{i(\vec{k}+m\vec{q}).\vec{n}}.$$

$$\tag{14}$$

26

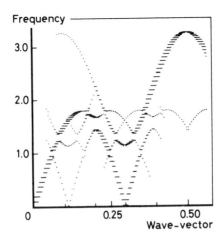

Fig. 2. As Fig. 1, but for other values of the parameters: $\alpha=0.95$, $\beta=-1$, $\delta=1.726$. For these values the structure shows discommensurations.

Because the function $\vec{f}_j$ is periodic one may expand

$$e^{i\vec{Q}.\vec{f}_j(\tau)} = \sum_\ell B_{\ell j}(\vec{Q})e^{i\ell\tau}. \tag{15}$$

Then the matrix element $H_{\vec{k}\nu}$ for a momentum transfer $\vec{Q}=\vec{K}-m\vec{q}-\vec{k}$ with $\vec{K}$ in the reciprocal lattice of the basic structure can be written as

$$H_{\vec{k}\nu} = \sum_{j=1}^{s} \bar{b}_j m_j^{-\frac{1}{2}} e^{-W_j(Q)} e^{i\vec{Q}.\vec{r}_j} \sum_\ell (\vec{Q}.\vec{A}_{\ell j}^{\vec{k}\nu}) \; B_{m-\ell j}(\vec{Q}) \tag{16}$$

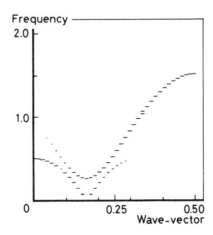

Fig. 3. Dispersion of the not translationally invariant model for $\alpha= 1.99, \beta= -1$, $\delta=0.2441$. Then the groundstate is sinusoidally modulated with wave vector $q=0.1622 \times 2\pi$.

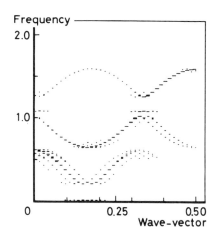

Fig. 4. As Figure 3, now for $\alpha=1.8$, $\beta=-1$, $\delta=0.2441$. The structure belongs to the discommensuration region.

So the contribution to the scattering is determined by terms B_{1j} that are given by the <u>static</u> deformation, and terms $A_{\ell j}^{\vec{k}\nu}$ given by the <u>dynamics</u>. Only those modes $\vec{k}\nu$ contribute that have a nonnegligible value. These modes are indicated in Figs. 1 to 4. So for small amplitude sinusoidal modulation there is still a clear dispersion curve. In the discommensuration region there is scattering for many values in the $\omega\vec{k}$-plane. This is even more so when there are more particles in the unit cell.

5. DEBYE-WALLER FACTOR

The geometric structure factor (2) is only valid for T=0. For nonzero temperature both the static and the dynamic structure factors contain a Debye-Waller factor. This factor has been calculated in the context of Landau theory, and for sinusoidally modulated IC phases with one particle per unit cell only by Overhauser [19], by Axe [20] and by Adlhardt [21]. Jaric [22] has discussed the long-wavelength excitations in quasi-crystals. They consider in particular the "new" modes, the phasons and amplitudons. Although, of course, for low temperature the fact that phasons have a very low frequency is important, it is, from a principal point of view, not correct to treat these excitations completely separate from the other modes. As a matter of fact all modes are influenced by the modulation, not only the phasons and amplitudons. This is certainly the case in the discommensuration region. In the present section we present an expression for the Debye-Waller factor based on the treatment of the excitations in section 3 which is rigorous in the harmonic approximation.

In general, the Debye-Waller function for the n-th particle is

$$W_n(\vec{Q}) = \frac{1}{2}\langle(\vec{Q}.\vec{u}_n)^2\rangle.$$ (17)

The displacements $\vec{u}(\vec{n}j\tau)$ are given in Eq. (11). Therefore,

$$W_{\vec{n}j\tau}(\vec{Q}) = W_j(\vec{Q};\tau+\vec{q}\cdot n) \qquad \text{with} \qquad (18)$$

$$W_j(\vec{Q};\tau) = \frac{h}{2Nm_j} \sum_{\vec{k}\nu} \omega_{\vec{k}\nu}^{-1} |\vec{Q}\cdot\vec{U}_j^{\vec{k}\nu}(\tau)|^2 (2\vec{n}_{\vec{k}\nu}+1).$$

For sufficiently high temperature the occupation numbers may be approximated and one obtains

$$W_j(\vec{Q};\tau) = \frac{hk_BT}{4Nm_j} \sum_{\vec{k}\nu} \omega_{\vec{k}\nu}^{-2} |\vec{Q}\cdot\vec{U}_j^{\vec{k}\nu}(\tau)|^2. \qquad (19)$$

This relation is analogous to the expression for ordinary crystals.

In order to get a view on the influence of the Debye-Waller factor one can calculate this factor in the DIFFFOUR model. Since that model is one-dimensional, the fluctuations diverge. Therefore, the linear chains are embedded in a three-dimensional array with harmonic interchain coupling. For a given mode $k\nu$ of the one-dimensional model and given two-vector (k_1,k_2) the frequency of the corresponding mode of the 3-dimensional model is

$$\omega_{\vec{k}\nu}^2 = c^2(\sin^2(k_1/2)+\sin^2(k_2/2)) + \omega_{k\nu}^2. \qquad (20)$$

As already found by Axe the DW factor consists of a constant part and a modulation. In the sinusoidal region this modulation has wave vector $2\vec{q}$: $W_j(\vec{Q};\tau) = W_{0j}+W_{2j}\cos(2\tau)$. The reason is that the fluctuations are maximal at the zero points of the sinusoidal modulation function and that W depends on the absolute value of the displacements only. In the discommensuration region the number of harmonics increases.

In the sinusoidal region the static structure factor becomes

$$F(\vec{Q}) = \sum_j \frac{1}{2\pi} \int_0^{2\pi} e^{i\vec{Q}\cdot\vec{r}_j+i\vec{Q}\cdot\vec{f}_j(\tau)-W_j(\vec{Q},\tau)+im\tau} \, d\tau \qquad (21)$$

$$= \sum_j e^{i\vec{Q}\cdot\vec{r}_j-W_{0j}} \frac{1}{2\pi} \int_0^{2\pi} e^{i\vec{Q}\cdot\vec{A}_j\sin(\tau)-W_{2j}\cos(2\tau)+im\tau}$$

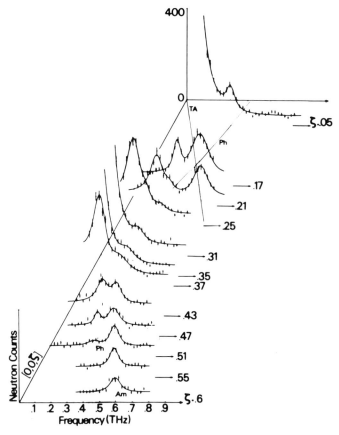

Fig. 5. Constant Q scan for inelastic neutron scattering from β-ThBr$_4$
for $Q=(2,3,1)-\vec{c}^*$ at $T=81K$ [24].

Using the Jacobi-Anger equality this gives (for one particle per unit cell)

$$F(\vec{Q}) = \frac{1}{2\pi} \int_0^{2\pi} \exp[i\vec{Q}.\vec{A}\sin(\tau)-Q^2W_0-Q^2W_2\cos(2\tau)+im\tau] \, d\tau \qquad (22)$$

$$= \delta(\vec{Q}-\vec{K}-m\vec{q}) \sum_\ell J_{-2\ell-m}(\vec{Q}.\vec{A}) \, I_\ell(Q^2W_2),$$

where J_ℓ and I_m are (modified) Bessel functions.

6. EXPERIMENTS

The soft mode leading to the normal-incommensurate phase transition
and the resulting special modes predicted by the theory have, till now,
only been seen for a rather restricted number of compounds: biphenyl [23]

30

and $ThBr_4$ [24-25], where phason branches have been observed with inelastic neutron scattering, and K_2SeO_4 [26], where an overdamped phason has been found by the same technique. The fact that phasons have not been found more frequently is believed to be due to the damping. Since the phason frequency is very low, the phason is generally overdamped because the damping does not go to zero when the phason branch tends to zero, as is the case for the acoustic branch [27]. The amplitudon has been observed for many systems by means of Raman spectroscopy.

As example we shall discuss here the case of $ThBr_4$. For temperatures above 95K β-$ThBr_4$ is in the tetragonal phase with space group $I4_1/amd$. An optic mode with wave vector $\gamma\vec{c}^*$ and belonging to the τ^4 representation (in Kovalev's notation) becomes unstable at 95K. Below T_i=95K $ThBr_4$ is in a phase which is modulated with the wave vector of the soft mode. The value of $\gamma(0.31\pm0.005)$ remains constant upon cooling to very low temperatures. No lock-in transition has been found.

Inelastic neutron scattering spectra have been measured near the strong satellite $(2,3,1-\gamma)$ at $T = 81K$ (Fig. 5). Three branches may be distinguished: i) A branch originating at $\omega\simeq0$, wave vector $\vec{q}$, interpreted as the phason branch. ii) A branch with minimum frequency at wave vector $\vec{q}$, which is the amplitudon branch, the minimum value of the frequency agrees with the frequency of a strongly temperature-dependent Raman active mode [28]. iii) A branch that originates at the main reflection $(2,3,1)$ and is interpreted as a transverse acoustic branch.

One may fit the spectra with quasi-harmonic oscillators and determine in this way the quasi-harmonic frequencies and oscillator strengths (Fig. 6).

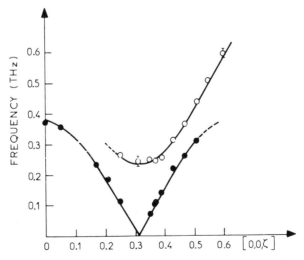

Fig. 6. Quasi-harmonic frequencies of the phason and amplitudon branches of $ThBr_4$ [24].

In this way one finds the phason and amplitude branches which resemble the dispersion in Fig. 3. As in that model calculation the oscillator strength ($|A_m^{k\nu}|$) is asymmetric with respect to $\vec{q}$: For $k<q$ the phason strength is greater, for $k>q$ it is the inverse.

The Raman activity depends in the first place on the values of $A_\ell^{k\nu} B_{m-\ell}$ for which $k+mq=0$ (effective wave vector zero because of the long wave length of the incoming light). In the model these modes can easily be read off from the figures. For the amplitudon $A_{\ell j}^{\vec{k}\nu}$ is nonvanishing for $\vec{k}+\ell\vec{q}$ $=\vec{q}$ while for $m-\ell=-1$, B_{m-1} is different from zero too and $\vec{k}+m\vec{q}=\vec{k}+(\ell-1)\vec{q}=0$. A second condition is then one on the symmetry of the mode: it should belong to a symmetric representation. This condition is also satisfied for the amplitudon, but not for the phason which is antisymmetric.

Analogous results have been obtained for biphenyl [23].

7. ACKNOWLEDGMENT

This work was partly done during the stay of T.J. at the Institut Laue-Langevin. He thanks the ILL for hospitality and the Stichting voor Fundamenteel Onderzoek der Materie for financial support.

8. REFERENCES

1. P. Bak, Phys. Rev. Lett. 54, 1517-1519 (1985).
2. M. Duneau and A. Katz, Phys. Rev. Lett. 54, 2688-2691 (1985).
3. V. Elser, Phys. Rev. B. 32, 4892-4898 (1985).
4. P. Bak, Phys. Rev. B. 32, 5764-5772 (1985).
5. T. Janssen, in Proceedings Phonon Conference Budapest, 1985 (World Scientific Publ., Singapore, 1985), 260-265.
6. A. Katz and M. Duneau, J. Physique 47, 181-196 (1986).
7. V. Elser, Acta Cryst. A 42, 36-43 (1986).
8. T. Janssen, Acta Cryst. A 42 H 261-271 (1986).
9. P. M. de Wolf, Acta Cryst. A 33, 493-497 (1977).
10. A. Yamamoto, Acta Cryst. A 38, 87-92 (1982).
11. S. Hendricks and E. Teller, J. Chem. Phys. 10, 147-167 (1942).
12. J. D. Ade, M. Iizumi, and G. Shirane, Phys. Rev. B 22, 3408-3413 (1980).
13. M. B. Walker, Can. J. Phys. 56, 127-138 (1978).
14. T. Janssen J. Phys. C 12, 5381-5392 (1979).
15. T. Janssen and C. de Lange, Journal de Physique C 6, 737-739 (1981).
16. T. Janssen and J. A. Tjon, Phys. Rev. B 25, 3767-3785 (1982).
17. T. Janssen, Jap. J. Appl. Phys.; Supplement 24-2 24, 747-749 (1985).
18. W. Marshall and S. W. Lovesey, Theory of Thermal Neutron Scattering (Clarendon Press, Oxford, 1971).
19. A. W. Overhauser, Phys. Rev. B 3, 3173-3182 (1971).
20. J. D. Axe, Phys. Rev. B 21, 4181-4190 (1980).
21. W. Adlhart, Acta Cryst. 38, 498-504 (1982).
22. M. Jaric, J. de Physique Colloques H C3, 259-270 (1986).
23. H. Cailleau, J. C. Messager, F. Moussa, F. Bugaut, C. M. E. Zeyen and C. Vettier, Ferroelectrics 67, 3-14 (1986).

24. L. Bernard, R. Currat, P. Delamoye, C. M. E. Zeyen, S. Hubert and R. de Kouchkovsky, J. Phys. C. Solid State Phys. 16, 433-456 (1983).

25. R. Currat, L. Bernard, and P. Delamoye, in Incommensurate Phases in Dielectrics 2, R. Blinc and A. P. Evanyuk (eds.) (North-Holland, Amsterdam, 1986), 162-204.

26. M. Quilichini and R. Currat, Sol. State Comm. 48, 1011-1015 (1983).

27. V. A. Golovko and A. P. Levanyuk, in Light Scattering Near Phase Transitions, H. Z. Cummins and A. P. Levanyuk (eds.) (North-Holland, Amsterdam, 1983), 169-226.

28. S. Hubert, P. Delamoye, S. Lefrant, M. Lepostollec and M. Hussonnois, J. Sol. St. Chem. 36, 36-44 (1981).

ARE EXOTIC CONSEQUENCES OF INCOMMENSURABILITY IN SOLIDS EXPERIMENTALLY

OBSERVABLE?

J. B. Sokoloff

Physics Department
Northeastern University
Boston, Massachusetts 02115

Abstract   The exotic transport properties expected for incommensurate
crystals, due to the fragmented nature of their energy bands, are
shown to be usually unobservable because of intrinsic defects in the
crystal structure and the usually weak nature of the modulation potential. Two systems for which they might be observable, however, are
some artificially grown superlattices and quasi-crystals.

I. INTRODUCTION

The electron energy band structure of an incommensurate compound is
highly fragmented by gaps on all energy scales [1]. This is because any
commensurate approximation to an incommensurate structure will have gaps at
wave vectors given by

$$\vec{k} = \frac{1}{2} \sum_j n_j \vec{Q}_j + \frac{1}{2} \sum_j m_j \vec{G}_j, \tag{1}$$

where $\{\vec{Q}_j\}$ and $\{\vec{G}_j\}$ are the reciprocal lattice vectors of the two period
structures which become incommensurate in the incommensurate limit and $\{n_j\}$
and $\{m_j\}$ are integers. In the incommensurate limit, these $\vec{k}$ values become
dense. The Q's and G's can be incommensurate because the ratios of their
magnitudes are irrational numbers or because the vectors have irrational
projections on each other. Indeed, Ostlund and Pandit [2] have obtained
such a band structure with gaps on all energy scales for a one-dimensional
tight binding model with a sinusoidal potential incommensurate with the

35

lattice. For strong potentials, they find that the gaps remain large as one looks on all energy scales; for weak potentials, however, the gaps become smaller and smaller as we look on smaller energy scales. This is expected on the basis of perturbation theory because the multitude of gaps that occur on small energy scales come from high orders in perturbation theory and hence become quite small for weak potentials. One might expect the presence of a multitude of gaps on all energy scales to make it possible to observe interesting phenomena in incommensurate crystals, such as Zener tunneling and Stark oscillations in the electrical conduction, interesting magnetic breakdown effects and interesting optical properties.

2. THEORY

In order to estimate the conditions that must be satisfied in order for Zener tunneling and Stark oscillation to be observable, let us consider a high order commensurate approximation, for which there exist minibands separating the gaps. (Remember that for a true incommensurate crystal with weak modulation potential, the gaps on arbitrarily small energy scales are negligibly small.) The distance traveled in a time interval Δt for an electron in a miniband of wave vector dependent energy $\varepsilon(\vec{k})$ is easily shown to be given by

$$\Delta x = \frac{1}{e\varepsilon} \left| \varepsilon(\vec{k}_o + \frac{e\vec{E}}{\hbar} \Delta t) - \varepsilon(\vec{k}_o) \right| = \text{(band width)}/e\varepsilon, \tag{2}$$

using the relations

$$\hbar\dot{\vec{k}} = e\vec{E}$$

and

$$\Delta x = \int \frac{1}{\hbar} \left| \nabla_{\vec{k}} \varepsilon(\vec{k}) \right| dt \ ,$$

where E is the applied electric field and $\vec{k}_o$ is the initial wave vector. Since the observability of Zener tunneling and Stark oscillations requires that Δx be much less than a mean free path for an electron at the Fermi level to be able to reach the gap edge, equation (2) shows for example that for an electric field of 0.1 V/cm and a mean free path of 100 Å, the miniband width must be $\ll 10^{-7}$ eV. The probability of Zener tunneling is given by

$$\exp\left[-\frac{\pi^2}{4}\frac{g^2}{eEa\varepsilon_F}\right] = \exp\left[-\frac{\pi^2}{4}\frac{g}{\varepsilon_F}\frac{g}{W}\frac{\Delta x}{a}\right] , \qquad (3)$$

where g is the gap, $a=2\pi/G$ where $\frac{1}{2}$ G is wave vector at which the gap occurs, and W is the miniband width. Since a = 1 Å and Δx = 100 Å, we may conclude that if g $\ll$ W, the electron will Zener tunnel quite readily through the gap, almost as if it were not present. Thus, we may conclude that for an incommensurate system, if as we look on smaller and smaller energy scales, the gaps become negligibly small, the electrons in an applied electric field will tunnel readily through these very small gaps, as if they were not present. Although Eq. 3 indicates that the tunneling probability will be reduced as E is reduced (making the small gaps more important) once Δx, given by Eq. 2, becomes larger than a mean free path, the electron will again not feel the effect of the small gaps because it will get scattered before it reaches a gap.

For the case of magnetic breakdown in the deHaas-van Alphen and other galvanometric properties, one might expect that the small gaps which fragment the energy versus wave vector relations, which are broken down for larger magnetic fields might become no longer broken down for smaller fields. This could lead to new (smaller) orbits of the electron around the edge of the Fermi surface (i.e., "lens" orbits. The other orbits caused by the gaps are open orbits). In practice, however, this may be impossible to observe. This is because in order to observe an orbit we must have the cyclotron frequency ω_C much larger than the inverse electron scattering time, which requires that ω_C be much greater than $\cong 10^{12}$ sec^{-1}, which for free electrons gives a magnetic field of about 6T. Thus, if lower order (i.e., larger) gaps are broken down for a field of 15 to 20T, if the next order gaps are less than 1/3 the size of these, they will always be broken down for any field greater than 6T (the minimum field for which the de Haas-van Alphen frequencies are observable) and hence will be unobservable. If the modulation potential causing the gaps is small compared to the Fermi energy (for example), this will always be the case (because the higher order gaps occur in high order perturbation theory in this potential). A cyclotron frequency of greater than 10^{12} sec^{-1} can be obtained for smaller fields when we consider that the orbits obtained when the smaller gaps are not broken down are quite small, and hence, the time for an electron to traverse them is reduced over the time to traverse the full Fermi surface. This can only reduce the permissible field by a factor proportional to the orbit circumference, however, whereas the higher order gaps drop off exponentially with the order.

The optical absorption due to an optically induced electronic transition, calculated in a model in which two Bloch states are mixed by the modulation potential is approximately proportional to

$$\frac{g}{\varepsilon_F} \left[\left(\frac{\hbar\omega}{g}\right)^2 - 1 \right]^{-\frac{1}{2}} \tag{4}$$

The g/ε_F factor represents the fraction of the section of the Fermi surface sliced by the Bragg plane for which the gap g straddles the Fermi level and the square root factor is due to the usual Van Hove singularity at a gap. Although in principle one should observe a hierarchy of transitions caused by the many gaps that straddle the Fermi surface, we see that if these gaps rapidly go to zero as we consider higher and higher order gaps, the g/ε_F factor will make their contribution negligible.

There are several ways of getting around these difficulties. First of all, one can consider artificially grown (by molecular beam epitaxy) incommensurate crystals with a non-sinusoidal modulation. For example, Kohmoto, Kadanoff and Tang consider a particular step function modulation in a one-dimensional tight binding model [4]. This model gives the totally fragmented band structure for all strength potentials which only occurs for potentials comparable to the bandwidth in the case of a sinusoidal modulation. Such step function modulations can be produced by growing artificial incommensurate superlattices with a step function modulation. For example, such a structure could be described by the following model potential:

$$V(\vec{r}) = \sum_n \left[v_1(\vec{r}-\vec{R}_n) + (v_2(\vec{r}-\vec{R}_n) - v_1(\vec{r}-\vec{R}_n))\theta(c-\cos\vec{Q}\cdot\vec{R}_n) \right] \tag{5}$$

where $\theta(x) = 1$ if $x > 0$ and 0 if $x < 0$, and $v_1(\vec{r}-\vec{R}_n)$ and $v_2(\vec{r}-\vec{R}_n)$ are respectively the atomic potentials of the two different types of atoms which occur in the alternating layers (i.e., we alternate layers of atoms of type 1 and of type 2 in an almost periodic way). The constant c takes a value between 0 and 1 and determines the relative widths of the layers and $\vec{Q}$ is the wave vector of the modulation and is usually chosen to be along a crystallographic axis. The Fourier transform of $V(\vec{r})$, which in lowest order perturbation theory determines the gaps is given by

$$V(\vec{k}) = \lim_{\varepsilon\to 0} \sum_n e^{i\vec{k}\cdot\vec{R}_n} v_1(\vec{k}) + (v_2(\vec{k})-v_1(\vec{k}))$$

$$\times \frac{1}{2\pi i} \int_{-\infty}^{\infty} dq \frac{e^{iq(c-\cos\vec{Q}\cdot\vec{R}_n)}}{q-i\varepsilon} \bigg], \tag{6}$$

where we have used the following integral representation for $\theta(x)$:

$$\theta(x) = \frac{1}{2\pi i} \int_{-\infty}^{\infty} dq \frac{e^{iqx}}{q - i\varepsilon} \quad .$$

The sum over n gives

$$V(\vec{k}) = v_1(\vec{k}) \sum_{\vec{G}} \delta_{\vec{k},\vec{G}} + \sum_{m} \left[v_2(\vec{k}) - v_1(\vec{k}) \right] \frac{1}{2\pi i} \int_{-\infty}^{\infty} dq \ i^m J_m(q) e^{iqc}$$

$$\times \sum_{m\vec{G}} \delta_{\vec{k},m\vec{Q}+\vec{G}} \quad . \tag{7}$$

where $\vec{G}$ is a reciprocal lattice vector of the underlying crystal lattice. The integral over q may then be carried out to give

$$V(\vec{k}) = v_1(\vec{k}) \sum_{\vec{G}} \delta_{\vec{k},\vec{G}} + \left[v_2(\vec{k}) - v_1(\vec{k}) \right] \sum_{\vec{G},m} \frac{1}{\pi} \frac{1}{m} (1-c^2)^{\frac{1}{2}} U_{n-1}(c)$$

$$\times \delta_{\vec{k},m\vec{Q}+\vec{G}} \quad . \tag{8}$$

where $U_m(x)$ is the Chebychev polynomial of the second kind [5]. For sufficiently large $\vec{G}$ and m, we can find values of $m\vec{Q} + \vec{G}$ which are arbitrarily closely spaced. Now, however, the gap size falls off only as m^{-1}, and hence a hierarchy of reasonably sized gaps is expected. This behavior is a consequence of the step function variation of the potential as we go from a region containing one type of atom to a region containing the other type. There are several types of defects in this perfect incommensurate structure which must be accounted for which we will categorize as follows: First of all, there could be wiggles in the planes, such that the thicknesses of the individual layers are maintained. This can be accounted for by adding a position dependent phase $\phi(\vec{r})$ to $\vec{Q} \cdot \vec{R}_n$ in Eq. 6, which in this article will be assumed to be a slowly varying random function of position. If, for example, ϕ is treated as a Gaussian random variable, the second term in Eq. 8 gets multiplied by a factor $e^{-m^2 <\phi^2>}$, which suppresses the very high order (i.e., large m) diffraction peaks including the ones that are very closely spaced). As we shall see, this factor also occurs in quasicrystals [6]. A second type of randomness is a variation in the relative thickness of the layers, which is governed by the parameter c in Eq. 6. If the deviation in c, δc, is a Gaussian random variable, the integral in Eq. 7 will contain a factor $e^{-q^2(\delta c^2)}$. The integral can be performed in the large m limit to obtain a factor

$$\left[\frac{Ce}{4m<\delta C^2>} \right]^m \qquad (9)$$

as the asymptotic dependence of the integral for large m. Again, this tends to reduce drastically the size of the extremely large m gaps and X-ray scattering Bragg peaks. For the Fibonacci lattice considered by Clark et al. [6a] phase disorder will again wipe out the extremely high order closely spaced gaps and Bragg peaks. Small variations in the thicknesses of the layers, however, might not interfere with their Fibonacci sequence arrangement, and therefore might not wipe out the high order gaps and Bragg peaks. They will simply introduce some disorder scattering. The third type of randomness is randomness in $\vec{Q}$, which will broaden the Bragg peaks in X-ray scattering. In general, these three types of disorder will occur in combination with each other. In addition, there is the usual broadening due to finite crystal sizes, substituted impurities, etc., which should be the same as for conventional crystals.

3. QUASICRYSTALS

   Quasicrystals are substances which exhibit icosahedral symmetry [7]. Since such symmetry is not allowed in conventional crystallography, they are almost periodic, like incommensurate crystals. They are believed to have an arrangement of atoms based on the three-dimensional Penrose tilting [8], whose structure factor was shown to fall off relatively slowly [9]. Like the layered structures, however, they also possess a factor due to phase disorder which drastically reduces the intensity of the very high order Bragg reflections (including the interesting closely spaced peaks which makes the spectrum dense [6]). Still, one might expect to see interesting transport and critical properties due to Bragg reflections which remain. Unfortunately, those quasicrystals whose transport properties have been measured to date have mean free paths of the order of interatomic spacings [10]. This is likely to be due to the transiton metal ions, which can be strong scatterers because of the presence of localized d-states at the Fermi energy which can resonantly scatter conduction electrons, as is believed to occur in transitional metallic glasses [11]. Unlike glasses, however, the atoms in quasi-crystals are almost periodic rather than random, and as we shall see, the almost periodic atomic arrangement of atoms

will not scatter the electrons. To see this, let us represent the manganese atom distribution by a three-dimensional Penrose lattice with hard
spheres at the vertices to represent manganese atoms. Then let us apply the
method of Lax to this problem [12]. Lax gives an expression for the total
scattering cross-section of an electron in any arrangement of scatterers,
which depends on their structure factor. For a periodic atomic arrangement, he shows that the scatterers' cross-section is zero unless the incident electron's wave vector satisfies the Bragg condition. For the quasi-
crystal, the Bragg condition is satisfied for a dense set of wave vectors,
but for most of them, the structure factor is negligibly small [9]. Thus,
to account for the short mean free paths, we consider the analogue of phase
defects in the quasicrystal, which consist in the projection technique of
random atomic displacements in the hyperplane perpendicular to the physical
hyperplane [6]. To accomplish this, we use the method of Zia and Dallas
[13], but we add a displacement $\vec{u}$ in the perpendicular hyperplane to each
lattice point in the six-dimensional hyperspace. Treating $\vec{u}$ as a Gaussian
random variable, we obtain a mean square structure factor

$$
<S^2> = \int d^3p \, d^3p' \sum_{jj'} ie^{i\vec{k}^6 \cdot (\vec{R}^6_j - \vec{R}^6_{j'})} R(-\vec{p})R(-\vec{p}')
$$

$$
<e^{i\vec{p}\cdot\vec{u}}><e^{-i\vec{p}'\cdot\vec{u}}> + \int d^3p \, |R(-\vec{p})|^2 \left[1 - |<e^{i\vec{p}\cdot\vec{u}}>|^2\right], \tag{10}
$$

where superscript 6 signifies a vector in the six-dimensional hyperspace,
$R(-\vec{p})$ is the Fourier transform of Zia and Dallas's "window function" and
$<e^{i\vec{p}\cdot\vec{u}}>$ is the Debye-Waller-like factor introduced by Elser to study the
effects of such disorder. The first term in Eq. 10 gives the Bragg scattering and the last term gives the incoherent scattering. For example, if
the distribution function of the u's is Gaussian $(\alpha^3/\pi^{3/2}) \, e^{-\alpha|\vec{u}|^2}$, where
$\alpha^{-2} = \langle u^2 \rangle$, we find that if $[\langle u^2 \rangle]^{\frac{1}{2}}$ is 0.2 of the mean radius of the tri-
acontrahedron cell, which contains the projection in the perpendicular
hyperplane of all the points whose projection in the physical hyperplane
gives the quasicrystal, the incoherent part of the structure is $\simeq 0.6$, but
the Bragg peaks are only reduced by 50% by the Debye-Waller-like factor
due to disorder [6]. Since the total scattering cross-section for a single
hard sphere is 4π times the square of the radius [14], we conclude from
Lax's formula that the total scattering cross section for the quasicrystal
σ_t is comparable to the square of this radius. Using the expression
$(n\sigma_t)^{-1}$ for the decay distance of the electric wavefunction, where n is the
number density of manganese atoms [12], we conclude that the wave function

decay distance could easily be of the order of the atomic separation. Thus, if we wish to observe effects of fragmented band structure in quasi-crystals, we must either improve the quality of the quasicrystals or to make them out of atoms which are weaker scatterers than manganese.

One possible way of producing materials which show exotic experimental consequences of almost periodicity is to use the gap transfer idea of Avron and Simon [15] and Azbel et al. [16]. The idea is that we have two incommensurate periodicate potentials, one of very long period and one of short period. Gaps which are higher order in perturbation theory in the long period potential but low order in the short period potential will not

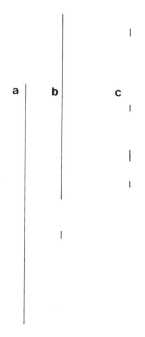

Fig. 1. Energy spectrum for a one-dimensional tight binding model (with two sinusoidal potentials of strengths 2.5% of the tight binding band width and wavelengths of 2π and 20π lattice constants. (a) is the spectrum for the short period potential only, (b) is the spectrum for the long period potential only and (c) is the spectrum for both potentials. The vertical lines represent energies in the band, and breaks in the lines represent gaps. The vertical axis is the energy, and the height of the figure runs from the lower edge of the tight binding model at the bottom of the figure to a point 6% of the way across the band.

be significantly smaller than the gap which is low order in both potentials because of the small energy denominators which occur. Thus, the band structure will be highly fragmented. For example, Fig. 1 shows the results of calculating the band structure of a one-dimensional tight binding model with two sinusoidal potentials, one of wavelength $2\pi a$ and one of wavelength $50\pi a$, where a is the lattice constant of the tight binding lattice. We see that with both potentials the band structure is highly fragmented, but with only one present it is not.

REFERENCES

1. D. Simon, Advances in Applied Mathematics $\underline{3}$, 463 (1982); J. B. Sokoloff, Physics Reports $\underline{126}$, 189 (1985).
2. S. Ostlund and R. Pandit, Phys. Rev. $\underline{B29}$, 1394 (1984).
3. J. M. Ziman, "Principles of the Theory of Solids," 2nd ed. (Cambridge University Press, Cambridge, 1972), pp. 190-196.
4. M. Kohmoto, L. P. Kadanoff and C. Tang, Phys. Rev. Lett. $\underline{50}$, 1870 (1983).
5. M. Abromowitz and I. A. Stagun (New York: Dover, 1965).
6. V. Elser, Phys. Rev. Lett. $\underline{54}$, 1730 (1985).
6a. R. Merlin, K. Bajema, R. Clarke, F. Y. Juang and P. K. Bhattacharya, Phys. Rev. Lett. $\underline{55}$, 1768 (1985); R. Clarke, this volume.
7. D. Schechtman, I. Blech, D. Gratias and J. W. Cahn, Phys. Rev. Lett. $\underline{53}$, 1951 (1984); R. D. Field and H. L. Fraser, Mat. Sc. and Eng. $\underline{68}$, L17 (1984).
8. D. Levine and P. J. Steinhardt, Phys. Rev. Lett. $\underline{53}$, 2477 (1984).
9. V. Elser, Phys. Rev. $\underline{B32}$, 4892 (1985); M. Duneau and A. Katz, Phys. Rev. Lett. $\underline{54}$, 2688 (1985).
10. S. J. Poon, A. J. Drehman, K. R. Lawless, Phys. Rev. Lett. $\underline{55}$, 2324 (1985); M. J. Burns, A. Behrooz, X. Yan, P. M. Chaikin, P. Bancel and P. Heiney, Bull. Am. Phys. Soc. $\underline{31}$, 268 (1985); D. Pavuna, C. Berger, F. Cyrot-Lackmann, P. Germi and A. Pasturel, Solid State Comm. $\underline{59}$,11 (1986); J.-L. Verger-Gaugry and P. Gyyot, J. de Physique (colloques) (in press); R. Markiewicz (unpublished).
11. G. Busch and H. J. Gunthrodt, "Solid State Physics", eds. H. Ehrenreich, F. Seitz and D. Turnbull (Academic Press, New York, 1974), p. 235.
12. M. Lax, Rev. Mod. Phys. $\underline{23}$, 287 (1951), see particularly p. 30.
13. R. K. P. Zia and W. J. Dallas, J. Phys. $\underline{A18}$, L341 (1985).
14. A. Messiah, "Quantum Mechanics" (Wiley and sons, New York, 1961).
15. J. Avron and B. Simon, Phys. Rev. Lett. $\underline{46}$, 1166 (1981).
16. M. Ya Azbel, P. Bak, P. M. Chaikin, Physics Letters $\underline{A117}$, 92 (1986); Phys. Rev. $\underline{A34}$, 1392 (1986).

THE APPLICATION OF AXIAL ISING MODELS TO THE DESCRIPTION OF MODULATED ORDER

Julia Yeomans

Department of Theoretical Physics
1, Keble Road
Oxford OX1 3NP, England

I. INTRODUCTION

In this paper I aim to review work on simple Ising models, the axial Ising models, which provide a useful phenomenological representation of modulated order in many compounds. The prototype of these systems is the axial next nearest neighbor Ising or ANNNI model and the first part of the article is devoted to a description of its phase diagram with emphasis on the features which might be observed experimentally. I then discuss the extent to which this behavior is in fact seen in polytypes, binary alloys and a ferrimagnet, cerium antimonide.

The ANNNI model [1] is described by the Hamiltonian

$$H = J_0 \sum_{nn}^{\perp} S_i S_j - J_1 \sum_{nn}^{\parallel} S_i S_j - J_2 \sum_{nnn}^{\parallel} S_i S_j \tag{1}$$

where the spins $S_i = \pm 1$ are Ising variables lying on the sites i of a cubic lattice. The second two terms denote sums over first (nn) and second (nnn) neighbors along a singled-out axial direction, $\parallel$, whereas the first term, $\perp$, is an interaction between nearest neighbor spins in the layers perpendicular to the axial direction.

We shall take $J_0 > 0$ so that the ordering within the layers is ferromagnetic. For $J_2 > 0$ the ground state along the axial direction is a simple ferro- or antiferromagnet for $J_1 > 0$ and $J_1 < 0$ respectively. For $J_2 < 0$, however, the first and second neighbor interactions compete and the axial order in the ground state depends on the value of $\kappa = J_2/J_1$. For $0 < \kappa < 1/2$ the ground state is ferromagnetic and for $-1/2 < \kappa < 0$ it is antiferromagnetic as before. However, for $|\kappa| > 1/2$, the antiferromagnetic second neighbor interaction is sufficiently strong to dominate the ordering, resulting in a ground state ...++--++.... The points $|\kappa|$ =1/2 are multiphase points [1] at which the ground state is infinitely degenerate.

Because there is an infinite degeneracy in energy at zero temperature, entropic contributions to the free energy are important at finite temperatures. This results in an infinity of stable phases in a region around the multiphase points [2-7]. To distinguish between different phases we use a notation introduced by Fisher and Selke [2,3]. Let a band be a sequence of consecutive layers of the same spin value terminated by layers of opposite spin. Then $\langle n_1 n_2 ... n_m \rangle$ is used to describe a phase where the repeating spin sequence comprises bands of length $n_1, n_2 ... n_m$. For example

$$...++--++---++--++---...\qquad(2)$$

is denoted $\langle 2223 \rangle$ or $\langle 2^3 3 \rangle$.

The mean-field solution of the ANNNI model [4-7] is expected to provide a good approximation to the phase diagram in three dimensions [8]. Those phases which are stable over the largest range of temperature and are shown in Figure 1 which mirrors much of what one would expect to see in an experiment. Note that, for $J_1 < 0$, only phases comprising 1- and 2-bands are stable whereas for $J_1 > 0$, only phases with band length ≥ 2 appear. At higher resolutions the following features would become apparent:

1. At sufficiently low temperatures series expansions indicate that there are infinite sequences of phases $\langle 2^k 3 \rangle$ and $\langle 1 2^{k+1} \rangle$, k=0,1,2..., for $J_1 > 0$ and $J_1 < 0$ respectively springing from the multiphase points [2,3,7]. The phases become exponentially narrower with increasing k.

2. As the temperature is increased, new phases appear through what has been termed in the literature "structure combination branching processes" [6,7]. The first instability of the boundary between two phases $\langle \nu_1 \rangle$ and $\langle \nu_2 \rangle$ is to the mixed phase $\langle \nu_1 \nu_2 \rangle$. For example, in Figure 1, $\langle 23223 \rangle$ appears between $\langle 23 \rangle$ and $\langle 223 \rangle$. It is likely that, for high enough temperatures, phases with wave vectors corresponding to every rational fraction within a given interval are stable.

3. The free energy differences between neighbouring long-period phases are, in general, extremely small. Moreover, the stable phases persist as metastable states when they cease to provide the global minimum in the free energy.

4. Incommensurate phases appear between the commensurate phases for temperatures $T \gtrsim T_c / 2$, where T_c is the transition temperature to the paramagnetic phase. These occupy an increasing fraction of the phase diagram as $T \to T_c$ [9,10].

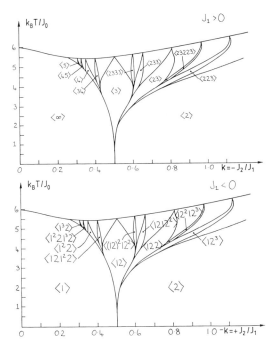

Fig. 1. Mean-field phase diagram of the three-dimensional ANNNI model.[4-7]

A fruitful way of understanding the physics leading to the modulated phases is in terms of the interactions between domain walls [11,8]. The definition of a domain wall is to some extent arbitrary, but in the vicinity of ⟨2⟩ they are usefully thought of as 3-bands or 1-bands for $J_1>0$ and $J_1<0$ respectively and near ⟨∞⟩ as the boundaries between bands of + and - spins. The domain walls fluctuate at finite temperatures and hence interact. The balance between the interaction, which decays exponentially with distance, and the wall self-energy determines the wall spacing and hence the wavelength of the modulated phase. It is important to remember that the effective long-range interaction between domain walls is an entropic effect that arises because of fluctuations in the spin ordering at finite temperatures. We shall emphasize later that other mechanisms, for example long-range elastic or electronic interactions, can also result in modulated ordering [12,13].

I shall now discuss the relevance of the ANNNI model to several experimental systems. All the compounds considered can be usefully thought of as being built-up of structural units, magnetic spins or groups of one or more atoms, which can lie in two different orientations. Because the structural units correspond to two-state variables it is useful to map a given compound onto an Ising model. If the interactions are short-range, the first guess would be to consider couplings between nearest neighbour spins only. If, however, for some reason (for example, near a crossover between ferro- and antiferromagnetic ordering) the first neighbor interaction is small, then the second neighbor interaction may be important and, if it is negative, modulated structures can result.

The interactions J_0, J_1 and J_2 are phenomenological representations of the difference in energy between different orientations of first or second neighbor structural units. They are expected to vary with pressure, temperature and composition, moving the compound through the phase space of the ANNNI model and sampling different modulated phases.

2. BINARY ALLOYS

Some of the best examples of systems with modulated ordering which fit well into the ANNNI framework [14] are binary alloys like Cu_3Pd,[15] $TiAl_3$ [16] or CuAu [17]. Below the ordering temperature many binary alloys lock-in to a face-centered cubic structure where, taking $TiAl_3$ as an example, planes of Al alternate with mixed planes of Ti and Al along the (100) direction. Within the mixed planes the two atomic species are ordered with each Al being surrounded by four Ti as nearest neighbors as shown in Fig. 2. Modulation is introduced into the crystal structure through antiphase boundaries which correspond to a displacement of the Ti sublattice through [1/2,1/2,0]. Hence each face-centered cube of atoms can lie in one of two positions which can be represented by an Ising spin variable and a given arrangement of antiphase boundaries can be described using notation analogous to that introduced for the ANNNI model. For example, the phase in Figure 2 is the ⟨2⟩ structure.

Modulated phases have been seen in many different binary alloys with the wavelength of the stable phase varying with temperature and composition. The compounds which have been studied most thoroughly using electron diffraction and high resolution electron microscopy are $TiAl_3$ [16], Cu_3Pd [15] and Ag_3Mg [18]. In $TiAl_3$ Louiseau et al. [16] identified 15

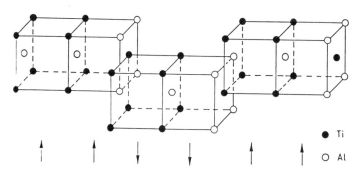

Fig. 2. Atomic configurations in the modulated phase, ⟨2⟩, in the binary alloy $TiAl_3$.

different long period structures containing 1- and 2-bands in specimens annealed at different temperatures. All the observed structures could be formed through structure combination branching processes and, if they were ordered as a function of increasing annealing temperature, they fell into the same sequence as the stable phases in the ANNNI model. Loiseau et al. [16] pointed out that there were often several different stacking sequences observed in the same crystal showing the sensitivity of the period to small differences in experimental conditions. Another interesting feature was the observation of jogs of the antiphase boundaries which increased in frequency with increasing temperature. These can be identified with fluctuations in the domain walls in the ANNNI model and are strong evidence that entropic effects are responsible for stabilizing the long period phases in TiAl3.

A new feature seen in Cu3Pd [15], which is discussed more fully in another paper in this volume [19], is a crossover with increasing concentration from long period structures with predominantly straight antiphase boundaries to phases where the boundaries are more wavy and not obviously pinned to the underlying lattice. This suggests that an increase in concentration corresponds to an increase in effective temperature (or, equivalently, a decrease in J_0) and that the compound crosses over from the low temperature region of the ANNNI-model phase diagram to higher temperatures where there is incommensurate ordering with the domain walls no longer pinned to the lattice [20]. Similar wavy antiphase boundaries have been observed in CuAu [17].

In Ag3Mg recent work [18] has led to the observation of $\langle 12^j \rangle$ with j=2,3...7,8,12 together with the mixed phases $\langle 12^2 12^3 \rangle$ and $\langle 12^3 12^4 \rangle$. Attempts to determine the temperature and concentration dependence of the period for this compound proved to be very difficult because of sluggish kinetics.

The conventional theory of the long period phases in binary alloys argues that incommensurate order can be stabilized by energy gained from the interaction between the Fermi surface and the new Brillouin Zone boundaries introduced by the periodic modulation [21,22]. These ideas explain the concentration dependence of the period of the modulated order but do not account for temperature effects nor predict lock-in to commensurate phases. The ANNNI picture allows the latter to be investigated but does not relate J_1 and J_2 to microscopic interactions. An amalgam of the two approaches, in which band theory calculations are used to estimate values for these interaction parameters and assess the effects of further neighbor interactions, would be very interesting.

3. CLASSICAL POLYTYPES

The classical polytypes, characterized by SiC and CdI_2, are comprised of close-packed layers of atoms which stack in varying sequences to form over 100 different long-period structures [23]. A useful way of distinguishing between the different phases, which emphasizes that one is dealing with a two-state system, is to use Zdhanov notation [24] which replaces the cyclic stacking sequences, AB, BC and CA, by + and anticyclic sequences, BA, CB and AC, by -. A phase

```
A   B   C   B   A   C   A   B   A   C   B   C   A   C   B
+   +   -   -   -   +   +   -   -   -   +   +   -   -         (3)
```

is then denoted ⟨23⟩ in a way similar to the ANNNI notation used above.

The long-period phases observed in SiC and CdI_2 [25] are strikingly reminiscent of those appearing in the phase diagram of the ANNNI model. In particular, in SiC, no 1-bands appear: 2- and 3-bands predominate and the occasional 4- and longer bands are stable. In CdI_2, on the other hand, 1- and 2-bands dominate the observed structures. Moreover, almost all the observed phases can be constructed from shorter wavelength phases through structure combination branching processes. Hence it is very tempting to identify SiC and CdI_2 as ANNNI systems with $J_1 > 0$ and $J_1 < 0$ respectively [26-29].

Previous theories of polytypism have assumed that the long-period structures are not equilibrium phases but result from growth around screw dislocations with the period of the structure reflecting the step height of the dislocation [25]. The growth theories are, however, not able to explain convincingly why only a very specific set of long-period phases are stable. Nevertheless, this is undoubtedly the mechanism through which many SiC crystals do grow. The resulting states are highly metastable because of the tiny free energy differences between the long period phases and the substantial atomic rearrangements needed to transform the material [30].

The importance of the ANNNI approach is that it shows that long--period structures can exist as stable phases even if only short-range atomic interactions are present. Hence, it may encourage experimentalists to try to construct phase diagrams and understand transformation mechanisms in the classical polytypes [30].

4. THE SPINELLOIDS

The spinelloids, AB_2O_4, where A and B represent cations such as Ni and Al, are one of the many polytypic minerals [31]. The basic structural unit for these compounds is an O cage containing cations which can lie in two different positions. The order within two-dimensional layers is normally ferromagnetic in one direction and antiferromagnetic in the other, but along the axial direction perpendicular to the layers several different stacking sequences have been observed.

Akaogi et al. [32] have published an experimental phase diagram for the spinelloids as a function of temperature and pressure. They observed that, although the phase boundaries showed no strong temperature dependence, as the pressure was increased, the phase sequence $\langle 3 \rangle$ - $\langle 2 \rangle$ - $\langle 12^2 \rangle$ - $\langle 12 \rangle$ - $\langle 1 \rangle$ became stable. This phase sequence corresponds exactly to that seen in Figure 1 as J_1 changes sign from positive to negative. The longer period phases that are not observed are expected to lie outside the resolution of the experiment.

Price et al. [33] have attempted to obtain quantitative values for J_1 and J_2 from models of the atomic interactions in the spinelloids. The results are model dependent but suggest $J_1 < 0$, $J_2/J_1 \sim 0.5$, $|J_3| < |J_2|$, $|J_4| < |J_2|$, consistent with using an ANNNI picture to represent these compounds.

5. CERIUM ANTIMONIDE

The best candidate for a magnetic system which shows ANNNI-like properties is the ferrimagnet cerium antimonide [4,34,35]. In this compound strong uniaxial spin anisotropy constrains the spins to point along [100]. Within the (100) planes the ordering is ferromagnetic with most planes having magnetization close to the saturation value pointing along or antiparallel to the axial direction. However, at first sight rather surprisingly, planes with zero net magnetization also appear. The phase diagram of cerium antimonide, together with the spin sequence in each phase is shown in Figure 3 [35].

A possible clue to the existence of the zero magnetization layers is that they lie on the boundaries of 1-bands. This corresponds to the position of the domain walls in the ANNNI model that can fluctuate most easily and suggests that one is seeing local fluctuations in the spin ordering that are fast on the time scale of the experiment. If this interpretation is correct, the zero-field phases can be identified as $\langle 12^k \rangle$, k=1-6 (but with $\langle 1212^2 \rangle$ replacing $\langle 12^2 \rangle$).

6. CONCLUSION

If the ANNNI model is to provide a useful representation of real systems, it is important to assess the effect of perturbations such as further neighbor interactions [36,37] and defects [38,39], which must inevitably be present, on the phase diagram. Adding an interaction between third neighbours in the axial direction leads to multiphase lines near which

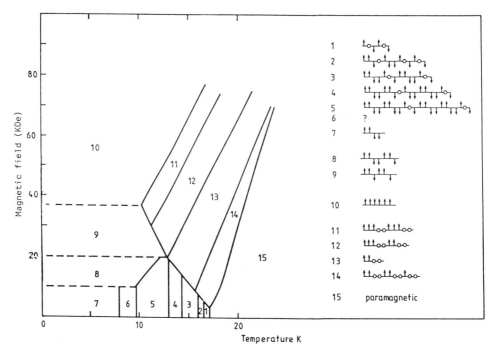

Fig. 3. Phase diagram of cerium antimonide (after Rossat-Mignod et al. [35]).

there are qualitatively similar sequences of long wavelength phases. However, there are quantitative differences with sequences like $\langle 23^k \rangle$ and $\langle 3^k 4 \rangle$, in which 3-bands predominate, being stable at the lowest temperatures [36,37].

It has been shown that the introduction of non-interacting annealed vacancies does not affect the phase sequences observed in the ANNNI model [38]. The vacancy concentration in each layer forms a wave which reflects the band sequence of the underlying spin structure. This behavior has been seen in experiments on thiourea [40]. Quenched defects, on the other hand, are thought to destroy phases of sufficiently long period at low temperatures, although it is likely that metastability will mask this effect in real systems [39].

It is important to emphasize two areas where more research is needed. Firstly, the J_i, i=0,1,2..., have been introduced as phenomenological parameters and an important test of the theory would be more calculations which attempt to calculate these from first principles [33]. Secondly, as was pointed out earlier, the ANNNI phases are stabilized by effective long-range interactions arising from entropic fluctuations of the domain walls [8,11]. Long-range interactions which result from elastic or electronic terms in the Hamiltonian can also stabilize long-period phases [12,13]. The dominant force will vary from compound to compound and it is of interest to determine the relevant physical mechanism in each case.

The experiments described above give convincing evidence for the applicability of the ANNNI model to natural phenomena. The model provides a mechanism through which polytypes can exist as equilibrium or highly metastable states and therefore challenges conventional growth theories and encourages experiments on the stability and kinetics of phase transformations in these compounds. It provides an explanation for commensurate modulated order in binary alloys and ferrimagnets and for the phase sequences observed and their dependence on temperature. The great advantage of such a phenomenological description is that the formalism is sufficiently simple that the effects of defects and changes in the Hamiltonian can be assessed.

REFERENCES

1. R.J. Elliott, Phys. Rev. 124, 346 (1961).
2. M.E. Fisher and W. Selke, Phys. Rev. Lett. 44, 1502 (1980).
3. M.E. Fisher and W. Selke, Phil. Trans. Roy. Soc. 302, 1 (1981).
4. J. von Boehm, and P. Bak, Phys. Rev. Lett. 42, 122 (1979).
5. P. Bak and J. von Boehm, Phys. Rev. B21, 5297 (1980).
6. P. M. Duxbury and W. Selke, J. Phys. A. Math. Gen. 16, L741 (1983).
7. W. Selke and P. M. Duxbury, Z. Phys. B57, 49 (1984).
8. A. Szpilka and M. E. Fisher, Phys. Rev. Lett. 57 1044 (1986)
9. M. H. Jensen and P. Bak, Phys. Rev. B27, 6853 (1983).
10. A. Aharony and P. Bak, Phys. Rev. B23, 4770 (1981).
11. J. Villain and M. Gordon, J. Phys. C. Solid St. Phys. 13, 3117 (1980).
12. R. Bruinsma and A. Zangwill, Phys. Rev. Lett. 55, 214 (1985).
13. A. Zangwill and R. Bruinsma, Comments on Condensed Matter Physics. In press.
14. D. de Fontaine and J. Kulik, Acta. Metall. 33, 145 (1985).
15. D. Broddin, G. van Tendeloo, J. van Landuyt, S. Amelinckx, R. Portier, M. Guymont and A. Loiseau. Phil. Mag. A54 395 (1986).
16. A. Loiseau, G. van Tendeloo, R. Portier and F. Ducastelle, J. de Physique, 46, 595 (1985).
17. M. Guymont and D. Gratias, Acta Cyst. A35, 181 (1979).
18. J. Kulik, S. Takeda and D. de Fontaine, Acta Metall. 35, 1137 (1987). preprint.
19. G. van Tendeloo, this volume.
20. D. de Fontaine, A. Finel, S. Takeda and J. Kulik in "Noble Metals Symposium," New York AIME meeting, eds. Massalski, Bennett, Pearson and Chang (1985).
21. H. Saro and R. S. Toth, Phys. Rev. 127, 469 (1962).
22. B. L. Gyorffy and G. M. Stocks, Phys. Rev. Lett. 50, 374 (1983).
23. A. R. Verma and P. Krishna, "Polymorphism and Polytypism in Crystals," Wiley, New York.
24. G. S. Zdhanov and Z. Minervina, J. Phys. (Moscow) 9, 151 (1945).
25. D. Pandey and P. Krishna, J. Cyst. Growth Charact. 7, 213 (1984).
26. S. Ramasesha, Pramana 23, 745 (1984).
27. J. Smith, J. M. Yeomans and V. Heine in Proc. N.A.T.O.A.S.I. on "Modulated Structure Materials," ed. T. Tsakalakos (Dordrecht, Nijhoff), p. 23.
28. G. D. Price and J. M. Yeomans, Acta Cryst. B40, 448 (1984).
29. J. M. Yeomans and G. D. Price, Bull. Min. 109, 3 (1986).

30. N. W. Jepps and T. F. Page, J. Cryst. Growth Charact. $\underline{7}$, 259 (1984).

31. H. Horiuchi, K. Horioka and N. Morimoto, J. Mineral. Soc. Japan $\underline{2}$, 253 (1980).

32. M. Akaogi, S. Akimoto, K. Horioka, K. Takahashi and H. Horiuchi, J. Solid St. Chem. $\underline{44}$, 257 (1982).

33. G. D. Price, S. C. Parker and J. M. Yeomans, Acta Cryst. B$\underline{41}$, 231 (1985).

34. G. Meier, P. Fischer, W. Halg, B. Lebech, B. D. Rainford and O. Vogt, J. Phys. C. Solid St. Phys. $\underline{11}$, 1173 (1978).

35. J. Rossat-Mignod, P. Burlet, H. Bartholin, O. Vogt and R. Langier, J. Phys. C. Solid St. Phys. $\underline{13}$, 6381 (1980).

36. W. Selke, M. N. Barreto and J. M. Yeomans, J. Phys. C. Solid. St. Phys.$\underline{18}$, L393 (1985).

37. M. N. Barreto and J. M. Yeomans, Physica $\underline{134A}$, 84 (1985).

38. H. Roeder and J. M. Yeoman, J. Phys. C. Solid St. Phys. $\underline{18}$, L163 (1985).

39. P. Bak, S. Coppersmith, Y. Shapir, S. Fishman and J. M. Yeomans, J. Phys. C. Solid St. Phys. $\underline{18}$, 3911 (1985).

40. J. P. Jamet and P. Lederer, J. Physique Lett. $\underline{44}$, L257 (1983).

TWO-DIMENSIONAL MODELS OF COMMENSURATE-INCOMMENSURATE PHASE TRANSITIONS

Paul D. Beale

Department of Physics, Campus Box 390
University of Colorado
Boulder, CO 80309

This paper presents a brief review of the phase diagram and phase transition properties of the two-dimensional Axial Next Nearest Neighbor Ising (ANNNI) model. The model exhibits a two-dimensional commensurate-incommensurate phase transition to a striped incommensurate phase. The phenomenological finite-size scaling method is able to correctly locate and identify all the phase transition lines in the phase diagram.

The ANNNI model is a very simple short-range spin model which displays a continuous commensurate-incommensurate phase transition line in two dimensions. The model is described by the Hamiltonian [1]

$$H = -J_1 \sum_{(ij)}^{nn} s_i s_j + J_2 \sum_{<ij>}^{nnn_x} s_1 s_j \tag{1}$$

where the spins take on the Ising values ± 1 and $J_1 > 0$, $J_2 > 0$ are the nearest neighbor and next-nearest neighbor coupling constants (see Fig. 1). The first sum is over all nearest neighbor pairs on a two-dimensional square lattice whereas the second sum is over next-nearest neighbor pairs in the x (axial) direction. The ground state of the model depends on the ratio $\kappa = J_2/J_1$. If $\kappa < 1/2$ the ground state configuration of spins parallel to the axial direction is ferromagnetic ($\uparrow\uparrow\uparrow\uparrow\uparrow\uparrow\uparrow\uparrow\uparrow\uparrow\uparrow\uparrow$) whereas for $\kappa > 1/2$ the in-ground state is antiphase ($\uparrow\uparrow\downarrow\downarrow\uparrow\uparrow\downarrow\downarrow\uparrow\uparrow\downarrow\downarrow$). At $\kappa \sim 1/2$ the ground state is infinitely degenerate. The commensurate phases ($\uparrow\uparrow\uparrow\uparrow\uparrow\uparrow\uparrow\uparrow\uparrow\uparrow\uparrow\uparrow$) and ($\uparrow\uparrow\downarrow\downarrow\uparrow\uparrow\downarrow\downarrow\uparrow\uparrow\downarrow\downarrow$) can be described by the wavevector of modulation, q=1/(wavelength). Hence the ferromagnetic phase has a wavevector q=0 and the two up, two down phase is denoted q=1/4.

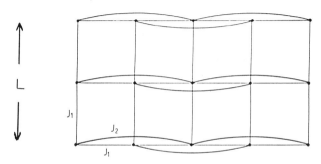

Fig. 1. Interactions in the ANNNI model. The nearest-neighbor couplings
are ferromagnetic and the next-nearest-neighbor couplings are
antiferromagnetic.

In three dimensions at nonzero temperature, T, mean field theory calculations [2] show that an infinity of stable commensurate phases grow out of
the multiphase point $\kappa=\frac{1}{2}$, T=0. (See Julia Yeoman's article in this proceedings for details.) However, the larger thermal fluctuations present in two
dimensions destabilize all of the commensurate phases except for q=0 and
q=1/4.

The two-dimensional phase diagram for the ANNNI model is shown in Fig.
2. The phases labeled q=0 and q=1/4 are the commensurate phases. They
exhibit long-range order of the spin field. The correlation function
$G(x)=\langle s(0,y)s(x,y)\rangle$ tends to a constant value as $x\to\infty$ in the q=0 phase.
(The brackets $\langle.\rangle$ denote a thermal average.) The correlation function in
the q=1/4 commensurate phase behaves like $G(x)\sim f(2\pi x/4)$ as $x\to\infty$. The
function f is periodic with period 1.

The two paramagnetic phases P_0,P_1 exhibit short range order. They
differ only in that in the P_1 phase the correlation function is modulated
with wavevector 0<q<1/4 while the correlation function is monotonic in the
P_0 phase. In each case the correlation function decays exponentially with
distance with a characteristic parallel correlation length, $\xi^{\parallel}$. The correlation length is finite throughout the paramagnetic phases and along
the disorder line which separates the P_0 phase from the P_1 phase [3]. The
disorder line is the line where the wavevector of modulation tends to zero
as κ is decreased in the P_1 phase.

The incommensurate phase, I, which lies between P_1 and the q=1/4 commensurate phase displays quasi-long range order. This means that the correlation function decays <u>algebraically</u> with distance along the x-axis. The
correlation length is infinite throughout the incommensurate phase. In
addition the correlation function is modulated with wavevector $(1-(2)^{-1/2})$
< q < 1/4.

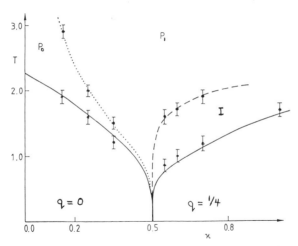

Fig. 2. Phase diagram of the two-dimensional ANNNI model. The phases labeled P_0 and P_1 are paramagnetic phases. The phases labelled q=0 and q=1/4 are ordered commensurate phases and I is a quasi-long-range ordered incommensurate phase. The error bars are estimates of the transition temperatures from the finite-size scaling analysis. The solid lines are the melting temperatures of the commensurate phases from surface free energy calculations. The dotted and dashed lines are guides to the eye.

$$G(x) \sim \frac{f(2\pi qx)}{|x|^\eta}, \text{ as } x \to \infty \quad . \tag{2}$$

The function f is again periodic with period 1 and the power law decay exponent η and the modulation wavevector q are continuous functions of T and κ.

The phase transitions between the different phases fall into very different universality classes. The $(q=0)$-P_0 commensurate-paramagnetic transition is continuous and falls within the two-dimensional Ising universality class. The $(q=1/4)$-I commensurate-incommensurate phase transition line is more interesting. Several critical effects are going on simultaneously as the transition line is approached. The wavevector deviation from the commensurate value, $\delta q = q-1/4$, tends to zero with a power law in the distance from the commensurate phase

$$\delta q \sim |t|^{\overline{\beta}} \tag{3}$$

The reduced temperature t is $t=(T-T_c)/T_c$. The exponent is $\overline{\beta} = 1/2$ for striped commensurate-incommensurate phase transitions of this type [5].

Within the q=1/4 commensurate phase near the commensurate-incommensurate transition line the system has two different correlation lengths, the usual one parallel to the axial direction, $\xi^{\parallel}$ and a different one perpendicular to the axial direction, $\xi^\perp$. Near the commensurate-incommensurate transition line within the commensurate phase the parallel and perpendicular correlation lengths diverge with the power law behaviors:

$$\xi^\| \sim |t|^{\nu_\|} \tag{4}$$

$$\xi^\perp \sim |t|^{\nu_\perp} \tag{5}$$

where t is the reduced temperature. The parallel correlation length exponent takes on the values $\nu_\| = 1/2$. The exponent $\nu_\perp$ is thought to be unity. Therefore, the phase transition is very anisotropic.

The incommensurate-paramagnetic transition falls within the Kosterlitz-Thouless universality class [4]. The correlation length in the paramagnetic P_1 phase diverges at the transition line with the behavior

$$\xi \sim \exp \left[\frac{\alpha}{|t|^{\frac{1}{2}}} \right] \tag{6}$$

as the reduced temperature t goes to zero. The transition is isotropic since the parallel and perpendicular correlation lengths diverge in the same way.

Now I would like to describe a single theoretical calculation [6] which is able to locate all the phase transition lines in the ANNNI model. The technique is based on the phenomenological finite-size scaling method developed by Nightingale [7]. For the two-dimensional strip geometry pictured in Fig. 1 in which the lattice has a width L, the system will have a phase transition only in the limit L→∞. However, a signature of the phase transition is observable for finite, even relatively small, L. The parallel correlation length and the wavevector of modulation can be determined numerically for finite width strips by finding the leading eigenvalues of the transfer matrix of the model. The finite-size scaling hypothesis [8] gives the following predictions for the functional dependence for the parallel correlation length and wavevector deviation from the commensurate value:

$$\delta q_L \approx t^{\bar{\beta}} \tilde{Q}(L/\xi^\perp) \approx L^{-\bar{\beta}/\nu_\perp} Q(L^{1/\nu_\perp} t), \tag{7}$$

$$\xi^\|_L \approx t^{-\nu_\|} \tilde{S}(L/\xi^\perp) \approx L^{\nu_\|/\nu_\perp} S(L^{1/\nu_\perp} t). \tag{8}$$

The functions Q and S are universal scaling functions. We can then locate

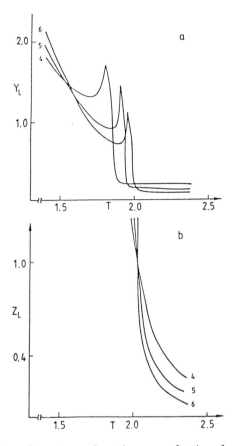

Fig. 3. The scaling functions for the correlation length and wave-vector for $\kappa = 0.25$ and L=4,5,6. The temperature T is in units of J_1/k_B.

the phase transition lines by defining the following logarithmic derivatives:

$$Y_L = \frac{d(\log(\xi_L^{\parallel}))}{d(\log(L))} \approx \frac{\nu_{\parallel}}{\nu_{\perp}} + a\, L^{1/\nu_{\perp}} t + \dots \, , \qquad (9)$$

$$Z_L = \frac{-d(\log(\delta q_L))}{d(\log(L))} \approx \frac{\bar{\beta}}{\nu_{\perp}} + b\, L^{1/\nu_{\perp}} t + \dots, \text{ as } t \to 0 \qquad (10)$$

Notice that for reduced temperature t=0 both Y_L and Z_L are asymptotically independent of L. The correlation length scaling function Y_L is independent of L when the correlation length is infinite in the L→∞ limit. The wave-vector scaling function Z_L is independent of L when $\delta q \to 0$ in the L→∞ limit.

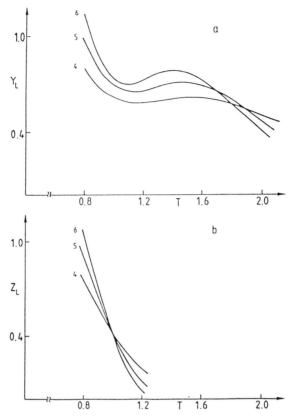

Fig. 4. The scaling functions for the w correlation length and wavevector
for κ=0.60 and L=4,5,6.

To illustrate this, the scaling functions are shown as functions of T
and L at fixed κ in Figs. 3 and 4. Figure 3 displays a cut through the
phase diagram of Fig. 2 made at κ=0.25. This cut intersects the commensurate-paramagnetic transition line and the disorder line. In Fig. 4 the cut
is made at κ=0.70 and intersects the commensurate-incommensurate transition
line and the incommensurate-paramagnetic transition line. Note that the behavior of the scaling functions is qualitatively different in the two figures.
In Fig. 3 the wavevector scaling function is independent of L at a higher
temperature than the temperature where the correlation length scaling function is independent of L. This indicates that the wavevector becomes zero
before the correlation length diverges as temperature is reduced. This
is the correct signature of a disorder line lying above the commensurate
phase. In Fig. 4 the correlation length scaling function is independent of
L at a higher temperature than the point where the wavevector scaling function is L-independent. This means in the L→∞ limit that the correlation
length has diverged at a higher temperature than the temperature where the
wavevector becomes commensurate. This is the correct signature for a phase

60

with quasi-long range order and incommensurate modulation lying above the commensurate q=1/4 phase. The phase diagram that results from making a number of fixed κ scans is shown in Fig. 2. See reference 6 for more details of the finite-size scaling method as applied to this model.

There are a number of other simple two-dimensional spin models which exhibit commensurate-incommensurate transitions. However, space does not permit me to elaborate here. The interested reader can consult reference 9.

In conclusion, simple spin models which exhibit commensurate-incommensurate transitions in two dimensions can be examined by a variety of theoretical techniques. For example, the finite-size scaling method is able to locate all of the phase transition lines in the ANNNI model. Commensurate-incommensurate transitions of the type found in these models have been observed experimentally. [10] The theoretical work on these models gives a number of predictions for universal quantities, such as the critical exponents which can be tested by the experimental work.

REFERENCES

1. R. J. Elliot, Phys. Rev. 124, 346 (1961).
2. P. Bak and J. von Boehm, Phys. Rev. B 21, 5287 (1980).
3. I. Pechel and V. J. Emery, Z. Phys. B 43, 241 (1981).
4. J. Villain and P. Bak, J. Phys. (Paris) 42, 657 (1980).
5. V. L. Prokrovski and A. L. Talapov, Zh. Eksp. Teor. Fiz. 78, 269 (1980).
6. P. D. Beale, P. M. Duxbury and J. Yeomans, Phys. Rev. B 31, 7166 (1985).
7. P. Nightingale, J. Appl. Phys. 53, 7927 (1982).
8. M. N. Barber, in "Phase Transitions and Critical Phenomena," vol. 7, C. Domb and J. Lebowitz (eds.) (Academic, London, 1984).
9. P. Bak, Rep. Prog. Phys. 45, 578 (1982).
10. S. G. J. Mochrie, Bull. Am. Phys. 30, 345 (1985).

GROWTH KINETICS IN A FRUSTRATED SYSTEM: THE QUENCHED AXIAL NEXT-NEAREST-
NEIGHBOR ISING MODEL

D. J. Srolovitz and G. N. Hassold

Los Alamos National Laboratory
Los Alamos, New Mexico 87545

Abstract   We have investigated the influence of the frustration
parameter, κ , in Monte Carlo quench simulations of the axial next-
nearest neighbor Ising (ANNNI) model. Quenches were performed from
the high temperature paramagnetic state into both the ferromagnetic
and modulated phase fields. For $0<\kappa<1$, the quenches lead to the formation of a glassy phase. For large κ, domains of the modulated phase
form and coarsen by the elimination of domain walls which meet at
four-fold vertices. Quenches were also performed by discontinuous
change of κ from the ferromagnetic regime to values in the modulated
phase field at fixed temperature. For $1<\kappa<2$, the modulated phase
forms by nucleation and growth. At larger values of κ and long times
in the nucleation and growth regime, the defected modulated structure
coarsens by the motion and annihilation of domain walls and vertices.

1.   INTRODUCTION

The axial next-nearest-neighbor Ising (ANNNI) model is among the simplest frustrated systems and as such provides a ready vehicle for the study
of incommensurate and modulated phases. The latter phenomenon is found in
a variety of physical systems and is discussed in detail in the paper by
Yeomans [1] in this volume. Recent studies [2] have investigated the kinetics of quenches in the ANNNI model from the paramagnetic (P) state. We
extend these studies by considering the role of the frustration parameter
in the growth kinetics of quenches from the paramagnetic state to the modulated and ferromagnetic (F) states. Additionally, we have examined the
kinetics of the growth of the modulated phase from the F state following a
discontinuous change in the frustration parameter. The study of the

kinetics of these phase changes following a quench are particularly important in the ANNNI model since the structure of such modulated materials as
silicon carbide (which has been described in terms of the ANNNI model [3])
are known [4] to be very sensitive to the conditions under which it was
synthesized (i.e., how strongly it was quenched).

The two-dimensional ANNNI model Hamiltonian is

$$H = -(1/2) \sum_{i,j} J_0 S_{i,j} S_{i,j\pm1} + J_1 S_{i,j} S_{i\pm1,j} + J_2 S_{i,j} S_{i\pm2,j} \tag{1}$$

We set $J_0 = J_1$, and define $\kappa = -J_2/J_1 \underline{>} 0$. κ is often referred to as the frustration parameter, as it governs the relative strength of the competing
ferromagnetic nearest-neighbor interactions and the anti-ferromagnetic
next-nearest-neighbor interactions along the x or i axis. The two-dimensional ANNNI phase diagram shown in Figure 1 has been confirmed by many
methods [5]. At high temperatures the system is paramagnetic for all κ.
For $\kappa > (1/2)$ the low temperature state is the $\langle 2 \rangle$ phase: stripes of width
two, oriented perpendicular to the frustration axis, and having the base
pattern $\dots \uparrow\uparrow\downarrow\downarrow\uparrow\uparrow\downarrow\downarrow\uparrow\uparrow\downarrow\downarrow \dots$. For the same range of κ, there exists an
incommensurate phase between the low temperature $\langle 2 \rangle$ and the high tempera-
ture P phases. The low temperature phase for $\kappa < (1/2)$ is ferromagnetic.

We have performed a series of quenches into the $\langle 2 \rangle$ and F phases using
standard Monte Carlo techniques with non-conserved (Glauber) dynamics. The
final temperature in all of the simulations was $kT/J_0 = 0.04$, which is only
slightly above zero. A large lattice size was deliberately chosen
(200X200) with periodic boundary conditions, so as to diminish the influence of finite size effects. However, only one or two simulations were
performed for each set of conditions. The transition probability, W, was
used for the Monte Carlo transitions

$$W(\delta E) = (1/2) [\tanh(\delta E/2kT) - 1] \tag{2}$$

where δE is the difference in energy between the trial and initial spin
orientations. The lattice energy was measured every 20 Monte Carlo steps
(MCS), where an MCS is defined as N (=200x200=40,000) spin flip attempts.
The configuration of the lattice was recorded every 50 MCS.1

2. QUENCHES FROM THE PARAMAGNETIC STATE

A. Glass Formation

In quenches from the P state, growth of the equilibrium $\langle 2 \rangle$ phase is
entirely suppressed for $(1/2) < \kappa < 1$ in favor of a metastable, glassy phase.

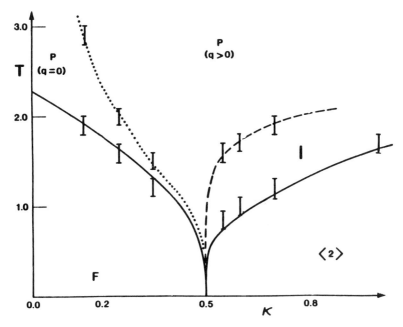

Fig. 1.   Two-dimensional ANNNI phase diagram (from Ref. 5), showing some of
the quench paths studied. I, P, F, and ⟨2⟩ correspond to the
incommensurate, paramagnetic, ferromagnetic, and modulated phase
fields, respectively.

The temporal evolution of the spin system following a quench from the
P state (T≫T_c) to T=0.04J_0/k at κ=0.8 is shown in Fig. 2. Clearly the ⟨2⟩
phase never forms during this 2000 MCS simulation. While no long range
order is observed, short range order is certainly present. The lack of
long range order and the presence of short range order (differing from
that of the equilibrium phase) is one of the signatures of a glassy phase.
The suppression of the ordered phase is due to the fact that the stripes,
characteristic of the ⟨2⟩ phase, once nucleated are not able to grow. In
the following diagram one of the two spins above any of the nucleated down
spin stripes must be flipped in order for that stripe to grow:

```
          . . . . . . . . . . . . . . . . . . .
          . . . . . . . . . . . . . . . . . . .
          . . . . . . . . . . . . . . . . . . .
     · · · ↑↑↑↑↑↑↑↑↑↑↑↑↑↑ · · ·
     · · · ↑↑↑↑↑↑↑↑↑↑↑↑↑↑ · · ·
     · · · ↑↑↑↑↑↑↑↑↑↑↑↑↑↑ · · ·
     · · · ↑↑↑↑↑↑↑↑↑↑↑↑↑↑ · · ·
     · · · ↑↑↓↓↑↑↓↑↓↑↑↑↓↓↑ · · ·
     · · · ↑↑↓↓↑↑↓↑↓↑↑↑↓↓↑ · · ·
     · · · ↑↑↓↓↑↑↓↑↓↑↑↑↓↓↑ · · ·
     · · · ↑↑↓↓↑↑↓↑↓↑↑↑↓↓↑ · · ·
          . . . . . . . . . . . . . . . . . . .
          . . . . . . . . . . . . . . . . . . .
          . . . . . . . . . . . . . . . . . . .
```

50MCS   200MCS   2000MCS

Fig. 2. Lattice configuration at $\kappa = 0.8$ subsequent to a quench to
T=0.04J$_0$/k.   The shaded and unshaded regions correspond
to spin up and down, respectively.

Such a spin flip entails an energy cost $\delta E=2(1-\kappa)$.   Thus for $\kappa<1$, $\delta E>0$ and
growth cannot be expected at low temperatures.

Quenches from the high temperature paramagnetic phase into the low
temperature ferromagnetic region of the phase diagram, $0<\kappa<(\frac{1}{2})$, produce
a glass phase which is in every way indistinguishable from the glass
phase found for $(1/2)<\kappa<1$.   The time evolution of such a quench is
shown in Fig. 3 for $\kappa=0.2$.   A spin configuration argument similar to that
above indicates that $\kappa=0$ delimits the lower end of the glass formation
regime.   Another indication of the presence of the glassy phase may be
found in the time evolution of the system energy.   Since the deviation of
the energy of the system from that of the ground state (ΔE) measures how
far out of equilibrium the system is, its time evolution indicates how
quickly the system is approaching its ground state.   In Fig. 4 we plot ΔE

50MCS   200MCS   2000MCS

Fig 3. Lattice configurations at $\kappa = 0.2$ subsequent to a quench to
T=-0.04J$_0$/k.

66

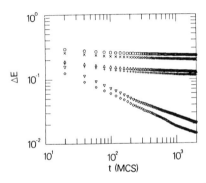

Fig. 4. The time dependence of the excess energy, ΔE, subsequent to a
quench to T=0.04J_0/k. The diamonds, circles, triangles, +'s,
x's, and inverted triangles correspond to κ= 0, 0.2, 0.4, 0.6,
0.8, and 1.0, respectively.

against time following quenches from the high temperature paramagnetic state
to T=0.04J_0/k for κ=0, 0.2, 0.4, 0.6, 0.8, and 1.0. Clearly, for 0<κ<1
ΔE is decaying in a sub-power law manner instead of the relatively fast
$t^{-1/2}$ manner observed [6] in the quenched Ising and other spin models. The
slow decrease in ΔE with time is attributable to short range ordering.
The nonzero values of ΔE to which these curves appear to be asymptotic
indicate that the system is approaching a metastable state.

B. Equilibrium Phase Formation

Figure 5 shows the time evolution of the spin configuration following
a quench from high temperature (T>>T_c) to T=0.04J_0/k for κ=3.0, i.e., the
<2> phase. Here we see that the striped pattern characteristic of the <2>
phase forms rapidly. However, since any of the four elements of the primi-
tive striped pattern (↑↑↓↓) can rest on any given site, there are four
degenerate states. Domains corresponding to these states are separated
from one another by domain walls. The domain patterns observed do bear
resemblance to those of a four-state clock model [7]. There is, however,
a significant difference. Simple calculations reveal that, for κ<1,
vertically oriented domain walls are more energetic than horizontal ones.
The energy density of vertical walls increases linearly with κ, while that
of the horizontal walls is independent of κ. Thus it is not surprising
that, at long times, the only domain walls in the system are horizontal.
The vertices at which four domain walls meet may be characterized in terms
of a vorticity or Burgers vector. There are two types of vertices in the
system (with equal density) corresponding to clockwise and counter-clock-
wise circulation. These two types of vertices may annihilate only on each
other. These two types of vertices may also be thought of as edge

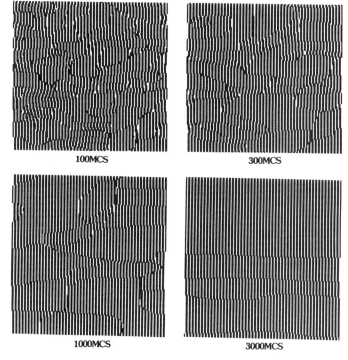

Fig. 5.   Lattice configurations at $\kappa=3.0$ subsequent to a quench to
T=0.04J_0/k.

dislocations with anti-parallel Burgers vectors.   Such a description is
useful in considering Fig. 5 in that each vertex corresponds to an extra
half-stripe of the <2> pattern which is pointing either up or down.

Although for $\kappa \geqslant 1$ the <2> phase does form it is, at least initially,
defected.   These defects are simply the domain walls which are described
above.   The rate at which these domain walls or, equivalently, vertices
anneal out is indicated in Fig. 4.   Here we see that for $\kappa=1$ the system
approaches equilibrium as $t^{-1/2}$.   This exponent is the same as for domain
growth in the simple Ising model [6] ($\kappa=0$).   In the simple Ising model the
correlation length grows as $t^{1/2}$ following the quench.   However, since
vertical walls and horizontal walls have different energies such a simple
scaling need not be found here.   In Fig. 6 we plot ΔE versus time for
quenches into the <2> state for four different values of κ.   Although the
anisotropy is increasing with increasing κ, all four curves remain parallel with a slope of approximately -(1/2).   This indicates that the system
as a whole is still behaving very much like the simple or isotropic Ising
model. The drop in these curves at approximately 3000 MCS is due to the
disappearance of all domain wall curvature when all of the vertices have
annealed out of the system, leaving only horizontal walls.   So while the

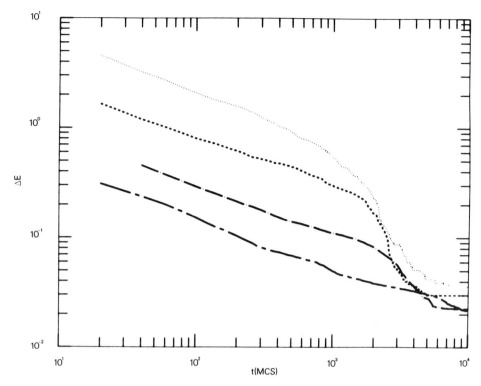

Fig. 6.   The time dependence of the excess energy, ΔE, subsequent to a quench to $T=0.04J_0/k$. The dotted, dashed, chain-dotted, and chain-dashed curves correspond to $\kappa = 30$, 10, 3, and 1, respectively.

energy is decaying uniformly as $t^{-1/2}$, the correlation length is indeed growing at different rates (although presumably still as $t^{1/2}$) in the axial and non-axial directions. This is in agreement with the observations of Kaski et al. [2].

3.   QUENCHES FROM THE FERROMAGNETIC STATE TO THE <2> STATE

Quenches from the F state to the <2> state at $T=0.04J_0/k$ were performed for $\kappa = 1.0$, 1.3, 1.5, 1.6, 1.7, 1.8, 1.9, and 2.0. Only for $\kappa \geqslant 1.7$ was the <2> phase observed during the simulation (2000 MCS). A simple energetic argument again establishes a critical value of κ for low temperature quenches. The energy cost involved in flipping a spin in a ferromagnetic environment is proportional to $(2-\kappa)$. For $T=0$ one expects the <2> phase to be formed only for $\kappa \geqslant 2$. However for $T \geqslant 0$, the <2> phase could be formed by nucleation and growth. The probability of nucleating the <2> phase increases with increasing κ or T.

Figure 7 shows a series of lattice configurations in a quench for the F state to the <2> state at $\kappa=1.7$. As expected from simple nucleation theory, it takes a relatively long time for the first <2> state nucleus to

appear. Following that, growth is rapid along the stripe (recall that
for $\kappa \geq 1$, growth of a stripe is energetically favorable). After a single
stripe has grown, it becomes necessary to nucleate additional stripes three
lattice spacings away. This is aided by the fact that the first stripe is
not completely straight and hence lowers the activation barrier for the
nucleation of a neighboring stripe. This leads to fast growth of a <2>
phase nucleus along the stripes and a much slower growth in the direction
perpendicular to the stripes.

By 500 MCS, the entire lattice has been filled with the defected <2>
phase. From this time on, the system attempts to "heal" itself by the
elimination of the defects (domain walls). Due to the relatively rapid
growth along the stripe, there are fewer horizontal domain walls than in
the quenches from the P state. The horizontal domain walls that do appear

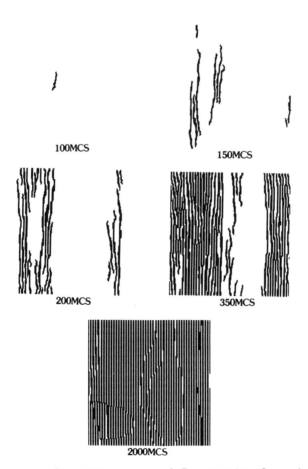

<table>
<tr><td>100MCS</td><td>150MCS</td></tr>
<tr><td>200MCS</td><td>350MCS</td></tr>
<tr><td colspan="2">2000MCS</td></tr>
</table>

Fig. 7. Lattice configurations at $\kappa = 1.7$, starting from the ferromagnetic
state at 0 MCS.

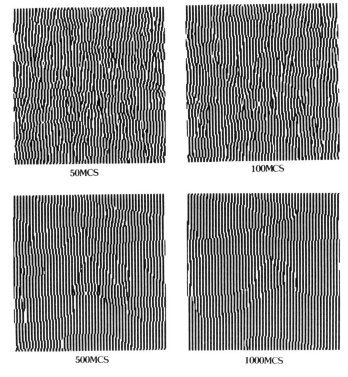

50MCS

100MCS

500MCS

1000MCS

Fig. 8. Lattice configurations at κ = 2.0, starting from the ferromagnetic
state at 0 MCS.

result mainly from positional mismatch as the two ends of a stripe grow
together (across the periodic boundary counditions).

As κ nears the critical value of 2.0, stripes nucleate with a much
greater frequency and at a smaller average distance from each other.
Figure 8 shows a temporal sequence of lattice configurations in a quench
from the F phase up to κ=2. These configurations are similar to the P to
⟨2⟩ phase quenches, with the same preponderance of horizontal domain walls.

The time dependence of ΔE (Fig. 9) exhibits an extreme sensitivity to
the frustration parameter κ. For simulations at or very near the critical
value (κ = 1.9, 2.0), behavior similar to the quenches from the P state is
observed. For quenches with κ further from the critical value (κ =1.7,
1.8), the early time behavior is dominated by the nucleation and growth of
stripes (for κ = 1.6 no nucleation was observed over the entire 2000 MCS
simulation). For κ = 1.8 this process appears complete by 200 MCS, while
for κ = 1.7 this process is not complete until approximately 600 MCS. Once
the nucleation and growth phases are complete, ΔE appears to scale as
$t^{-1/2}$ for all values of κ.

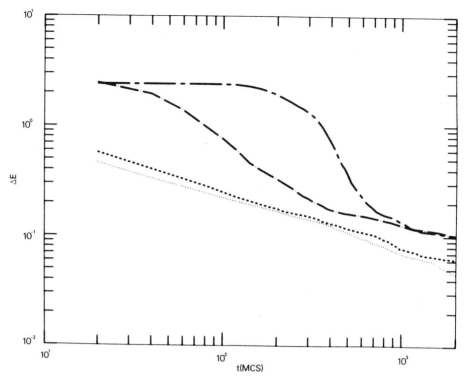

Fig. 9. The time dependence of the excess energy, ΔE, starting from the ferromagnetic state at 0 MCS. The dotted, dashed, chain-dotted, and chain-dashed curves correspond to quenches in κ to $\kappa = 2.0$, 1.9, 1.8, and 1.7, respectively.

4. CONCLUSIONS

For a large range of values of the frustration parameter κ quenches from the high temperature paramagnetic state to low temperature lead to the formation of a glass-like state. Unlike in other systems such as the antiferromagnetic Ising model on an FCC lattice, the disordered or glassy state in the ANNNI model is not the lowest energy state of the system. Furthermore, the frustration that leads to this glassy state does not have its source in random terms in the Hamiltonian of the system as does the random field Ising model. The glassy state we observe is strictly a result of competing interactions in the Hamiltonian which lead to a situation in which the phase space of the system has very many, truly metastable states. To our knowledge this is the first observation of a true glassy (metastable, disordered) phase in a spin model which contains no random terms in the Hamiltonian.

While metastabilities are observed in quenches from high to low temperature, they are also observed on quenches or rapid changes in the frustration parameter. A system originally in the ferromagnetic state upon having the frustration parameter raised into the modulated phase region of the phase diagram remains in the ferromagnetic state over certain ranges of the frustration parameter. In some cases, this is due to the difficulty involved in growing the modulated phase once nucleated. In others, it is due to the difficulty in nucleating the modulated phase in the first place.

The existence of the variety of metastabilities found in this model bears on the question of whether experimentally observed frustrated systems are truly in equilibrium. For example, it is known [4] that varying the method by which certain polytypic materials (such as SiC) are produced affects the crystal structure in which they are found. Barring kinetic factors, the structure of such materials should only be determined by the instantaneous value of the temperature and stoichiometry. While the present model is two-dimensional, most materials of interest are three-dimensional. Since the phase diagram of the ANNNI model becomes more complicated and shows more modulated phases with increasing dimensionality, we expect metastability to be even more common in three dimensions than in two.

5. REFERENCES

1. J. Yeomans, this volume.
2. K. Kaski, T. Ala-Nissila, and J. D. Gunton, Phys. Rev. B31, 310 (1985); T. Ala-Nissila, J. D. Gunton, and K. Kaski, Phys. Rev. B33, 11 (1986).
3. G. D. Price and J. Yeomans, Acta Cryst. B40, 448 (1984).
4. S. S. Sinozaki and H. Sato, J. Am. Ceram. Soc. 61, 425 (1978).
5. P. D. Beale, P. M. Duxbury, and J. Yeomans, Phys. Rev. B31, 11 (1985) and references therein.
6. I. M. Lifschitz, Zh. Eksp. Teor. Fiz. 42, 1354 (1962); M. K. Phani, J. L. Lebowitz, M. H. Kalos, and O. Penrose, Phys. Rev. Lett. 45, 366 (1980).

ELECTRON MICROSCOPY OF INCOMMENSURATE STRUCTURES

G. Van Tendeloo, J. Landuyt and S. Amelinckx

Universiteit Antwerpen, R.U.C.A.
Groenenborgerlaan 171
B-2020 Antwerp, Belgium

Abstract   Some of the different electron microscopy techniques
useful for the study of incommensurate modulated structures are
briefly discussed and applied to two different systems: the incommensurate phase in barium sodium niobate and the long period superstructures in Cu_3Pd.

1.   INTRODUCTION

Electron microscopy is a unique technique; it is able to combine
information from reciprocal space through the electron diffraction pattern
with information from direct space through electron micrographic images.
The real space information can be obtained using different imaging modes:

a) if only a single reflection is selected in the back focal plane (i.e.,
the diffraction pattern) and allowed to contribute to the image formation process, we obtain a conventional "bright field" or "dark field"
image depending on whether the selected reflection is the transmitted
beam or one of the diffracted beams.   This technique is particularly
well suited to study crystal defects up to a resolution of 1 nm [1,2]
but imaging on an atomic scale is excluded since no interference
between different beams is possible.

b) if several reflections are allowed to contribute to the image formation, one obtains an interference image or usually called a high resolution image.   If the objective aperture chosen is large enough, the
obtainable resolution is now essentially limited by the instrument.
For recent microscopes this resolution limit is around 0.2 nm.   More
information on the technique can be found in ref. [4,5].

The applications of electron microscopy are mainly limited by the material itself.

a) Because of absorption and other inelastic processes the thickness of the material is limited to a few hundred nm for conventional microscopy and to only 20-30 nm for high resolution electron microscopy. Standard methods to obtain such thin materials are found in [6].

b) The material should resist hundred (or more) kV electron irradiation; this makes it e.g. impossible to study the $NaNO_2$ incommensurate phase and a number of other incommensurate insulators.

Selected area electron diffraction patterns can be obtained from regions as small as 500 nm (with convergent beam electron diffraction even from areas smaller than 2 nm) and thanks to the strong electron-specimen interaction (compared to X-rays) it is possible to detect weak periodicities associated with slight deformations or modulations. Such periodic modulations will introduce extra reflections, i.e., satellites, of which the positions are related to the basic reflections of the unmodulated structure. From the positions of these satellite reflections one can easily deduce the direction and the wavelength of the modulation. Whenever the structure can be described as an interface modulated structure characterized by a displacement vector $\bar{R}$, this displacement can be determined from the geometry of the diffraction pattern only. Consider Fig. 1 which represents a direct high resolution image of a long period superstructure in $Cu_{3+x}Pd$, together with its diffraction pattern oriented in the correct way. The basic underlying structure is the cubic LI_2 (Cu_3Au-type) structure. The modulation with average half wavelength M gives rise in the corresponding diffraction pattern to a splitting of some of the LI_2-type reflections. This splitting (see e.g., 011) is inversely proportional to the average period M and the direction of splitting immediately determines the defect plane. The position of the reflections $\bar{H}$ with respect to their corresponding basic reflection g allows one to determine the displacement vector $\bar{R}$ at the periodic interface. It is easily shown [7] that

$$\bar{H} = \bar{g} + \frac{1}{M} (m - \bar{g}.\bar{R}) \bar{e}_n$$

where m is an integer and $\bar{e}_n$ is the normal to the interface. By considering three independent reflections $\bar{g}_1$, $\bar{g}_2$ and $\bar{g}_3$ the displacement vector $\bar{R}$ can be completely determined.

If properly used, high resolution electron microscopy is a powerful technique to extract atomic scale information on the crystal structure and particularly concerning defects in the structure. Care should be taken

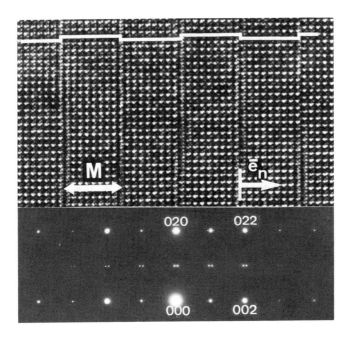

Fig. 1. High resolution electron microscopy image of a long period superstructure and its corresponding electron diffraction pattern.
The LPS has an average spacing M=9.

that the material is extremely thin, the crystal is oriented exactly along
the zone axis and that the electron microscope (especially the beam tilt
[8] and the objective astigmatism) is properly aligned. Under these circumstances a useful resolution of around 0.2 nm can be achieved which is
certainly beyond the interatomic distances in a number of materials. We
may, however, not forget that the observed image is always a projection
along the electron beam and that due to multiple diffraction effects and
the inevitable spherical aberration, chromatic aberration and beam divergence in the electron microscope, the final observed image is not always a
direct enlargement of the projected crystal potential. In order to allow
detailed interpretations, computer simulations are mostly required; they
are to be compared with the experimental observed images. Such computer
simulations start from a well known structure and calculate the wave function at the exit surface of the specimen. This wave function is then twice
Fourier transformed, including the effects introduced by different aberra-
tions and microscope limitations. Details about the technique, the approximations and the limitations involved can be found in refs. [5, 9];
examples on the applications are numerous and go back to the initial work
of Iijima [10] and Allpress and Sanders [11].

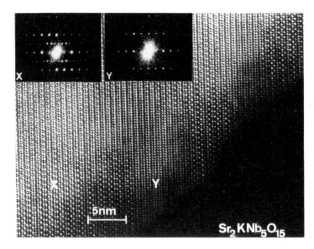

Fig. 2. High resolution image of $Sr_2KNb_5O_{15}$ showing different micro-do-
mains X and Y. The incommensurate modulation is present in X
but not in Y. The optical diffraction patterns from both areas
are shown as an inset.

With respect to the study of the incommensurate modulated structures,
HREM can mainly contribute to:

a) identify the origin of the incommensurability (see further when discussing the discommensurations in $Ba_2NaNb_5O_{15}$).

b) determine the shape of the modulation, i.e., whether it is a step
function, a sinusoidal modulation or a more complex behaviour (see
further when discussing the long period superstructure in Cu_3Pd).

c) locate, with the help of optical diffraction, the modulated areas in
the crystal. An example is illustrated in Fig. 2 for $Sr_2KNb_5O_{15}$ which
in the electron diffraction pattern taken along b^* shows an incommensurate modulation along a^* similar to the one observed in $Ba_2NaNb_5O_{15}$
close to the transition at 300°C [12]. The high resolution image of
Fig. 2 clearly reveals the modulation only in restricted areas (indicated X) while areas indicated Y do not show any modulation. Optical
diffraction of the high resolution image is an excellent help to treat
such problems [13]. The electron microscope negative taken at a mag-
nification of around 5.10^5x is used as a grating for a laser beam and
an area of the order of a few nm wide can be selected to obtain an
optical diffraction pattern. The technique allows:

    1) to detect weak periodicities which might be difficult to
detect in the real space image

    2) to obtain diffraction information from very small areas
(down to 2 nm)

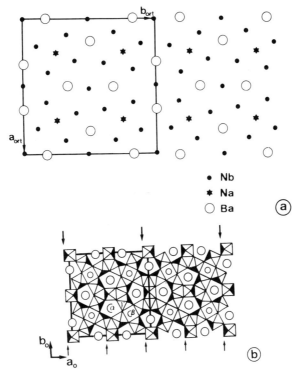

● Nb
✳ Na
○ Ba

(a)

(b)

Fig. 3. Schematic structure representation of BSN projected along the
c-axis:
a) only the metal atom positions are represented; the pseudo
unit cell indicated is the one defined by Jamieson et al.
[17]
b) the true unit cell is doubled with respect to a) due to a
shear of the tops of the NbO_6 octahedra. These shears are
indicated by shaded sides of the octahedra.

In the present case optical diffraction patterns from areas X and Y
are represented as an inset; they clearly indicate the absence of any
modulation in the area Y (the reflections present correspond to the
undistorted tetragonal phase) and the presence of a modulation in the
area X. The observed satellites are similar to those in $Ba_2NaNb_5O_{15}$
[14] and will be explained further when considering this material.
With the help of the techniques just mentioned: electron diffraction, conventional electron microscopy and high resolution electron microscopy combined with optical diffraction, we will analyze incommensurate modulations
occurring in two completely different structures
a) $Ba_2NaNb_5O_{15}$ (BSN) and its homologues
b) $Cu_{3\pm x}Pd$.
Details on the incommensurate phase in SiO_2 (quartz) and $AlPO_4$ (berlinite)
were also revealed by electron microscopy [15]. They will be discussed in
this volume by M. Walker.

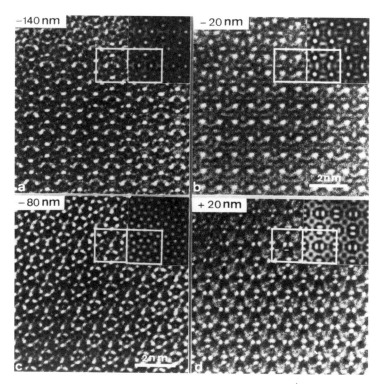

Fig. 4.   High resolution images of BSN imaged along the c-axis. For each
          different defocus value, indicated in the upper left corner, the
          calculated image is represented in all images. In Figs. a, b
          and c one images respectively the Ba, Na and Nb configuration;
          in Fig. d no direct relationship with the projected crystal
          structure can be made.

2.   INCOMMENSURATE PHASE IN $Ba_2NaNb_5O_{15}$ (BSN) AND RELATED MATERIALS

BSN has a tetragonal high temperature phase (a = 1.24 nm; c = 0.4 nm)
and undergoes a paraelectric-ferroelectric transition at 580°C [16]. At
300°C the material transforms from paraelastic to ferroelastic lowering
hereby its symmetry to orthorhombic (2mm) with lattice parameter $a_o = 2\sqrt{2}$
a_t; $b_o = \sqrt{2} a_t$ and $c_o = 2 a_t$ [12]. The structure is rather complex but is
extremely well suited for HREM for the following reasons:
- along the c axis the structure is only one NbO_6-layer (i.e.,0.4 nm) thick.
  (apart from small atom shifts which double the c-axis [17]).
- projected along [001] only atoms of the same chemical species superimpose
  (i.e., form columns) and the different metal sublattices (Ba, Na, Nb) are
  well separated and each have a particular geometrical pattern (see Fig.
  3). Na columns (stars in Fig. 3) form a square lattice and are separated
  by 0.88 nm; Ba atoms (open circles) form a regular arrangement of squares
  and lozenges separated by 0.62 nm; Nb atoms (black dots) form a regular
  arrangement of pentagons, squares and triangles separated on the average
  by 0.37 nm [17].

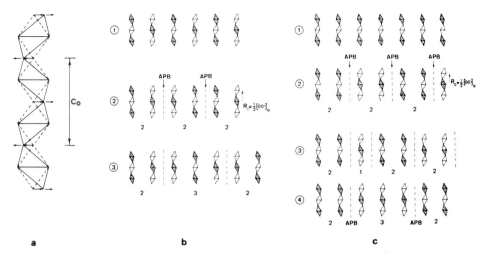

Fig. 5. a) Deformation of the chains of NbO$_6$ octahedra by collective
shearing results in a doubling of the c-axis.

① b,c) schematic representation of two hypothetical
structures
② b,c) commensurate orthorhombic superstructure formed by
introducing periodic antiphase boundaries in ①
③ b,c) ④ if a singular strip of "1" or "3" is present in
the sequence, the structure becomes incommensurate.

Under bright field high resolution conditions (Fig. 4), the different sublattices can be imaged separately for different defocus values (i.e.,
different strengths of the objective lens); the defocus values have been
indicated in the upper left corner of each picture. The experimental
images have for each image been compared with the calculated images (upper
right corner) and the unit cell which is chosen to have the same origin as
in Fig. 3 is indicated on both the experimental and calculated images.
The thickness of the crystal for the calculated images was only 3.6 nm
(i.e., 9 unit cells). In Fig. 4a, b and c, respectively, the Ba, Na and
Nb configurations are imaged as bright dots (compare Fig. 4a,b,c with
Fig. 3) but in Fig. 4d a bright dot configuration is observed which has no
direct relationship with any of the projected atom configurations. This
example clearly indicates that care must be taken when interpreting HREM
images in terms of atom configurations. Along this [001] section, however,
no evidence for any modulation is present either in the direct image or in
the diffraction pattern.

At the paraelastic-ferroelastic transition Ba-ions slightly shift
place and break the tetragonal symmetry to make a ($\sqrt{2}$a, $\sqrt{2}$b, 2c) unit
cell. Superimposed on this is a collective shear of the tops of the NbO$_6$
octahedra parallel with (001) [17]. Since the NbO$_6$ octahedra form chains
along [001] (see Fig. 5a), the displacements in the next layer are necessarily in the opposite sense and double the c-parameter (Fig. 5a). In the

81

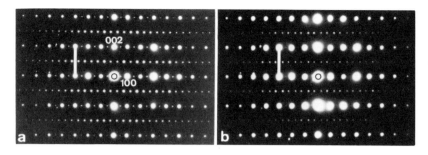

Fig. 6. a) [010] electron diffraction pattern of BSN taken at:
         a) room temperature
         b) close to 300°C

electron diffraction pattern this is translated by the presence of an extra
row of weak reflections at ℓ = 2n+1 (Fig. 6). At room temperature an
(almost) commensurate structure is formed and in the electron diffraction
pattern the satellites at ℓ = 2n+1 are all equally spaced. The room temperature BSN structure can be looked at as a stacking of columns as in
Fig. 5a with deformed octahedra, the Nb atoms are situated in the center
of such octahedra, whereas Ba and Na atoms sit in between columns of octahedra. If we just concentrate on the stacking along the a-axis, we can
describe the room temperature commensurate phase of BSN as represented in
Fig. 5b. ② i.e., a sequence of two left-sheared octahedra (hatched) followed by two right-sheared octahedra which can be described as a $2\bar{2}2\bar{2}$
sequence, using the Fujiwara notation [18]. Such a structure can be considered as a modulated structure where all octahedra are sheared in the
same direction (Fig. 5b ①) or alternatively a structure where octahedra are sheared alternatingly left and right (Fig. 5c ①).

When the temperature is raised close to the transition temperature of
300°C, the satellites in the ℓ =2n+1 row no longer remain equally spaced
and an incommensurate structure is formed (Fig. 6b). In view of the
previous description, this indicates that antiphase boundaries no longer
appear every 2 layers but that random defects (discommensurations) appear,
locally changing a "2" configuration into "3" or "1" (see Fig. 5b,c ③).
Both possibilities can be distinguished on the electron diffraction pattern
from the shift of satellites towards or away from a given position (e.g.,
the position 001). A bright or dark field image of BSN directly reveals
the discommensurations as wavy, spaghetti-like defects (Fig. 7) coalescing
together in groups of four. This indicates that their displacement vector
must be one fourth of a lattice vector. From high resolution observations
one can further deduce that the displacement vector is actually 1/4 [102]
with respect to the unit cell described above. When the temperature is
raised, the discommensurations become very mobile and their number quickly

increases. They never form regular arrangements (i.e., periodic defects) such as in the long period superstructures (see further) or in the different forms of ZnP_2 and CdP_2 [19].

In BSN the incommensurability never becomes very large, i.e., never deviates far from its commensurate value M = 2. However, when replacing Ba by Pb and substituting partially Na by K, a number of different incommensurate long period superstructures were found at room temperature [20]. In particular, the hypothetical structures in Fig. 5.b ① and 5.c ① corresponding to M = ∞ and M = 1 respectively are apparently realized in the lead homologue of BSN. Also in this case the long periods are attributed to a modulation as a result of the deformation by the shearing of NbO_6-octahedra. Figure 8.a-e shows a series of [010] electron diffraction patterns with respectively the M values: M = ∞, M = 2.78; M = 2, M = 1.62,

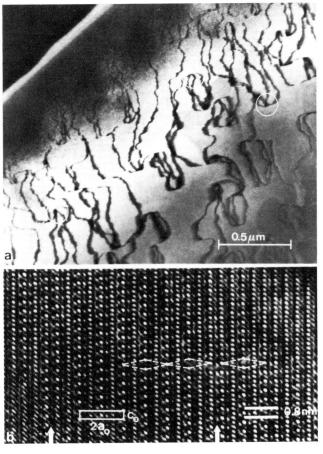

Fig. 7. a) Conventional dark field image showing discommensurations in BSN. Note the fourfold coalescence of the discommensurations.
b) High resolution image in $Sr_2KNb_5O_{15}$ where deviations from the period M = 2 are observed (see arrows).

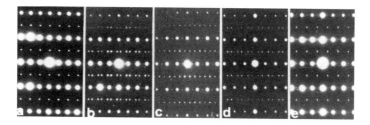

Fig. 8. Electron diffraction patterns along the [010] zone of Pb_2 $(K_xNa_{1-x})Nb_5O_{15}$ exhibiting different incommensurabilities. a) $M = \infty$; b) $M = 2.77$; c) $M = 2$; d) $M = 1.62$; e) $M = 1$.

$M = 1$. Figure 8a represents a structure where all octahedra in a given (001) layer are sheared in the same sense. At the other extreme, Fig. 8e is the diffraction pattern of a structure in which in successive rows on proceeding along the a-axis the sense of shearing changes its sign, causing antiphasing after each unit cell $M = 1$ (see representation in fig. 4c 1). The diffraction pattern of the commensurate BSN structure is represented in fig. 8c; it corresponds to $M = 2$. If strips with a width equal to 3 are introduced as in fig. 5b ③ or fig. 5c ④ the diffraction pattern evolves towards that of fig. 8b. The structure giving rise to this diffraction pattern can be approximated by the stacking symbol 33323332 which corresponds to an M value of $\frac{11}{4} = 2.77$. Fig. 8d finally represents the diffraction pattern of a structure where "1" blocks are introduced into a "2" sequence; the structure is approximated by $M = \frac{5}{3} = 1.67$ i.e. 221221.

3. INCOMMENSURATE LONG PERIOD SUPERSTRUCTURES IN $Cu_{3\pm x}$ Pd

Around the Cu_3Pd composition the Cu-Pd phase diagram shows three phases which are very much related to each other:

a) the basic LI_2 (Cu_3Au-type) structure which is only stable below ~ 21 at % Pd

b) a one-dimensional long period superstructure (LPS) which is deduced from the LI_2 structure by periodically introducing conservative antiphase boundaries along (001) with an average separation M. The structure is stable between approximately 21 and 30 at % Pd.

c) a two-dimensional LPS which is deduced from the LI_2 by introducing periodically conservative APB's along one of the cube planes and non-conservative APB's along another cube plane. This structure, however, is only stable in a narrow temperature interval between the one dimensional LPS and the solid solution (disordered phase).

84

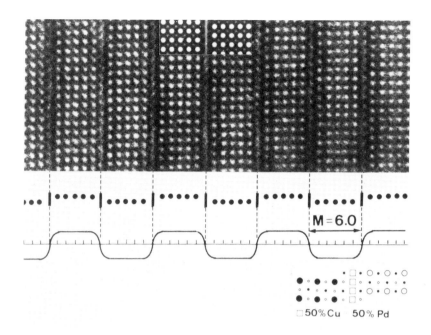

Fig. 9.   High resolution image of a LPS in $Cu_{3+x}Pd$ with M = 6. Considerable disorder is present along the interfaces; this is observed
in the image as a diffuse line. A calculated image using the
model represented in the lower right corner is inserted at the
top of the figure. This model assumes disorder (open squares)
along one plane on the Pd-sites (large circles) as well as on
the Cu-sites (small circles).

It is well known that the M value of the one-dimensional LPS changes with
temperature and composition [21] and this has been confirmed by a recent
electron microscopy investigation [22]. At lower Pd concentration
(around 20 at % Pd) the M value is temperature dependent and varies for a
given composition between ∞ (LI_2) at lower temperatures and M $\approx$ 10 at
higher temperatures (see further). At higher concentrations the M value is
no longer temperature dependent and decreases with increasing composition
along commensurate values M = 8, M = 6, M = 5, M = 9/2, M = 4, M = 7/2,
until finally M = 3 is obtained for the composition $Cu_{70}Pd_{30}$, This sug-
gests that for Cu-Pd the two ANNNI ground states may be $\langle\infty\rangle$ and $\langle3\rangle$ rather
than $\langle1\rangle$ and $\langle2\rangle$ for Ag_3Mg [23] or Al-Ti [24]. A general survey about the
ANNNI model is given ·in [25], [26], [27] while the application of the model
to the Cu-Pd system is treated by de Fontaine et al. in [28]. This study
also revealed that the interfaces become sharper with increasing composition. For the M = 3, M = 7/2 or M = 4 the boundary plane is well defined

and ordering persists up to the boundary plane. For M = 6, however, which
was heat treated at the same temperature, one observes a diffuse line
(Fig. 9) every 6th layer. This is indicative of disorder of Cu and Pd
atoms along this plane, as can be confirmed from image simulations (see
upper inset) of a model introducing only disorder in one atomic plane (see
model below). The simulations also allow to conclude that the bright dots
indeed correspond to the projected Pd atom positions.

Fig. 10. Dark field (110) image in Cu-19.3 at % Pd annealed at 400°C.
The interfaces are highly confined to cube planes and reflect
a preferential spacing.

Particularly interesting is the evolution of the period as a function
of temperature for the low Pd-concentration alloys. For Cu-19.3 at % Pd
the LI$_2$ structure (M = ∞) is stable below 360°C. When the material is heat
treated below this temperature the LI$_2$ structure develops occasional antiphase boundaries, located preferentially along cube planes and separated
by several hundred nm. When the material is then heated above the 360°C
transition, the boundaries become unstable and develop a zig-zag-shaped
arrangement (see Fig. 10) which upon prolonged heating will slowly develop
a long period superstructure. The average period of this LPS decreases
from M = 14 at temperatures just above 360°C until M ≈ 10 at the highest
temperature (500°C), i.e., just below the order-disorder transition. Apart

from this variation in average spacing M, there is also a variation in the degree of order along the antiphase boundary. Compare in Fig. 11 the irregularly spaced interfaces annealed at 400°C which, however, are relatively sharp and well defined with the more regularly separated boundaries after annealing at 500°C. The latter become very wavy and the diffuseness spreads out over several atomic planes. These diffuse images indicate disorder, which is consistent with computer simulations. The modulation

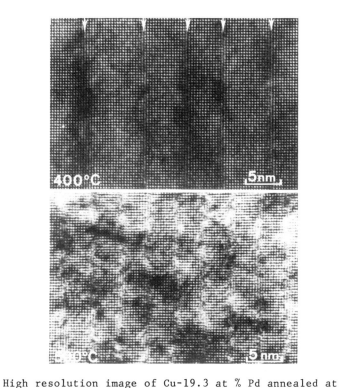

Fig. 11. High resolution image of Cu-19.3 at % Pd annealed at
a) 400°C
b) 500°C
In a) the boundaries are irregularly spaced but well defined;
in b) the antiphase boundaries are more regularly spaced but
with a large disorder along the interfaces.

function therefore is closer to a sinusoidal modulation as compared to the step function for the antiphase boundaries at lower temperatures (or higher concentrations). This effect, which has been predicted by Selke [26] but never really observed, is not so surprising; a similar phenomenon has been observed in LEED by Sundaran et al. [29] for the disordering at the surface of a Cu₃Au crystal. Due to the large disorder along the interfaces and the wavy character of the boundaries, these structures in Cu-Pd can be considered to be incommensurate long period superstructures; their behavior is very similar to the LPS in CuAu [30,31].

4. ACKNOWLEDGMENT

The subjects described here originate from joint work with D. Broddin, M. Guymont, A. Loiseau, C. Manolikas and R. Portier.

5. REFERENCES

1. P. B. Hirsch, A. Howie, R. B. Nicholson, D. W. Pashley and M. J. Whelan, "Electron Microscopy of Thin Crystals" (London, Butterworths, 1965).
2. S. Amelinckx, Solid State Physics, supplement 6 (1964).
3. J. M. Cowley, "Diffraction Physics" (North Holland, Amsterdam-Oxford, 1975).
4. J. C. H. Spence, "Experimental High Resolution Electron Microscopy" (Clarendon Press, Oxford, 1981).
5. D. Van Dyck, Advances in Electronics and Electron Physics 65, 295 (1983).
6. K. C. Thompson-Russell and J. W. Edington, "Practical Electron Microscopy in Materials Science 5" (The Macmillan Press, Ltd., 1977).
7. J. Van Landuyt, R. De Ridder, R. Gevers and S. Amelinckx, Mat. Res. Bull. 5, 353 (1970).
8. D. J. Smith, W. O. Saxton, M. A. O'Keefe, G. J. Wood and W. M. Stobbs, Ultramicroscopy 11, 263 (1983).
9. W. Coene and D. Van Dyck, Ultramicroscopy 15, 41 (1984).
10. S. Iijima, J. Applied Physics 42, 5891 (1971).
11. J. G. Allpress and J. V. Sanders, J. Appl. Cryst. 6, 165 (1973).
12. J. C. Toledano, Phys. Rev. $B12$, 943 (1975).
13. G. Van Tendeloo and J. Van Landuyt, J. Microsc. Spectrosc. Electron. 8, 461 (1983).
14. G. Van Tendeloo, S. Amelinckx, C. Manolikas and When Shulin, Phys. Stat. Sol. $A91$, 483 (1985).
15. G. Van Tendeloo, J. Van Landuyt and S. Amelinckx, Phys. Stat. Sol. $A(33)$, 723 (1976).
16. S. Singh, D. A. Draegert and J. E. Geusic, Phys. Rev. $B2$, 2709 (1970).
17. P. B. Jamieson, S. C. Abrahams and J. L. Bernstein, J. Chem. Physics 50, 4352 (1969).
18. K. Fujiwara, J. Phys. Soc. Japan 12, 7 (1957).
19. A. V. Sheleg and V. P. Novikow, Sov. Phys. Solid State 24, 2000 (1982). C. Manolikas, G. Van Tendeloo and S. Amelinckx, Phys. State. Sol. A (1986).
20. C. Manolikas, G. Van Tendeloo and S. Amelinckx, Sol. State Comm. 58, 845 (1986).
21. K. Schubert, B. Kiefer, M. Wilkens and B. Haufler, Z. Metallk. 46, 692 (1956).
22. D. Broddin, G. Van Tendeloo, J. Van Landuyt, S. Amelinckx, R. Portier, M. Guymont and A. Loiseau, Phil. Mag. 53 (1986).
23. R. Portier, D. Gratias, M. Guymont and W. M. Stobbs, Acta Cryst. $A36$, 190 (1980).
24. A. Loiseau, G. Van Tendeloo, R. Portier and F. Ducastelle, J., Physique (Paris) 46, 595 (1985).
25. D. de Fontaine and J. Kulik, Acta Met. 33, 145 (1985).
26. W. Selke in "Modulated Structure Materials," T. Tsakalakos (ed.), (Martinus Nijhoff, The Hague, 1984), p. 23.
27. J. Yeomans, this volume.
28. D de Fontaine, A. Finel, S. Takeda and J. Kulik, Nobel Metal Symposium, Massalski, Bennett, Pearson and Chang (Eds.), Met. Soc. AIME, p. (1986), p. 49.

29. V. S. Sundaram, Brian Farrell, R. S. Alben and W. D. Robertson, Phys. Rev. Lett. $\underline{31}$, 1136 (1973).

30. M. Guymont, R. Portier and D. Gratias, Acta Crystl. $\underline{A36}$, 792 (1980).

31. K. Yasuda, M. Nakagawa, G. Van Tendeloo and S. Amelinckx, J. Less Common Metals (1986).

EFFECT OF TRANSVERSE ELECTRIC FIELDS ON THE INCOMMENSURATE PHASE OF

NaNO$_2$

H. Z. Cummins, S. L. Qiu and Mitra Dutta

Department of Physics
City College
City University of New York
New York, New York 10031

J. P. Wicksted and S. M. Shapiro
Department of Physics, Brookhaven National Laboratory
Upton, New York 11973

Abstract   Elastic neutron scattering experiments on sodium nitrite in
transverse electric fields are reported. The results are interpreted
with the Ishibashi phenomenological free energy. Although no true
Lifshitz point is possible, the analysis shows that the incommensurate
phase can be narrowed by a suitable perturbation and can disappear at a
triple point.

Sodium nitrite (NaNO$_2$) was the first dielectric crystal found to have
an incommensurate phase [1]. In the 25 years since it was discovered many
other examples have been found but NaNO$_2$ has remained nearly unique (with
the recent exception of quartz) for the very small temperature range of its
incommensurate phase.

NaNO$_2$ is ferroelectric at room temperature with polarization along the
b-axis. The ferroelectric ordering process involves reorientation of the
NO$_2$ groups, which can flip between two stable (+b-b) orientations [2]. The
incommensurate phase, which exists between the C$_{2V}$ ferroelectric and the
D$_{2H}$ paraelectric phases, is only 1.4° wide in zero field, extending from
T$_L$ = 162.5°C to T$_I$ = 163.9°C. The narrowness of the incommensurate phase
has invited speculation about the possible existence of a Lifshitz point
(LP) in this system, accessible by a perturbation such as a transverse
electric field [3]. Previous dielectric [4,5] and neutron scattering [6,7]
experiments have explored the effect of a longitudinal electric field E$_b$
which leads to a triple point at E$_b$=3.0 kV/cm. This is not an LP, however,

since E_b is the ordering field. Furthermore, the modulation wavevector $q_0 = \delta a^*$ does not decrease significantly near this point.

In order to explore the possible existence of a Lifshitz point in transverse electric fields, elastic neutron scattering experiments on $NaNO_2$ were performed at the High Flux Beam Reactor at Brookhaven National Laboratory. The samples were grown from the melt at City College. They were typically 0.34 cm x 0.62 cm x 0.42 cm and had gold films evaporated on the a-faces or c-faces for the application of electric fields parallel to the a − or c − axes, respectively. The room temperature lattice parameters of $NaNO_2$ are a = 3.556 Å, b = 5.563 Å and c = 5.384 Å [8]. The samples were installed in a temperature controlled oven mounted on a triple-axis spectrometer with the c-axis vertical. The incident neutrons were fixed at an energy of 14.7 meV and were collimated to yield a FWHM q-resolution limited by the crystal mosaic of 0.006a* (0.01 $Å^{-1}$) where a* = $2\pi/a$. Scans were taken along the a-axis with q extending to 0.25 a* on either side of the (020) Bragg peak, to measure the wavevector and intensity of the satellites that appear in the incommensurate phase. Typical scales are shown in Fig. 1. The temperature was measured by a thermistor mounted inside the oven touching the sample, and the temperature of the oven was controlled by an Omega proportional temperature controller. The temperature was maintained constant to better than 0.01°C and our measurements were completely reproducible. This is an important consideration since at temperatures above 130°C there is a significant current through the sample when a field is applied, producing ohmic heating of the sample [9]. The oven was evacuated to about 5 x 10^{-3} Torr to minimize the water vapor attacking the sample, as $NaNO_2$ is hygroscopic. The fields were applied to a maximum of 1.3 kV/cm for the E-field parallel to the a-axis, and to 2.2 kV/cm for the E-field parallel to the c-axis, the maximum fields possible before electrical breakdown of the sample occurred.

In Fig. 1, we show the temperature dependence of the scans for a field of E = 2.2 kV/cm applied along the c direction. There is no change of the satellite width with temperature and the widths are resolution limited. The inset in Fig. 1 shows q scans at zero field over a much wider range of temperature and q. It reveals that, in addition to the incommensurate resolution-limited peaks, there is broad, diffuse intensity which persists well into the paralelectric regime, but disappears when the sample enters the ferroelectric phase. Figure 2 shows the intensity of the satellites versus temperature, while Fig. 3 shows the modulation wavevector versus temperature for different transverse fields parallel to the a-axis. The range of the incommensurate region clearly decreases with increasing field

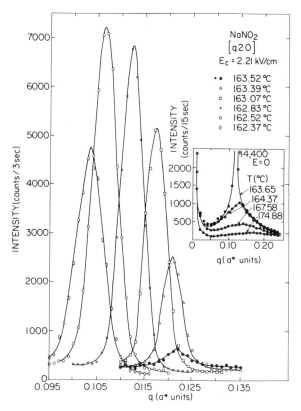

Fig. 1. Temperature dependent scans for $NaNO_2$ within the incommensurate
phase for a fixed field $E_c = 2.2$ kV/cm applied along the c-
axis. The inset shows scans over a wider range of temperature
and q.

while the end-points (T and T_L) both move towards lower temperatures. The
wavevector of the satellites also decreases with increasing field.

In Fig. 4 we plot the transition temperatures T_I and T_L found from the
extrapolated endpoints of the curves in Fig. 2, and the corresponding modulation wavevectors δ_I and δ_L found from the curves in Fig. 3. In Fig. 4,
we also show theoretical curves for δ_I, δ_L, T_I and T_L obtained by adjusting
the coefficients α_1 and T_{01} in Eq. (2) discussed below to obtain optimum
fits to the data. The other coefficients used in the free energy were
obtained from fits to our dielectric constant data [7] and from the zero
field neutron experiments.

The Lifshitz point (LP), first proposed by Hornreich, Luban and
Shtrikman in 1975, is a special multicritical point in the phase diagram
of systems with incommensurate phases [10]. At the LP, the disordered
(para), modulated (incommensurate) and ordered (ferro) phases meet, and
the three phase boundary lines join with common tangents, as shown in

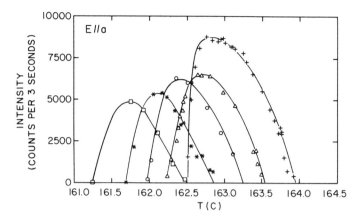

Fig. 2. Intensity of the neutron diffraction satellites at δa^* vs T
for different values of the transverse electric field E_a in
kV/cm: 0(+), 0.55($\triangle$), 0.82(0), 1.09(*), 1.36($\square$). The solid
lines are guides for the eye.

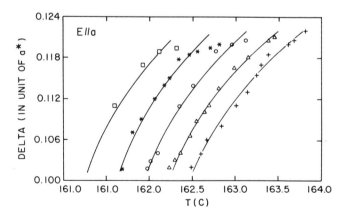

Fig. 3. Modulation wavevector in units of a* versus temperature for
different values of the transverse electric field E_a in kV/cm:
0(+), 0.55($\triangle$), 0.82(0), 1.09(*), 1.36($\square$). The solid lines
are guides for the eye.

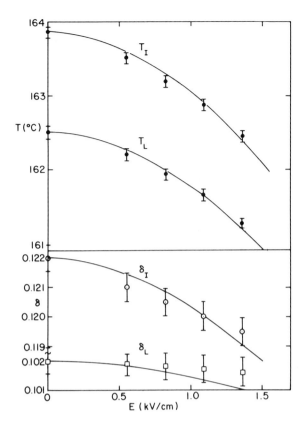

Fig. 4. T_I, T_L, $\delta_I(0)$ and $\delta_I(\square)$ in NaNaO$_2$ with E along the a-axis. The solid lines are theoretical predictions discussed in the text, with the parameters α_1 and T_{01} adjusted to give a best fit to the experimental data.

Fig. 5. The modulation wavevector q_0 of the incommensurate phase goes to zero continuously at the Lifshitz point, and critical fluctuations can extend through a large region in k space.

The usual LP phenomenology which leads to the phase diagram shown in Fig. 5 is based on a mean field analysis of the free energy density functional

$$F(x) = \frac{1}{2} AP^2 + \frac{1}{4} BP^4 + \frac{1}{2} \alpha |\nabla P|^2 + \frac{1}{4} \beta |\nabla^2 P|^2 \tag{1}$$

where $P = P(x)$ is the order parameter, $A = A_0(T-T_0)$, $B > 0$, and $\beta > 0$. The coefficient α is negative, but is assumed to increase monotonically with some perturbation (e.g., pressure, field or composition), passing through zero at the LP. As α increases, both the width of the incommensurate phase and the modulation wavevector δ decrease, both reaching zero at the LP. With the free energy of Eq. 1, the direct para-ferro transition line $T_0(\alpha)$ in the phase diagram shown in Fig. 5 is second order. The second order

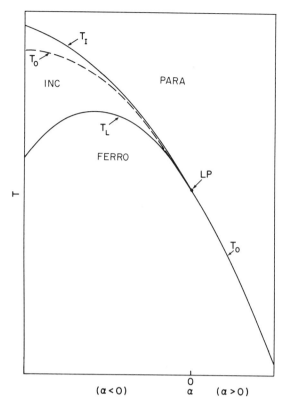

Fig. 5. Schematic mean-field phase diagram of an incommensurate syste
with free energy given by Eq. (1). The direct para-ferro
transition temperature T_0 is arbitrarily assumed to depend
quadratically on α as shown. At the Lifshitz point LP,
passes through zero, the three phases meet, and the modulation
wavevector k reaches zero. ——— (physical transitions);
- - - - - (virtual transitions).

para-inc transition line is indicated by T_I, and the first order inc-ferro
lock-in transition line by T_L.

In systems which exhibit a first order direct para-ferro transition,
such as $NaNO_2$, a LP, as defined above cannot exist. Nevertheless, the
width of the inc phase and the modulation wavevector k may decrease under
some perturbation producing behavior which closely resembles that of systems possessing a LP.

We now consider systems in which the direct para-ferro transition is
first order. The free energy functional of Eq. (1) must be extended to
include a sixth-power term since B is negative; we also include a term in
$P^2|\nabla P|^2$ which is required to make the modulation wavevector temperature
dependent [11,12,6]. Since we are concerned with a ferroelectric, we take
the order parameter P to be the electric polarization, and the perturbing
field E to be an electric field. The specific free energy density functional we consider is [12]

$$F(X) = (\tfrac{1}{2})AP^2 + (\tfrac{1}{4})BP^4 + (\tfrac{1}{6})CP^6 + (\tfrac{1}{2})\alpha|\nabla P|^2 + (\tfrac{1}{4})\beta|\nabla^2 P|^2 + (\tfrac{1}{4})\eta P^2|\nabla P|^2 - E\cdot P \quad (2)$$

with $A = A_0(T-T_0)$, $B < 0$, $C > 0$, $\beta > 0$ and $\eta > 0$. All coefficients except A are assumed to be independent of temperature. The perturbing field E (which cannot be the ordering field) will be an electric field perpendicular to the polar axis, so $E\cdot P = 0$. Both α and T_0 are taken as functions of E. Note that with suitable redefinitions of P and E, this free energy functional could apply to incommensurate systems other than ferroelectrics.

We employ the simplest forms for α and T_0 allowed by symmetry which are:

$$\alpha = \alpha_o + \alpha_1 E^2 \quad (3a)$$

$$T_0 = T_{00} + T_{01}E^2 \quad (3b)$$

The free energy density F is computed from the functional $F(x)$ by $F = (\tfrac{1}{V}) \int F(x)dV$ where V is the volume of the crystal. For the ferroelectric phase we assume that the order parameter is homogenous, so $P(x) = P_o$; for the incommensurate phase, we use the incommensurate plane wave approximation $P = P_k\cos(kx)$.

The resulting free energy densities are:

(a) (ferro): $F_o(P_o) = \dfrac{1}{2} AP_o^2 + \dfrac{1}{4} BP_o^4 + \dfrac{1}{6} CP_o^6$ \quad (4)

(b) (inc): $F_k(P_k) = \dfrac{1}{4} A_k P_k^2 + \dfrac{3}{32} B_k P_k^4 + \dfrac{5}{96} CP_k^6$ \quad (5)

where

$$A_k = A + \alpha k^2 + \tfrac{1}{2} \beta k^4 \quad (6)$$

$$B_k = B + (\tfrac{1}{3})\eta k^2 \quad (7)$$

The description of the incommensurate modulation by a single plane wave (IPW approximation) is justified for $NaNO_2$ by the fact that no higher harmonics of the satellites are observed experimentally. We note that in the presence of a <u>longitudinal</u> electric field, we would need to consider the mixed form $P(x) = P_o + P_k\cos kx$ since in the inc phase, a longitudinal electric field induces homogeneous polarization. The resulting free energy would contain cross-terms in P_o and P_k which would, in turn, induce harmonics in the modulation [11].

In general, the values of P_o, P_k and k that minimize the free energy can be found by solving the simultaneous equations:

$$\frac{\partial F}{\partial P_o} = \frac{\partial F}{\partial P_k} = \frac{\partial F}{\partial k} = 0 \quad (8)$$

97

Alternatively, the free energy can be minimized by computer with a non-linear least-squares program that varies P_o, P_k and k, a procedure which avoids the approximations required by the algebraic method.

We now turn to the construction of the phase diagram shown in Fig. 6 which is implied by Eqs. (2)-(8). The T_o line where $A(E,T) = 0$ is a line of virtual direct second-order para-ferro transitions. With $B < 0$, a first order transition always intervenes while A is still positive, at

$$T_1 = T_o + \frac{3B^2}{16A_o C} \tag{9}$$

i.e., at a constant temperature interval above T_0. (Details of the derivation from the free energy of this and other results are given in reference (13)). If $B_k > 0$, the para-inc transition is second order and occurs at

$$T_I = T_o + \frac{\alpha^2}{2A_o \beta} \tag{10}$$

just as in the second-order case. The inc-ferro lock-in transition occurs at a temperature T_L determined by the crossing of $F(P_k)$ and $F_o(P_o)$ given by Eqs. (4) and (5), each evaluated at their respective minima. We have evaluated $T_L(E)$ numerically.

The T_o and T_I lines meet at the value of E for which $\alpha = 0$ which is the virtual Lifshitz point (VLP) of the system. It does not correspond to a physical phase transition, however, since the T_I line crosses the first-order direct transition line T_1 at the crossing point CP where

$$\alpha^2 = 3\beta B^2/8C \tag{11}$$

To the right of this point, there is no incommensurate phase since the direct first order para-ferro transition intervenes; the lock-in transition line T_L also terminates at CP which is, therefore, a triple point. The modulation wavevector k, found by minimizing F_k [Eq. (5)] with respect to k, is given by

$$k = \left[-\frac{\alpha}{\beta} - \frac{1}{8} \frac{\eta}{\beta} P_k^2 \right]^{\frac{1}{2}} \tag{12}$$

Since $P_k = 0$ at T_I and increases monotonically as T is lowered, k will decrease with decreasing T through the incommensurate phase. The modulation wavevector k (or equivalently $\delta = k/a^*$ where a^* is the reciprocal lattice vector in the direction of the a-axis) at T_I and T_L can be found from Eq. (12). In Fig. 7 we show the electric field dependence of T_I-T_L (the width of the incommensurate phase) and of δ_I and δ_L, the limiting values of the modulation wavevectors (in units of a^*) at T_I and T_L,

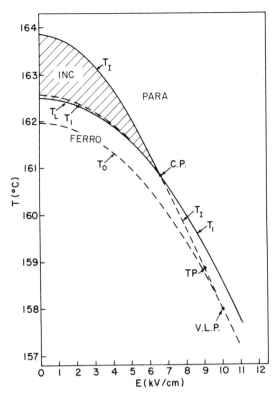

Fig. 6. Schematic mean-field phase diagram of an incommensurate system
with the free energy given by Eq. (2) for which the direct para-
ferro transition at T_1 is first order. T_0 and   are assumed
to depend quadratically on the perturbing field E as shown
in Eq. (3). All parameters in the free energy were taken
from column 2 of Table 1. The plot was constructed with the
arbitrarily chosen values $\alpha_1 = 1.952 \times 10^{-19}$ cm^4/statvolt^2 and
$T_{01} = -3.97 \times 10^{-2}$ °C.cm^4/statvolt^2 which puts the virtual
Lifshitz point LP at T = 158°C and E = 10 kV/cm.
————— (physical transition); ------ (virtual transition).

respectively. Note that δ_I and $\delta_L \to 0$ at the LP, while $T_I-T_L \to 0$ at the

crossing point CP.

    There is a further complexity in the phase diagram associated with the

para-inc transition line. Since the modulation wavevector at T = T_I is

$k = [-\alpha/\beta]^{1/2}$, B_k in Eq. (7) must eventually become negative as α approaches

zero, since B is negative. This occurs when $k^2 = -3B/\eta$, or $\alpha = 3\beta B/\eta$.

Beyond this tricritical point (TP in Fig. 6), the para-inc transition is

first order. In Fig. 6, the tricritical point is shown in the physically

inaccessible region below CP which turns out to be appropriate for NaNO$_2$.

The relative positions of CP and TP depend on the values of the coeffic-

ients in the free energy, however, and the tricritical point could in

principle be physically accessible. Finally, we note that if the term

$(1/4)\eta P^2 |\nabla P|^2$ were not included in the free energy functional [Eq. (2)], i.e., if we take $\eta = 0$, then B_k in Eq. (7) would be negative <u>regardless</u> of the value of k, and the para- inc transition would always be first order, even at $E = 0$. (This point was also discussed in Ref. 6).

We now turn to the relevance to $NaNO_2$ of the free energy functional of Eq. (2) and the analysis of it presented in the theory section. The identification of the direct para-ferro transition in $NaNO_2$ as first-order rests on two observations [6]. First, when the inverse dielectric constant ε_b^{-1} is plotted against temperature, extrapolations of the two straight lines representing the para and ferro phases cross rather than both extrapolating to zero at T_0 as they would if the direct transition were second order. Second, in the presence of a longitudinal electric field E_b, there would be no transition at fields greater than that of the triple point if the direct transition were second order. However, the neutron scattering experiments of Durand et al [6] as well as earlier work they cite indicated that the direct transition line continues above the triple point, terminating at a critical end point at E=6kV/cm.

It is therefore reasonable to assume that $NaNO_2$ should exhibit the type of behavior shown in Fig. 6. In fact, Fig. 6 was constructed using all the free energy coefficients for $NaNO_2$ listed in Table 1. The coefficients $T_{01} = -3.97 \times 10^{-2}$ °C.cm^2/statvolt^2 and $\alpha_1 = 1.952 \times 10^{-19}$ cm^4/statvolt^2 were selected arbitrarily to provide a convenient range in T and E for the plot.

Since the data in Fig. 4 shows the expected decrease with increasing field of δ_1, δ_L, T_I and T_L, we fit the theoretical predictions to the data to obtain:

(E_a): $\alpha_1 = 5 \times 10^{-19}$ cm^4/statvolt^2, $T_{01} = -0.7$ °C.cm^2/statvolt^2

(E_c): $\alpha_1 = 7 \times 10^{-20}$ cm^4/statvolt^2, $T_{01} = -0.04$ °C.cm^2/statvolt^2

The phase diagrams constructed using these values look essentially like Fig. 6, uniformly deformed by moving LP to lower temperature and different fields. The inferred (E, T) position of LP, CP and TP for the two cases, with E in kV/cm, and T in °C are:

(E_a): LP = (6.75, 134), CP = 4.13, 150), TP = (5.57, 140)

(E_c): LP = (16.7, 151), CP = (11, 158), TP = (14.9, 153)

The correctness of these phase diagrams has not been fully demonstrated in the present experiments, of course, since the maximum fields that can be applied before electrical breakdown occurs are far below those required to reach the interesting part of the ET plane. Nevertheless, we suggest that the phase diagram of Fig. 6 is qualitatively correct for $NaNO_2$ in transverse electric fields.

Finally, we consider a question about the free energy functional of Eq. (2) raised by Levanyuk [14]. The cross-term with coefficient η was

Table 1. The Values of the Coefficients in the Phenomenological Free Energy (Eq. 2)

	Present work		W. Buchheit et al.[a]		D. Durand et al.[b]	
	MKS	CGS	MKS	CGS	MKS*	CGS
A_0	2.16×10^7	2.40×10^{-3}	1.9×10^7	2.1×10^{-3}	2.2×10^7	2.4×10^{-3}
B	-3.30×10^{10}	-4.07×10^{-11}	-2.2×10^{10}	-2.7×10^{-11}	-7.0×10^{10}	-8.6×10^{-11}
C	1.57×10^{13}	2.157×10^{-19}	4.3×10^{12}	5.9×10^{-20}	4.3×10^{13}	5.9×10^{-19}
α	-1.757×10^{-11}	-1.952×10^{-17}			-2.3×10^{-11}	-2.6×10^{-17}
β	3.78×10^{-30}	4.20×10^{-32}			-5.0×10^{-30}	5.5×10^{-32}
η	1.04×10^{-7}	1.28×10^{-24}			1.5×10^{-7}	1.8×10^{-24}

$T_{00} = 161.97°C$

[a]Ref. 19.

[b]Ref. 6.

*The values of the coefficients α, β and η determined by D. Durand et al. have been converted to mks units with one reduced unit (r.u.) = $2\pi/a = 1.767 \times 10^{10}$ m.

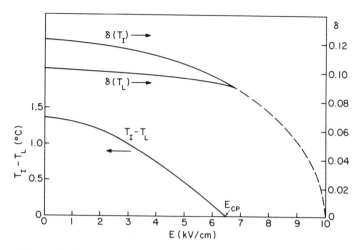

Figure 7.   Electric field dependence of $T_I - T_L$, δ_I and δ_L for the phase diagram of Fig. 6.

introduced primarily to make the modulation wavevector q_0 temperature dependent, but this could also be due to an intrinsic temperature dependence of α.

In the paralectric phase, there are no satellites near the Bragg peaks, but a broad diffuse (or critical) scattering is observed near the incommensurate phase transition indicating enhanced critical fluctuations in that region of q-space (see inset in Fig. 1). If the q-dependence of diffuse scattering in the paralectric phase is computed from Eq. (2), one obtains

$$I(q) \propto (A(T) + \alpha q^2 + \beta q^4)^{-1} \tag{13}$$

which is independent of η since $P^2 = 0$. As first noted by Yamada and Yamada [15], the value of q at the peak of the diffuse x-ray scattering shifts with temperature, suggesting that α $\underline{is}$ temperature dependent, a point also noted by Michel in connection with his microscopic model of $NaNO_2$ [16]. This diffuse scattering was also observed in neutron scattering by Sakurai, Cowley and Dolling [17], and was further investigated by Durand et al. [18].

Levanyuk suggested that the temperature dependence of the satellite position $\delta(T)$ and the position of the peak of the diffuse scattering q(IMax) should be plotted together; if there is no change of slope at T_I, then the temperature dependence of δ could be attributed entirely to the temperature dependence of α and the η term would be unnecessary. Our result, shown in Fig. 8, shows that there is a marked change in slope at T_I indicating that both η and $\alpha(T)$ should be included. The η term is also

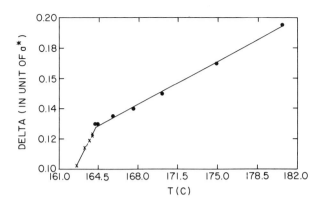

Fig. 8. Temperature dependence of the satellite wavevector δ_I for $T_L \leq T \leq T_I$ (x) and of the peak of the diffuse scattering (o) for $T > T_I$. Note the marked change of slope at T_I.

necessary, as mentioned above, to make the para-incommensurate transition second order. We have not extended the phenomenological free energy by adding additional terms, however. Instead, we have recently turned to the microscopic theory and, in collaboration with K. H. Michel, are attempting to connect the thermodynamic properties of the incommensurate phase of NaNO2 directly with the microscopic properties of the crystal (see the chapter of Michel in these proceedings).

ACKNOWLEDGMENTS

We thank Y. Ishibashi, R. A. Cowley, J. D. Axe, G. Shirane, Y. Yamada, R. Currat, R. M. Hornreich, A. P. Levanyuk, K. H. Michel, K. Hamano, K. Ema, L. Garland and J. Steiner for discussions and for their interest in this work. We also thank Wen Li for participation in the latter phases of these experiments. This research was supported by the National Science Foundation Division of Materials Research under Grant No. DMR83-14498 and, at Brookhaven, by the Division of materials Science, U.S. Department of Energy under Contract No. DE-AC02-76CHO0016. The figures appearing in this chapter were reprinted from Ref. 13 which contains a more extensive report of the work described here.

REFERENCES

1. S. Tanisaki, J. Phys. Soc. Jpn. _16_, 579 (1961).
2. S. Sawada, S. Nomura, S. Fujii and I. Yoshida, Phys. Rev. Lett. _1_, 302 (1958); Y. Yamada, I. Shibuya and S. Hoshino, J. Phys. Soc. Jpn. _18_, 1594 (1963).

3. D. G. Sannikov., Sov. Phys. Solid State $\underline{23}$, 1827 (1982).

4. K. Hamano, J. Phys. Soc. Jpn. $\underline{19}$, 945 (1964).

5. K. Gesi, J. Jpn. App. Phys. $\underline{4}$, 818 (1965).

6. D. Durand, F. Denoyer, D. Lefur, R. Currat and L. Bernard, J. Phys. (Paris) Lett. $\underline{44}$, L207 (1983).

7. S. L. Qiu, H. Z. Cummins, S. M. Shapiro and J. C. Steiner, Ferro-electrics $\underline{52}$, 181 (1983).

8. G. E. Zeigler, Phys. Rev. $\underline{38}$, 1040 (1931); M. I. Kay and B. C. Frazer, Acta Cryst. $\underline{14}$, 56 (1961).

9. Y. Asao, I. Yoshida, R. Ando and S. Sawada, J. Phys. Soc. Jpn. $\underline{17}$, 442 (1962).

10. R. M. Hornreich, M. Luban and S. Shtrikman, Phys. Rev. Lett. $\underline{32}$, 1678 (1975).

11. Y. Ishibashi and H. Shiba, J. Phys. Soc. Jpn. $\underline{45}$, 409 (1978).

12. Y. Ishibashi, W. Buchheit and J. Petersson, Solid State Commun. $\underline{38}$, 1277 (1981).

13. S. L. Qiu, M. Dutta, H. Z. Cummins, J. P. Wicksted & S. M. Shapiro, Phys. Rev. B (in press); S. L. Qiu, Incommensurate Phase Transitions in NaNO$_2$, Ph.D. Thesis, The City University of New York (1985). (Several computational errors have been corrected in this manuscript.)

14. A. P. Levanyuk - private communication.

15. Y. Yamada and T. Yamada, J. Phys. Soc. Jpn. $\underline{21}$, 2167 (1966).

16. K. H. Michel, Phys. Rev. $\underline{B24}$, 3998 (1981).

17. J. Sakurai, R. A. Cowley and G. Dolling, J. Phys. Soc. Jpn. $\underline{28}$, 1426 (1970).

18. D. Durand, F. Denoyer, M. Lambert, L. Bernard and R. Currat, J. Phys. (Paris), $\underline{43}$, 149 (1982).

19. W. Buchheit and J. Petersson, Solid State Commun. $\underline{34}$, 649 (1980).

LIGHT SCATTERING FROM INCOMMENSURATE INSULATORS;   MAINLY BaMnF4

D. J. Lockwood

Physics Division
National Research Council
Ottawa, Canada K1A OR6

Abstract   The light scattering work on pyroelectric BaMnF4 is reasses-
sed here in view of new structural information on the incommensurate
phase. The dispersions in the acoustic phonon velocities V_{cb} and V_{cc}
are ascribed to coupling to the phason and amplitudon, respectively;
the failure to observe a true soft mode in the Raman spectrum for
$T < T_i \simeq 247$ K is attributed to the overdamped nature of the amplitudon even for $T \simeq 85$ K; the static central peak is associated with
symmetry breaking frozen defects; and the dynamic central peak previously thought to be due to the phason is now assigned to LA-phonon -
amplitudon coupling. Thus the light scattering experiments have provided no direct evidence for phasons or amplitudons. It is suggested
that Brillouin scattering experiments should be performed on incommensurate ThBr4, which is a very simple system with small phason and amplitudon damping, to search for the overdamped phason and predicted
amplitudon - acoustic-phonon coupling. Finally, it is demonstrated that
quasiperiodic superlattices provide exciting possibilities for detailed
comparisons between theory and experiment.

1.   INTRODUCTION

Like the classic ferroelectric materials (rochelle salt, KH_2PO_4 and
$BaTiO_3$) before them, $NaNO_2$, thiourea and K_2SeO_4 form the original three incommensurate ferroelectrics. Now, of course, there are many known examples
of all kinds of ferroelectrics, but a theoretical understanding of the
microscopic details of the transitions in many commensurate and incommensurate materials has proved elusive. A prime example is KH_2PO_4, which has been
studied for 50 years [1]!

From an experimental viewpoint, light scattering techniques have been widely used during the last decade to investigate the dynamics of incommensurate transitions in insulators [2,3]. The main objective in such work has been the detection of the Goldstone phase mode (or phason) that is predicted to occur in the incommensurate phase. Efforts to find an underdamped phason have been singularly unsuccessful, and it is now known that this mode is overdamped near zero wave vector [4-6]. In the next section we reassess the recent light scattering work on $BaMnF_4$ in view of new structural information [7-9]. In Section 3 we give details of results for $ThBr_4$ that point to interesting Brillouin experiments, while in Section 4 we discuss exciting new results obtained by Raman scattering from incommensurate superlattices.

2. $BaMnF_4$

Pyroelectric $BaMnF_4$ has become one of the classic incommensurate materials. Nevertheless, despite the scores of papers published since 1976 concerning the incommensurate phase transition at $T_i \simeq 247$ K, this improper ferroelastic remains enigmatic in many respects [10]. $BaMnF_4$ is a rare example of an incommensurate system possessing a four-component order parameter and having no low-temperature lock-in phase [11]. The star of the modulation wave vector contains $q_1 = (\alpha, 1/2, 1/2) = q_2$, $q_3 = (\alpha, -1/2, 1/2) = -q_4$, with $\alpha \simeq 0.39$. The incommensurate wave vector is weakly temperature - dependent and shows thermal hysteresis [12]; it is also sample dependent, indicating the importance of lattice defects [13]. The transition temperature is sample dependent too, and may be reduced by defects or impurities which pin the modulation [8,14]. The defects of importance in $BaMnF_4$ are thought to be fluorine vacancies [15].

Until recently the exact form of the low temperature phase was unknown and there were three possibilities predicted by theory [16] based on the high temperature space group ($A2_1am$ or C_{2v}^{12}) and the star of the soft mode. The point group of the incommensurate phase is now known to be C_2 (the average structure has $P2_1$ or C_2^2 symmetry) [7-9] due to condensation of q_1 (and q_2) with q_3 (and q_4) zero or vice versa. Thus monoclinic domains form below T_i with a "yz" strain, and it has been found that the twin fraction of these domains is both sample and temperature dependent [8]. The transition is now known to be slightly first order [7-9].

All of the light scattering work performed on $BaMnF_4$ to date [17-27] was done before the incommensurate structure was known, and, in some cases [17-19], even before $BaMnF_4$ was known to be incommensurate. It is informative to reassess the results of these experiments in view of the recently published structural information. The monoclinic structure below T_i implies that only the Goldstone phason and one amplitude mode (or amplitudon)

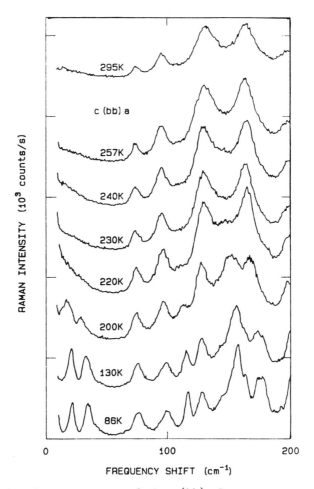

Fig. 1. The low-frequency part of the c(bb)a Raman spectrum of BaMnF$_4$.

are of long wavelength and are potentially detectable in a light scattering
experiment [6,11,25]. The phason is expected to be overdamped [6], but a
gap may exist in the phason dispersion due to defect or impurity induced
pinning [28].

A. <u>Raman Scattering</u>

The most detailed temperature-dependent Raman scattering studies of
BaMnF$_4$ have been reported by Lockwood et al. [25] and Murray et al. [26].
Typical diagonal-polarization spectra are shown in Fig. 1. Two temperature dependent modes are visible near 22 and 35 cm^{-1}, and a coupled oscillator model was required to reproduce this part of the low-frequency spectrum [23]. However, neither of the decoupled mode frequencies behaved as
expected for a soft amplitude mode [25,26]. Essentially similar results
were obtained in all diagonal polarizations [26]. Thus the earlier report
of soft mode behavior by Ryan and Scott [17] has not been reproduced in

subsequent experiments [29]; the differences are attributed to the scattering geometry used by Ryan and Scott, which resulted in a mixed A_1-B_2 (in C_{2v} symmetry) spectrum [25,30]. Submillimeter dielectric spectroscopy [31] also revealed two low-frequency modes in agreement with the Raman data of Reference 25. The mode softening reported for a Raman line of mixed polarization near 150 cm^{-1} by Popkov et al.[19] was not found in pure polarizations by Lockwood et al. [25] and it probably arose from the combined intensity variation of two overlapping modes, neither of which showed a strong temperature dependence in their frequencies [27].

The appearance of two temperature-dependent low-frequency modes below T_i lead to a tentative assignment [25,29,31] to the two amplitudons, or one amplitudon and one non-Goldstone phason, appropriate to the incommensurate structure that would result from the condensation of modes at all four of the symmetry-related wave vectors $q_1 \to q_4$ [6,11,25]. Lockwood et al. [25] did comment that if the monoclinic structure was appropriate, one of the modes could be the amplitudon. However, this conjecture must now be rejected, as neither of the two low-frequency modes has the expected amplitudon temperature dependence. We conclude that the amplitudon has not been observed directly in any experiment so far, and we suggest a different interpretation.

Although the neutron scattering experiment [11] failed to detect the phason or amplitudon for $T < T_i$, it did discover the soft mode expected for $T > T_i$. This mode was found to be overdamped at the incommensurate wave vector q_i for all temperatures up to at least 581 K. Thus it is most likely that <u>the amplitudon</u>, which is derived from the $T > T_i$ soft mode at $T = T_i$, <u>is also overdamped</u> below T_i (at least for temperatures close to T_i). This supposition fits in well with the negative results of the Raman and microwave experiments, which showed no underdamped soft mode even for temperatures as low as 85 K, and suggests further experiments to look specifically for an overdamped amplitudon at $q \simeq 0$.

For point group C_2, group theory [32] predicts $35A + 34B$ optic modes in the incommensurate phase, comprising $11A_1 + 6A_2 + 5B_1 + 11B_2$ zone-centre modes from the orthorhombic phase ($A_1, A_2 \to A$; $B_1, B_2 \to B$) plus $18A + 18B$ modes from q_1. A new analysis of the earlier mode assignments [25,26] shows that many modes that were previously attributed to polarization "leak through" are actually allowed in these polarizations in monoclinic symmetry. A revised symmetry assignment of the modes is given in Table 1, where it can be seen that $9A_1 + 5A_2 + 3B_1 + 11B_2$ modes are observed for $T > T_i$ and at least $20A + 20B$ modes are found for $T < T_i$. Thus the Raman results are in agreement with the new structural information.

Table 1. Symmetry Assignments of Raman Lines Observed in BaMn$_4$ at Temperatures Above and Below T$_i$ = 247 K[*]

| C$_{2v}$/C$_2$ Symmetry and Temperature (K) | | | | | | | | | | | |
| A$_1$/A | | | A$_2$/A | | | B$_1$/B | | | B$_2$/B | | |
295	257	86	295	257	86	295	257	86	295	257	86
		23			22					30	25
		36			34						37
									54⌉	53	60
									67⌋		67
74	71	75	73	71	73						79
91⌉	93	100							85⌉	83	87
97⌋						108⌉	110	111	90⌋		103
		116			115	114⌋					116
123⌉	130	127							137⌉	136	136
135⌋					148				140⌋		149
		156			156				160⌉	160	
156⌉	161	163	163	160	164				162⌋		169
175⌋		174			173						
					186			183	180	176	181
197	196	201				199					200
		224					221	224	225	221	223
232	229	236									
			238	233	244	238	238	241	239	238	239
243⌉					249						
247⌋											
288	287	289	288	287	289				310		287
		325							329⌉	327	332
		347							359⌋		
395⌉	399	407	391	397	406				380	378	389
399⌋											
									452	455	455

[*]The bracketed line frequencies (cm^{-1}) are transverse-longitudinal pairs.

B. Brillouin Scattering

The Brillouin studies by three groups have led to some puzzling and inconsistent results. The first measurements by Bechtle et al. [20,21] revealed a frequency dispersion in the transverse acoustic (TA) phonon velocity V_{cb} near T$_i$. A large anomaly in V_{cb} observed at T$_i$ by ultrasonic techniques [33] had nearly disappeared at Brillouin frequencies. The frequency dependence of the anomaly (see Fig. 2) could remarkably be characterized by a single relaxation time $2\pi\tau_2$ = 1.5 ns at T$_i$, and the temperature dependence at 4 MHz could be described by

$$\tau^{-1}(T) = \tau_1^{-1}\left(\frac{T_i - T}{T}\right) + \tau_2^{-1} \qquad (1)$$

with $2\pi\tau_1$ = 6 ps.

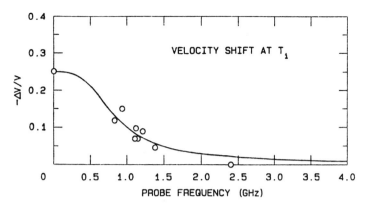

Fig. 2. Frequency dependence of the anomaly in V_{cb} at T_i for BaMnF$_4$. The curve is a least-squares fit of a single relaxation time to the experimental points [after Ref. 21].

Bechtle et al. [21] found an anomaly at T_i in the longitudinal acoustic (LA) phonon velocity V_{cc}, which agreed reasonably well with the ultrasonic data [33] (although it was noted that in some runs the Brillouin anomaly was smaller than the ultrasonic one). On the other hand, Lyons et al. [27] reported a dispersive effect in the V_{cc} anomaly. To date the differences between the results of these two groups have not been reconciled [34]. Lyons et al. found $2\pi\tau_2 \simeq 0.1$ ns at T_i, but they were unable to explain the origin of this relaxation.

The difference in relaxation times for the TA and LA phonons suggests different relaxation mechanisms. Scott [34] proposed from geometrical arguments that it is the phason that interacts with the TA mode, while it is the amplitudon that couples to the LA mode.

From measurements made in c(aa)b polarization, Murray and Lockwood [23] discovered an asymmetry in the LA mode lineshape for $T < T_i$, and an intensity variation in the TA$_1$ (quasi-shear) mode for $T > T_i$. The temperature dependence of the Brillouin spectrum is shown in Fig. 3, and the LA phonon lineshape asymmetry is clearly visible in Fig. 4. This asymmetry does not appear in either the a(cc)b or a(bb)c spectra. The LA-phonon integrated intensity is essentially temperature independent, whereas the TA$_1$ phonon exhibits a marked decrease in integrated intensity with $T \to T_i$ from above and then remains nearly constant for $T < T_i$ (see Fig. 5). The TA$_1$ mode strength has a critical exponent $\beta \simeq 0.55$ for $T > T_i$. Similar temperature-dependent acoustic mode behavior was seen in the a(cc)b and a(bb)c spectra. In all these cases neither the TA$_1$ nor the LA modes exhibited observable frequency anomalies at T_i [25]. The temperature dependent behavior of the TA$_1$ mode intensity for $T > T_i$ has yet to be interpreted. From Brillouin measurements at 295K, Lockwood et al. [25] were able to determine

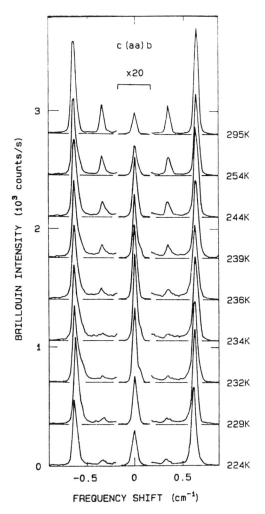

Fig. 3. Temperature dependence of the c(aa)b Brillouin spectrum of $BaMnF_4$.

all of the elastic and piezoelectric constants and most of the elasto-optic constants for commensurate $BaMnF_4$.

The Brillouin spectra of Fig. 3 also exhibit a central peak anomaly, which was not found in other polarizations; as shown in Fig. 5, the central peak intensity is a maximum at a temperature near the TA_1-phonon intensity discontinuity and has a critical exponent $\beta \approx 0.55$ for $T > T_i$. The observation that the critical central peak had no intrinsic width (within the resolution) and also manifested itself near T_i in the familiar "speckle" pattern of a static central peak [35] supported an assignment to a purely static effect [25]. The presence of the static central peak was later confirmed by Lyons et al. [27] who found that the intensity maximum occurred at $T_s \approx 241$ K $\approx T_i - 6$ K. Following Halperin and Varma [37] this strongly suggests that the static central peak is due to symmetry breaking frozen

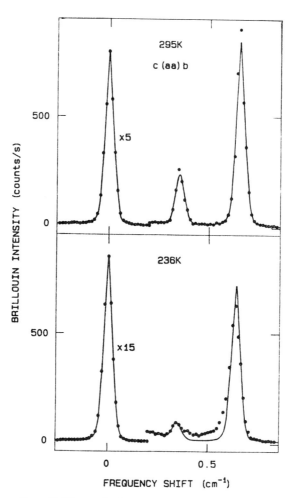

Fig. 4. The results of fits of the symmetric instrumental function [25]
to the c(aa)b Brillouin spectrum of $BaMnF_4$ at temperatures
above and below T_i.

defects, which could be the fluorine vacancies mentioned earlier. More
experimental work on a variety of samples is needed to prove this hypothesis. Another possibility is domain effects, as suggested by Lyons et al
[27]. The observation of a central peak could be correlated with the formation of large domains at or just below T_i, but the relative sizes of the
monoclinic domains continue to vary with temperature over a wide temperature
range [8].

Another central peak, this time a dynamic one, has been observed by
Brillouin scattering in (bb) polarization with wave vector q along the c
axis [22,24,27]. In this scattering configuration the central peak linewidth was shown to be proportional to q^2, but showed no appreciable temperature dependence, and the intensity maximum occurred at $T_d \simeq 247$ K [22].
The central peak linewidth, Γ_d, increased with q varying in the ac plane

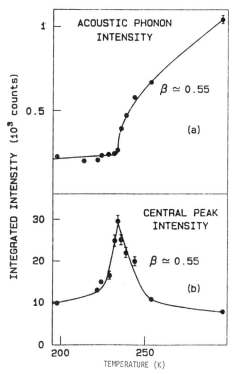

Fig. 5. The integrated intensity of (a) the TA_1 phonon at 0.34 cm^{-1} and
(b) the central peak in the c(aa)b Brillouin spectrum of BaMnF$_4$
[after Ref. 25].

away from the c axis, but the peak itself could not be observed for angles
between q and c greater than ~30° [24,27]. Lyons et al. [24,27] found that

$$\Gamma_d \sim q^2 (D_\parallel \cos^2\phi + D_\perp \sin^2\phi) \ , \tag{2}$$

where ϕ is the angle q makes with the incommensurate a axis in the ac plane,
and estimated that $D_\parallel$ = 0.98 cm^2 s^{-1} and $D_\perp$ = 0.14 cm^2 s^{-1}. These values of
D are too large by far to be associated with thermal diffusion [27], and
thus entropy fluctuations are not the cause of the inelastic central peak.
Lyons et al. [27] attribute this peak to overdamped phason scattering mediated by the LA- phonon – phason interaction.

Lyons et al. [27] have used their model for the dynamic central peak
in BaMnF$_4$ to interpret the asymmetric LA-phonon line profile observed by
Lockwood et al. [25]. The latter authors demonstrated [23,25] that the
difference in lineshapes for T < T_i, as shown for example in Fig. 4, could
be accommodated by an underdamped oscillator near 0.6 cm^{-1}; they attributed
it to the phason at a time when the phason was thought to be underdamped.
However, by assuming that the LA phonon is coupled to their overdamped
phason, Lyons et al. [27] obtained the remarkably good parameter-free fit to

113

the LA phonon lineshape shown in Fig. 6. Thus the LA phonon and the dynamic central peak are closely related.

From a theoretical viewpoint, Golovko and Levanyuk [6] pointed out that in BaMnF$_4$ the Goldstone phason should be observed in diagonal polarization with $q \parallel a$ only, which is not what was found experimentally by Lyons et al. [27]. Furthermore, the observed central peak linewidth is much greater than the theoretically predicted phason linewidth, which is expected to be $\sim 10^{-5}$ of the amplitudon linewidth [6]. Also, the intensity of the phason scattering is expected to be comparatively weak [6]. Thus it now appears that the central peak observed by Lyons et al. is not due to phason scattering.

As discussed previously, the amplitudon in BaMnF$_4$ is likely to be overdamped, as well as the phason, and is expected to have a greater width than the phason. This immediately suggests that the dynamic central peak of Lyons et al. is due to the overdamped amplitudon interacting with the LA phonon. This explanation is compatible with present theory for BaMnF$_4$ and is supported by the qualitative arguments of Scott [34].

C. Summary

From this review it is clear that the low frequency dynamics of BaMnF$_4$ for $T < T_i$ are not fully understood. The evidence for the existence of the phason and amplitudon is only secondary at best, appearing in the form of a dispersion of TA (for the phason) and LA (for the amplitudon) phonons, and also in a dynamic central peak and LA phonon lineshape anomaly (for the amplitudon), via the respective phason--or amplitudon--acoustic-phonon interactions. More experimental and theoretical work is needed to elucidate the forms of the phase and amplitude modes in defect containing BaMnF$_4$.

It is quite possible that BaMnF$_4$ does not behave like many of the other incommensurate insulators studied to date. It is a special case, because of defects pinning the phase of the incommensurate modulation and because it shows no lock-in transition. Indeed, BaMnF$_4$ may even prove to be an example of an incomplete "devil's staircase" [37]. The effects of the defect-induced pinning and monoclinic domain formation on the low frequency dynamics have yet to be investigated, although some predictions have been made [6] for light scattering from a devil's staircase. From a statics viewpoint, a recent X-ray diffraction study has shown that defects introduce random fields and destroy the long-range order of the incommensurate phase [38].

Finally, because of the sample dependence of some of the results obtained from various physical measurements on BaMnF$_4$ it is important to note in the future the source of the material used in the experiment. In this

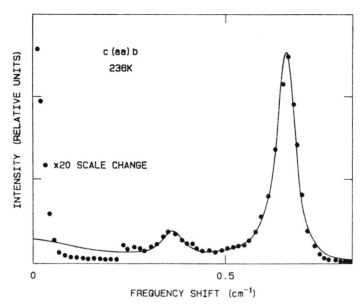

Fig. 6. Comparison of the calculated LA phonon profile in BaMnF$_4$ (solid
line)with the experimental data in c(aa)b polarization [after
Ref. 27]. (Copyright, 1982, Bell Telephone Laboratories In-
corporated, reprinted by permission).

regard we should mention that the sample used by Lockwood et al. [25] was
the same as (or very similar to) one of the crystals studied by Ryan [8] and
by Cox et al. [11]: Lyons et al. [27] obtained their crystals from this
same source [39].

3. ThBr$_4$

Thorium tetrabromide is one of the very few insulators where the phason dispersion has been measured directly (by neutron scattering) in the
incommensurate phase below T_i = 95 K [40,41]. The modulation wave vector
$q_i = 0.31c^*$ is practically independent of temperature and no lock-in is
observed down to 4 K. In the tetragonal phase (space group I4$_1$/amd or
D$_{4h}^{19}$) above T_i a soft mode is observed that becomes overdamped only very
close to T_i. Below T_i this soft branch splits into the expected phase and
amplitude branches, with the phason exhibiting linear dispersion.

Compared with BaMnF$_4$ this system is most favorable for light scattering studies, because the incommensurate transition is very simple [40] and
because the phason and amplitudon linewidths are very narrow at moderate q
values. Bernard et al. [40] measured linewidths in the range 50-100 GHz
for the soft modes above and below T_i. A Raman scattering study [42] of
ThBr$_4$ has revealed the temperature-dependent amplitudon for T < T_i and the
frequency of this mode has a critical exponent β = 1/3. It would be informative to carry out Brillouin scattering experiments on ThBr$_4$ to look for

the overdamped phason and for a predicted [40] strong coupling between the amplitudon and a LA phonon.

4. QUASIPERIODIC SUPERLATTICES

One of the most interesting recent developments in the field of incommensurate systems has been the first realization of a quasiperiodic superlattice [43]. The sample was grown by molecular beam epitaxy and comprised alternating layers of GaAs and AlAs, which were arranged to form a Fibonacci sequence. Such a sequence can be derived from two building blocks, layers A and B, which are stacked according to the following rules:

$$S_1 = |A|$$
$$S_2 = |AB|$$
$$S_3 = |AB:A|$$
$$S_4 = |ABA:AB|$$
$$S_n = |S_{n-1}:S_{n-2}| \; .$$

The physical properties of this quasi-crystal have been discussed elsewhere in these Proceedings [44] and need not be repeated here. From a light scattering point of view, it has been shown that resonance Raman scattering from LA phonons provides an excellent probe of the quasiperiodic nature of the superlattice [43]. The back-scattering Raman spectrum of the GaAs/AlAs Fibonacci superlattice [44] shows a continuum with clearly resolved "dips" that are ascribed to gaps in the density of states of LA modes [43].

We have studied another quasiperiodic superlattice grown on (100) Si in our laboratory using molecular beam epitaxy. The sample comprised a 10th generation Fibonacci sequence using 150 Å of Si for block A and 80 Å of $Si_{0.75}Ge_{0.25}$ for block B. The Raman spectrum of this superlattice was measured in a 90° scattering geometry, as described elsewhere [45], using a non-resonant excitation wavelength of 457.9 nm. The spectrum shown in Fig. 7(a) comprises a number of discrete lines in this case, rather than a continuum with dips in it, but still corresponds to scattering from LA phonons. The quasiperiodic nature of the superlattice is reflected in the frequencies and relative intensities of the lines, which do not show the behavior expected for a periodic superlattice. This can be seen by comparing Figs. 7(a) and (b).

Since their parameters can easily be adjusted to meet a specific requirement, superlattices grown by molecular beam epitaxy provide exciting new possibilities of performing light scattering experiments on samples specially tailored to compare with detailed theories of incommensurate

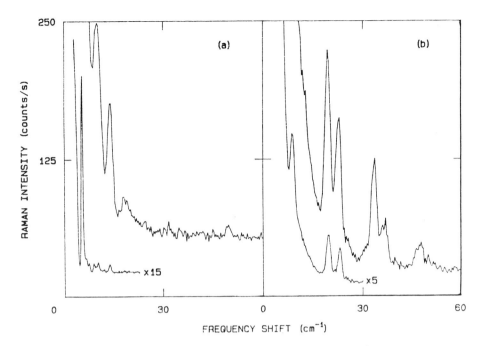

Fig. 7. The Raman spectrum of (a) the quasiperiodic $Si/Si_{0.75}Ge_{0.25}$ superlattice and (b) a periodic $Si/Si_{0.7}Ge_{0.3}$ superlattice (40 periods of 145 Å of Si plus 40 Å of alloy) recorded under identical conditions.

systems. For once, we do not have to just simply accept what nature offers.

5. REFERENCES

1. See the review by T. Matsubara, Jpn. J. Appl. Phys. 24, suppl. 24-2, 1 (1985).
2. J. Petzelt, Phase Transitions 2, 155 (1981).
3. H. Z. Cummins and A. P. Levanyuk, eds., "Light Scattering Near Phase Transitions" (North-Holland, Amsterdam, 1983).
4. V. A. Golovko and A. P. Levanyuk, Sov.-Phys. JETP 54, 1217 (1981).
5. R. Zeyher and W. Finger, Phys. Rev. Lett. 49, 1833 (1982).
6. V. A. Golovko and A. P. Levanyuk in Ref. 3, Chapt. 3.
7. R. V. Pisarev, B. B. Krichevtzov, P. A. Markovin, O. Yu. Korshunov and J. F. Scott, Phys. Rev. B28, 2667 (1983).
8. T. W. Ryan, J. Phys. C19, 1097 (1986).
9. P. Saint-Grégoire, R. Almairac, A. Freund and J. Y. Gesland, Ferroelectrics 67, 15 (1986).
10. See, for example, reviews by J. F. Scott in Rep. Prog. Phys. 12, 1055 (1979); Ferroelectrics 24, 127 (1980); in Ref. 3, Chapt. 5; and Ferroelectrics 66, 11 (1986).
11. D. E. Cox, S. M. Shapiro, R. A. Cowley, M. Eibschütz and H. J. Guggenheim, Phys. Rev. B19, 5754 (1979).
12. D. E. Cox, S. M. Shapiro, R. J. Nelmes, T. W. Ryan, H. J. Bleif, R. A. Cowley, M. Eibschütz and H. J. Guggenheim, Phys. Rev. B28, 1640 (1983).
13. M. Barthés-Régis, R. Almairac, P. St-Grégoire, C. Filippini, U. Steigenberger, J. Nouet and Y. Gesland, J. Physique Lett. 44, L829 (1983).

14. P. St-Grégoire, M. Barthes, R. Almairac, J. Nouet, J. Y. Gesland, C. Fillippini and U. Steigenberger, Ferroelectrics 53, 307 (1984).

15. B. B. Lavrencic and J. F. Scott, Phys. Rev. B24, 2711 (1981).

16. V. Dvorak and J. Fousek, Phys. Stat. Sol. A61, 99 (1980).

17. J. F. Ryan and J. F. Scott, Solid State Commun. 14, 5 (1974) and in "Light Scattering in Solids," M. Balkanski, R. C. C. Leite and S. P. S. Porto, eds. (Flammarion, Paris, 1976), p. 761

18. V. V. Eremenko, A. P. Mokhir, Yu. A. Popkov and O. L. Reznitskaya, Ukr. Fiz. Zh. 20, 146 (1975).

19. Yu. A. Popkov, S. V. Petrov and A. P. Mokhir, Sov. J. Low Temp. Phys. 1, 91 (1975).

20. D. W. Bechtle and J. F. Scott, J. Phys. C. 10, L209 (1977).

21. D. W. Bechtle, J. F. Scott and D. J. Lockwood, Phys. Rev. B18, 6213 (1978).

22. K. B. Lyons and H. J. Guggenheim, Solid State Commun. 31, 285 (1979).

23. A. F. Murray and D. J. Lockwood in "Proc. 7th Int. Conf. Raman Spectrosc.," W. F. Murphy, ed. (North-Holland, Amsterdam, 1980), p. 58.

24. K. B. Lyons, T. J. Negran and H. J. Guggenheim, J. Phys. C13, L415 (1980).

25. D. J. Lockwood, A. F. Murray and N. L. Rowell, J. Phys. C14, 753 (1981).

26. A. F. Murray, G. Brims and S. Sprunt, Solid State Commun. 39, 941 (1981).

27. K. B. Lyons, R. N. Bhatt, T. J. Negran and H. J. Guggenheim, Phys. Rev. B25, 1791 (1982).

28. A. D. Bruce and R. A. Cowley, J. Phys. C11, 3609 (1978).

29. See comments by J. F. Scott, in Ref. 3, Chapt. 5.

30. See the review by M. V. Klein, in Ref. 3, Chapt. 8.

31. A. A. Volkov, G. V. Kozlov, S. P. Lebedev, J. Petzelt and Y. Ishibashi, Ferroelectrics 45, 157 (1982).

32. J. Petzelt and V. Dvorak, J. Phys. C9, 1587 (1976).

33. I. J. Fritz, Phys. Lett. 51A, 219 (1975).

34. See J. F. Scott, Ferroelectrics 47, 33 (1983).

35. W. Taylor, D. J. Lockwood and H. Vass, Solid State Commun. 27, 547 (1978).

36. B. I. Halperin and C. M. Varma, Phys. Rev. B14, 4030 (1976).

37. J. F. Scott, private communication.

38. T. W. Ryan, R. A. Cowley and S. R. Andrews, J. Phys. C19, L113 (1986).

39. The crystals were grown from zone melted materials by H. J. Guggenheim at Bell Laboratories.

40. L. Bernard, R. Currat, P. Delamoye, C. M. E. Zeyen, S. Hubert and R. de Kouchkovsky, J. Phys. C16, 433 (1983).

41. C. M. E. Zeyen, Physica 120B, 283 (1983).

42. S. Hubert, P. Delamoye, S. Lefrant, M. Lepostollec and M. Hussonnois, J. Solid State Chem. 36, 36 (1981).

43. R. Merlin, K. Bajema, R. Clarke, F.-Y. Juang and P. K. Bhattacharya, Phys. Rev. Letters 55, 1768 (1985).

44. See the article by R. Clarke in these Proceedings, p. 357.

45. M. W. C. Dharma-wardana, D. J. Lockwood, J.-M. Baribeau and D. C. Houghton, Phys. Rev. B34, 3034 (1986).

BRILLOUIN SCATTERING STUDY OF INCOMMENSURATE CRYSTALS

Toshirou Yagi

Department of Physics
Kyushu University 33, Hakozaki 6-10-1
Fukuoka 812, Japan

Abstract   A sound velocity dispersion in the incommensurate phase
of $NaNO_2$ appears in the propagation of the longitudinal sound wave
along the [010] direction contrary to the other two directions [100]
and [001], where no dispersion appears. Brillouin scattering study
shows a negative result on the asymmetric propagation of the transverse wave in the incommensurate phase of $NaNO_2$. The lower incommensurate phase of $Ba_2NaNb_5O_{15}$ has similar relaxational mechanisms to
those in the upper incommensurate phase.

1. INTRODUCTION

Incommensurability between some excitations and a basic periodic
structure has been reported recently in a wide variety of condensed matter,
for example crystals, liquid-crystals and quasi-crystals. This phenomenon
has attracted intensive interest among many researchers by virtue of its
new qualitative aspects in condensed matter physics in addition to its
widespread occurrence. In particular, incommensurate crystals give us an
experimental stage suitable for the precise measurement of the static and
dynamic properties of the incommensurability. An atomic/ionic order incommensurate with respect to the original lattice periodicity appears in the
incommensurate phase of many dielectric crystals (incommensurate crystals). The incommensurate phase appears usually in a narrow temperature
region between the normal phase (high-symmetry phase) and the commensurate
phase (low-symmetry phase). We can examine the physical properties of the
incommensurate structure with good reproducibility only by careful regula-
tion of temperature of the crystal.

Using Raman and Brillouin scattering methods, we can observe dynamic properties of the incommensurate phase. Many studies have been reported in this field of structural phase transition in dielectric crystals by the use of these light scattering methods. In particular, the Brillouin scattering method is adequate for the study of the dynamic property of the physical quantities which couple to the elastic strains [1]. In Brillouin scattering we can effectively employ the sound wave as a probe to detect the dynamical instability in the incommensurate phase transition. This provides us a detailed understanding of incommensurability from the dynamical viewpoint. In the following sections we describe several topics dealing with the Brillouin scattering study of the incommensurate crystals.

2. SODIUM NITRITE (NaNO$_2$)

Sodium nitrite undergoes a normal-incommensurate phase transition at T_N = 437.7 K. The normal phase is paraelectric. In the incommensurate phase below T_N, an incommensurate order of the orientation of NO$_2$ ions appears with the wavevector k_0 along the orthorhombic [100] direction. The amplitude vector (displacement vector) of the order is parallel to the [010] direction. The incommensurate phase becomes a commensurate phase at T_C = 436.3 K. The commensurate phase is ferroelectric. The spontaneous polarization appears along the [010] direction. The incommensurate phase of sodium nitrite exists in a narrow temperature range of 1.4 K. We have observed the temperature dependence of the Brillouin spectra of both longitudinal and transverse sound waves in this narrow temperature region [3]. A sound velocity dispersion in the incommensurate phase is seen by comparison between the results of the longitudinal sound velocity obtained by the Brillouin scattering and ultrasonic experiments. The temperature dependence of the transverse waves is used to discuss the asymmetric propagation of the shear sound wave in the incommensurate phase.

Sound Velocity Dispersion

Figures 1(a) and (b) show the temperature dependence of the sound velocity of the longitudinal waves propagating along the three orthorhombic axes: [100], [001] and [010] in the incommensurate phase of sodium nitrite. The small dots are the ultrasonic data reported by Hatta et al. at 6.60 MHz [4]. In the [100] and [001] directions (Fig. 1(a)) we can see no significant difference between Brillouin scattering measurement and the ultrasonic result throughout the three phases near T_N and T_C. On the contrary, the longitudinal wave propagating along the [010] direction shows

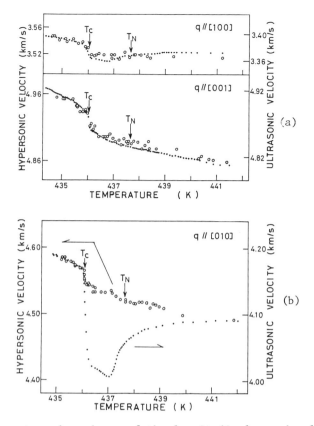

Fig. 1. Temperature dependence of the longitudinal sound velocity obtained
from the Brillouin frequency shift of $NaNO_2$ [3]. The wavevector q
is parallel to [100], [001](a), and to [010](b). Small dots are
the ultrasonic data reported by Hatta et al. at 6.60 MHz [4].

a clear sound velocity dispersion between the frequency regions of GHz and
MHz, as shown in Fig. 1(b). This means the longitudinal sound wave propagating along the [010] direction couples to some excitation in the incommensurate phase. The longitudinal sound wave couples to the amplitude
mode in the incommensurate phase. The present sound velocity dispersion
indicates that the amplitude mode has a frequency dependence with a characteristic frequency between few MHz and a few tens of GHz. The sound velocity dispersion appears gradually with lowering temperature above T_N and
takes its maximum value in the incommensurate phase. The dispersion disappears completely in the ferroelectric-commensurate phase below T_C. The
temperature dependence suggests that activation of the amplitude modes
begins above the phase transition temperature T_N of sodium nitrite. This
is a typical example of the detection of the dynamical instability in the
incommensurate phase by the observation of the elastic dispersion through
a comparison of the Brillouin and the ultrasonic experiment. In this case
we employed the sound waves as a probe to detect the dynamical property of
the amplitude mode in the incommensurate phase.

121

Asymmetric Sound Wave Propagation

Esayan et al. have reported an interesting asymmetry in the propagation of the transverse sound wave by the ultrasonic experiment of the
incommensurate $RbH_3(SeO_3)_2$ crystal [5]. Namely, a transverse sound wave
propagating along the [001] direction and polarized along the [010]
direction does not have the same velocity as the one propagating along the
[010] direction and polarized along [001] direction. The incommensurate
wave-vector is parallel to the [001] direction. The attenuation of the
sound wave was also asymmetric in those two sound waves. However, the two
sound waves should have the same velocity and attenuation with each other
because of the well-known thermodynamical relation in the elastic stiffness
constants $c_{ij} = c_{ji}$. Their study suggests an asymmetric relation $c_{ij} \neq$
c_{ji} making a striking contrast to the usually accepted symmetrical relation
$c_{ij} = c_{ji}$. Dvořák and Esayan gave a theoretical investigation for this
phenomenon [6]. Their theory implies that the asymmetry in sound velocities will vary approximately linearly with probe frequency:

$$\left| V_{ij}(\omega) - V_{ji}(\omega) \right| \propto \omega \tag{1a}$$

By comparing asymmetries in the MHz and GHz regime, we have shown that this
is not true. The asymetry we measure is nonzero in the IC phase but nearly
independent of frequency. Thus, we believe that the theory of Dvořák and
Esayan is qualitatively wrong: the asymmetry does not arise from finite
wave vector and finite frequency effects (dispersion), but from the fact
that $c_{ij}(\omega=0) \neq c_{ji}(\omega=0)$. This can arise, as hypothesized by Scott [5b],
from "local rotations," i.e., internal torques. We have tried to observe
the asymmetric propagation of the transverse sound wave by Brillouin scattering methods, which have several advantages in the measurement of the
sound velocity compared with the ultrasonic method (which usually requires
a transducer and adhesives).

In order to examine their hypothesis we have observed the Brillouin
spectra of the two kinds of transverse sound waves in the incommensurate
phase of $NaNO_2$ [7]. One propagates along the incommensurate wavevector k_0,
that is, along the orthorhombic [100], and the other along [010], which is
perpendicular to k_0. The former transverse mode has a polarization parallel to the [010] and the latter to the [100]. Both sound velocities belong
to the same elastic stiffness constant c_{66} in the well known thermodynamical notation. The result is shown in Figs. 2(a) and (b). In Fig. 2(a)
the temperature dependence of the sound velocities of the two transverse
modes is shown in a temperature range from room temperature to a region
far above T_N. The effect of the piezoelectricity induced below T_C appears

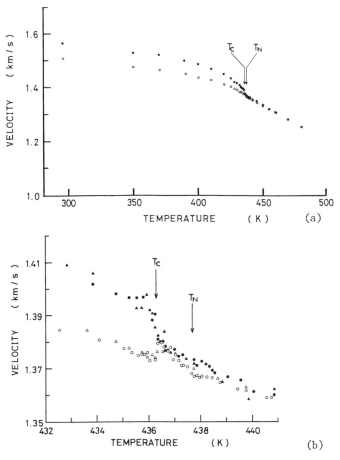

Fig. 2.   Temperature dependence of the c_{66} transverse sound velocity of
NaNO$_2$(a), and in magnified scales (b) [7]. Closed circles and
triangles are for the transverse wave propagating along [100]
polarized along [010]. Open circles and triangles are for the
sound wave propagating along [010] and polarized along [100].

clearly in the temperature dependence of the sound velocity in the ferroelectric phase as shown in Fig. 2(a).   In Fig. 2(b) both sound velocities
are compared with each other in a magnified scale of both vertical and
horizontal axes.   The difference between both sound velocities seems very
small compared with the experimental errors.   Just above and below T_N, the
difference seems to become comparable to the experimental errors.   However,
we cannot draw any definitive conclusions from the present result for the
asymmetric propagation of the transverse wave in the incommensurate phase
of this material.   In order to make this problem clearer, we need a new
method more accurate than the present system.   At the present stage it is
concluded that the difference between the velocities of the two sound
waves, even if it exists, is smaller than the present experimental errors
in the measurement of the Brillouin frequency shift.

3. BARIUM SODIUM NIOBATE ($Ba_2NaNb_5O_{15}$)

Barium sodium niobate (BSN) crystals undergo successive phase transitions, a paraelectric-ferroelectric transition at T_t = 833 K and then a paraelastic-ferroelastic phase transition at T_0^1 = 573 K. Below T^1, BSN crystal has an incommensurate phase with a wavevector K_0 = ± (1 + δ) (a^* + b^*)/4 + c^*/2 [8]. The incommensurate phase becomes a commensurate phase at T_c = 543 K. Many studies have been reported on the phase transition at T_0^1, including Brillouin scattering studies [9,10].

Recently it has been reported that a highly stoichiometric BSN crystal has a second incommensurate phase below 110 K [11]. This means that the commensurate phase below T_c becomes an incommensurate phase again at T_0^2 = 110 K. Few experimental studies have been reported on this second incommensurate phase. We have observed the temperature dependence of the Brillouin spectra near T_0^2 in order to investigate dynamics of the phase transition [12].

Relaxational Phenomena

We have observed anomalous propagation of the longitudinal sound wave which corresponds to the elastic stiffness constants c_{11} and c_{22}. Figure 3(a) shows the temperature dependence of c_{11} and the attenuation constant Γ_{11}, which were calculated from the Brillouin frequency shift and the spectral width. A clear anomaly is seen in the temperature dependence of both quantities near T_0^2 (105 K in our specimen). We analyzed the data assuming a single relaxational mechanism.

$$c_{11}(\omega, T) = c_{11}(\infty) - \frac{\Delta c_{11}}{1 - i\omega\tau(T)}, \tag{1b}$$

where Δc_{11} and $\tau(T)$ are the anomalous part of c_{11} and the relaxational time, respectively. The relaxational time $\tau(T)$ is determined by fitting the data. The temperature dependence of $\tau(T)$ is shown in Fig. 3(b) indicating a critical slowing down at T_0^2. The present result is well expressed by the relation

$$\tau^{-1}(T) = \left|\frac{T - T_0^2}{T_0^2}\right| \tau_1^{-1} + \tau_2^{-1}. \tag{2}$$

Here τ_1 characterize amplitudon-acoustic phonon interaction and τ_2 is a temperature-independent term which includes a diffusive term. $\tau^{-1}(T)$ calculated from Eq. (2) is shown with a straight line in Fig. 3(b) with

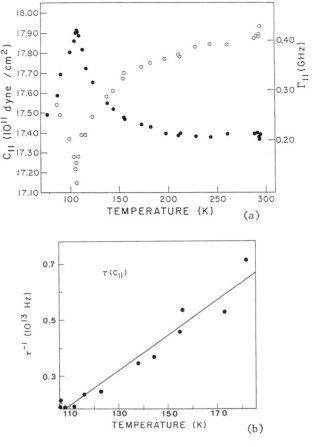

Fig. 3.   Temperature dependence of the elastic stiffness constant
c_{11} (o) and the attenuation constant Γ_{11} (●) of BSN (a), and the
inverse relaxational time τ^{-1} above T_0^2 (b) [12]. The straight line
is obtained by Eq. (2) in the literature.

$\tau_1 = (2.0 \pm 0.2) \times 10^{-13}$ s and $\tau_2 = (8.0 \pm 0.3) \times 10^{-12}$ s [1,12].   The
analysis of c_{22} gives consistent values of τ_1 and τ_2 with the result
obtained from c_{11}.   These values suggest to us similar mechanisms in the
incommensurate transition at T_0^2 to those at T_0^1.

4.   CONCLUSIONS

Brillouin scattering studies have an advantage in the observation of
the dynamical behavior in incommensurate crystals, as we briefly reviewed
here.   For the other group of the incommensurate materials, liquid crystal
or quasi-crystal, the method will bring much useful information on their
dynamical properties, though from the experimental viewpoint at present we
have many problems to be overcome.   We expect to continue the research on
the liquid crystal and quasi-crystals by the Brillouin scattering method.

5. REFERENCES

1. J. F. Scott, Ferroelectrics $\underline{47}$, 33 (1983).
2. M. V. Klein, "Light Scattering Near Phase Transitions," H. Z. Cummins and A. P. Levanyuk, eds. (North-Holland, New York), 1983, p. 503. Vol. 5, "Modern Problems in Condensed Matter Sciences," V. M. Agranovich and A. A. Maradudin, eds.
3. T. Yagi, Y. Hidaka, and K. Miura, J. Phys. Soc. Jpn $\underline{48}$, 2165 (1980); ibid. $\underline{51}$, 3562 (1982).
4. I. Hatta, Y. Shimizu and K. Hamano, J. Phys. Soc. Jpn $\underline{44}$, 1887 (1978).
5. S. K. Esayan, V. V. Lemanov, N. Mamatkulov and L.A. Suvalov, Sov. Phys. Crystallogr. $\underline{26}$, 619 (1981).
5b. J. F. Scott, Ferroelectrics, $\underline{66}$, 11 (1986).
6. V. Dvořák and S. K. Esayan, Solid-State Commun. $\underline{44}$, 901 (1982).
7. T. Yagi, N. Sugimoto, A. Sakai and J. F. Scott (to be published).
8. J. Schneck and F. Denoyer, Phys. Rev. $\underline{B23}$, 383 (1981).
9. P. W. Young and J. F. Scott, Bull. Am. Phys. Soc. $\underline{26}$, 303 (1981).
10. J. C. Toledano, M. Busch and J. Schneck, Ferroelectrics $\underline{13}$, 327 (1976).
11. J. Schneck and D. Paquet, Ferroelectrics $\underline{21}$, 577 (1978).
12. M. Zhang, T. Yagi, W. F. Oliver and J. F. Scott, Phys. Rev. $\underline{B15}$, 1381 (1986).

EXPERIMENTAL RESULTS RELATED TO THE NORMAL-INCOMMENSURATE PHASE

TRANSITION IN A_2MX_4 COMPOUNDS

M. Quilichini

L.P.M.T.M.
Université Paris XIII
93430 Villetaneuse, France

Abstract   Inelastic neutron scattering results taken in the incommensurate phase of K_2SeO_4 show that the phase mode response is overdamped at low q wave-vector. A rather complete set of data obtained
at 120 K, 110.5 K, 100 K, 96 K tends to demonstrate that the phase mode
dispersion curve exhibits a gap at q = 0 and is relatively insensitive
to temperature. The present results can be related to available Raman
data. A single crystal neutron diffraction experiment has been carried out at 588 K in the normal phase of K_2ZnCl_4. In refining the
data we tested two models: a two-position, split-atom harmonic model
(R = .026) and an ordered anharmonic model (R = .019). For Cl atoms
anharmonic temperature factors based on the Gram-Charlier expansion
have been refined up to fourth order.

1. INTRODUCTION

The experimental work presented here has been done in collaboration
with R. Currat for Inelastic Neutron Scattering (I.L.L., H.F. Reactor
Grenoble) and with G. Heger for Neutron Diffraction (ORPHEE Reactor
Saclay). This paper contains two main sections: one devoted to the study
of the phase mode in K_2SeO_4; the other, with the normal incommensurate
transition in K_2ZnCl_4.

Potassium selenate and potassium tetrachlorozincate belong to the
large A_2MX_4 family of insulating compounds in which most present the characteristic and well known sequence of two phase transitions sandwiching a
temperature range in which the material is incommensurate. The upper one
is second order and takes place at $T = T_i$ where the crystal transforms
from the Normal (N) paraelectric phase (space group Pnma) to the Incommen-

127

surate modulated phase (INC). The lower one is slightly first order and occurs at the lock-in point T_c where the crystal goes into the ferroelectric phase (space group (Pna21) with a tripling of the unit cell. The N phase is pseudo-hexagonal and indeed in K_2SeO_4 there exists an order-disorder phase transition at a higher temperature T_h = 745 K above which the crystal is in a disordered hexagonal phase (space group P6$_3$/mmc) with a smaller ($X^{\frac{1}{2}}$) unit cell. In K_2ZnCl_4 this hexagonal transition is only hypothetical, T_h being above the melting point. It may be interesting to start with a brief summary of the characteristic transition points in both materials. The melting point T_f is about 900 K in K_2SeO_4 and 635 K in K_2ZnCl_4. The T_i points are 130 K and 563 K, while the T_c points are 93 K and 403 K for K_2SeO_4 and K_2ZnCl_4, respectively. In potassium selenate the INC state extends over a 37 K temperature range 750 K below T_f, while in potassium tetrachlorozincate it ranges over 130 K only 70 K below T_f. This temperature range of the INC state is comparable to those observed in Rb_2ZnCl4 and R_2ZnBr_4 and is much greater than in K_2SeO_4. In this latter compound T_i is far away from T_f and so the ordering of the SeO_4 groups can be thought to be complete while in K_2ZnCl_4 T_i is not so far from T_f, and disorder is likely to manifest itself.

In both materials the wavevector of the static modulated distortion is given by $q_\sigma = (1 - \delta)\ \underset{\sim}{a}^*/3$ where $\underset{\sim}{a}^*$ stands for the reciprocal lattice parameter of the normal phase and δ measures the deviation from the commensurability. These static displacements give rise to satellite eleastic Bragg reflections which are the signature of the INC state. The INC parameter δ decreases monotonically with decreasing T and jumps to zero at the lock-in transition for both materials; the δ (T) curve is different for each and so far there is no satisfactory method to calculate it.

Since the very complete inelastic neutron scattering study by Iizumi et al. [1] on potassium selenate, an extensive effort, mainly experimental, has been devoted to this family of compounds. In the work by Iizumi it was clear that the dynamics at the transition point T_i are displacive; a transverse acoustic phonon branch of symmetry $\Sigma_3 - \Sigma_2$ (in an extended zone scheme) is shown to soften when the temperature is approached from above at a point $q_{\underset{\sim}{\delta}}$ in reciprocal space. The star at this point has two components $(q_{\underset{\sim}{\delta}}, - q_{\underset{\sim}{\delta}})$, and the order parameter (O.P.) of the transition is thus of dimension 2. This displacive mechanism can be described in the framework of Landau theory.

Let us call the normal mode coordinate of the non-degenerate soft mode $Q_{q_{\underset{\sim}{\delta}}} = \eta_{\underset{\sim}{\delta}}\ e^{i\phi}$. Then the theory predicts two specific excitations in the INC state: namely the amplitude and the phase modes which appear as two independent normal modes, originating from the renormalization of the $Q_{\underset{\sim}{q_\sigma} + \underset{\sim}{q}}$

and $Q_{-q_\delta+q}$ soft branches for small q (where the parabolic dispersion of normal mode is valid) via a fourth order anharmonic term V^4 $Q^2_{-q_\delta}$ $Q_{q_\delta+q}$ $Q_{-q_\delta+q}$. The amplitude mode is the modulation of the amplitude of the static distortion. It is shown to have a dispersion relation with a temperature dependent gap. It can be observed by either Raman spectroscopy or inelastic neutron scattering. The phase mode is the modulation of the phase of the static distortion. It has an acoustic-like dispersion which is due to the invariance of the free energy to an infinitesimal change in the phase of the O.P. in the INC phase.

Theoretical work by R. Pick and H. Poulet [2] predicted that the phase mode should be visible using Brillouin spectroscopy. In potassium selenate several Raman works have demonstrated the existence of the amplitude mode [3-4] but so far in this compound as in thallium tetrabromide or in biphenyl the phase mode has been seen only in the INC phase using inelastic coherent neutron scattering.

2. PHASE MODE IN K_2SeO_4

In potassium selenate we already reported [5] inelastic neutron data which revealed a low energy, overdamped inelastic feature located near the strong (1.312,0,2) satellite reflection at T = 120 K. We assigned this signal to the phase branch, since it is much too broad to be associated with the transverse acoustic branch emanating from the satellite reflection. Our argument was based on the work by Zehyer et al. [6], who showed that there are fluctuating forces acting on the O.P. if its phase is shifted rigidly. If anharmonicity is taken into account, and in the plane wave limit approximation, these forces are shown to be identical with the self energy of the soft mode in the hydrodynamic limit. For us the only way to extract values for the phase mode quasi-harmonic frequency was to fit the data with a damped harmonic oscillator (DHO) function where the damping coefficient Γ_ϕ was extrapolated from that of the soft mode and considered independent of q, following Golovko et al. [7].

Since this earlier work we obtained a more complete set of data. We completed our previous data at 120 K and collected more at 110.5 K, 100 K and 96 K in the INC state. As before, the measurements were recorded on the I N 12 cold-neutron triple-axis spectrometer at I.L.L. We summarize our results in Fig. 1, which gives the quasi-harmonic frequency ω_ϕ of the phase mode for different q in the a^* direction. The positive side refers to data taken near the (1/3[4- δ], 0,2) satellite, while the negative side refers to data taken near the (-1/3[4-δ], 0,2) satellite reflection for focalization setting convenience. This set of data tends to demonstrate

that the phase mode dispersion curve for q parallel to the modulation's
axis a^* is relatively insensitive to T in the whole INC phase and presents
a gap at q = 0. In this range of temperature δ varies from 0.055 to 0.025
in a^* reduced units. Even though this set of data is rather complete, we
do not have a quasi-harmonic frequency value for q < δa^* at any temperature. It is worthwhile to remark that q = δa^* represents the location of
the first gap which is opened in the phase branch when one takes into
account the anharmonic interaction with the long wavelength polarization
$P(3\delta)$ while diagonalizing the Q part of the deviation of the thermodynamic
potential, to obtain the amplitude and phase modes coordinates (8). Dvorak
and Petzelt have shown that the phason with vector q = K = δa^* is both
Raman and I.R. active; this has been confirmed experimentally using Raman
technique by Inoue et al. [9], whose results agree with ours, as demonstrated in Fig. 1.

During this experiment we worked with different incident neutron wavevector (k_i = 1.4 Å^{-1}, 1.55 Å^{-1}, 2 Å^{-1}) so the energy resolution ranged from
15 GHz to 60 GHz; the damping coefficient Γ_ϕ we used during the fitting procedure was 310 GHz at 120 K and 250 GHz at 96K. To answer the question
"does a gap exist or not?" we have to compare the instrumental width with
the line-width of the quasi-elastic Lorentz-like response we would get
for a small wavevector q in the regime where the phase mode is obviously
overdamped ($\omega_\phi \ll \Gamma_\phi$). This linewidth is given by the well known relation $\gamma = \omega_\phi^2/\Gamma_\phi$. If we used it loosely for our lower frequency data point, we get
γ = 58 GHz (q = δa^* at 96 K) which extrapolates to ω_ϕ = 57 GHz at q = 0 and
leads to γ (120 K) = 10.5 GHz and γ (96 K) = 13 GHz. In a typical setting
for a Brillouin experiment the free spectral range has an average value of

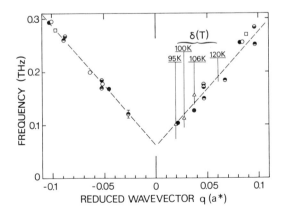

Fig. 1. Phase mode dispersion curve for $q \parallel a^*$: neutron data at 120 K ($\circ\bullet$);
110.5 K ($\square$); 100 K (o); 96 K ($\bullet$), Raman results from ref. 9
($\triangle$) corresponding to 95 K, 100 K and 106 K from left to right.

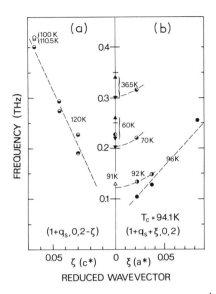

Fig. 2. a) Phase mode dispersion curve for $\underset{\sim}{q} \| \underset{\sim}{c}^{*}$.
         b) Commensurate phase $\underset{\sim}{q} \| \underset{\sim}{a}$. Raman data from refs. 3 (▲),
             4 (▼).

30 to 100 GHz and an instrumental resolution of 0.6 to 2 GHz. With this technique we could study the width of the quasi-elastic phase mode response, as long as it gives a signal that is not too weak to be detected. Nevertheless, if a gap exists, an easy and straightforward explanation would be that near T_i where the displacive model holds, there is pinning by impurities whose nature is not known and which gives rise to a gap; by comparison, near T_c where the so-called soliton regime can describe the INC phase, discommensurations would in any case give a gap for the phase mode dispersion curve.

In Fig. 2a we see that along c^{*} the dispersion of the phase mode is steeper than along $\underset{\sim}{a}^{*}$, which is consistent with the steep soft mode dispersion in that direction. The pseudo-phase mode quasi-harmonic frequency has been measured in the ferroelectric state at 92 K, 70 K, 60 K and 36.5 K. The results, shown in Fig. 2b, are in good agreement with Raman data and give the renormalization of the mode frequency.

Finally, in Fig. 3 we present estimations, for different values of q, of the phase mode dynamical structure factor $F^2 = H\Gamma_{\phi}\omega_{\phi}^{2}/kT$ where H is the fitted height of the damped harmonic oscillator response. Within the approximations of the fitting procedure F^2 appears to be constant, as predicted by theory. For one data point we check that the relation F^2 (ampl.) = F^2 (phas.) holds.

So for potassium selenate the dynamics at T_i seem quite well explained by the displacive model, and this compound remains the only incommensurate system presenting a lock-in transition where a phase mode has been

observed. But in contradiction with theory and as long as we are only
dealing with neutron data, a gap seems to exist at q = 0 in the dispersion
curve; by comparison, for $ThBr_4$ it is shown that the gap does not exist
[10] on the basis of the same quality data.

3.   THE DYNAMICS AT T_i IN K_2ZnCl_4

When one wants to study the other members of this family he is very
soon disappointed. Nothing works as predicted. We have studied both
rubidium and potassium tetrachlorozincate which, as already mentioned,
present the same sequence of isostructural phase transitions as potassium
selenate.

First, the soft mode could not be observed by inelastic neutron scattering in these two compounds. The data obtained in the normal phase of
K_2ZnCl_4 show that there is. no appreciable softening of the Σ_3 - Σ_2 phonon
branch (Fig. 4). Starting from the (4,0,0) Bragg peak, which has the
largest structure factor, we observed a well-formed phonon up to about 0.65
in reduced units. On the other hand, one observes a quasi-elastic diffuse
scattering which peaks at the position of satellite reflections in the INC.
phase (Fig. 5). Gesi et al. (11) have shown that this intensity obeys a
Curie-Weiss law and tends to diverge at T_i. The widths along $\underset{\sim}{b^*}$ and $\underset{\sim}{a^*}$
directions decrease as the temperature approaches T_i.

The critical behavior has not yet been studied, and the correlation
length is not known. Nevertheless, we can assume that this critical behavior is the same as in Rb_2ZnCl_2, which has been well described by a d = 3
XY Heisenberg model [12]. Such a behavior for $T > T_i$ is quite different

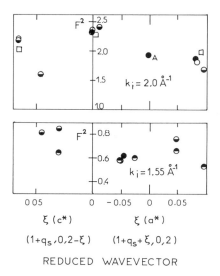

Fig. 3.   Dynamical structure factor F^2 for different value of q and for
two incident neutron wave-vectors.

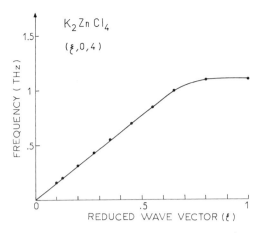

Fig. 4. $\Sigma_3 - \Sigma_2$ Transverse acoustic mode at 588 K (extended zone scheme) in K_2ZnCl_4.

from the one observed by Axe in K_2SeO_4, who sees only the condensation of the soft phonon. This diffuse scattering is clearly a precursor effect of the second order phase transition at T_i, showing the existence of correlated movements, but it does not favor either the displacive model or the order-disorder mechanism. Secondly, when one compares the Raman results obtained below T_i, one sees a very nice amplitude mode in K_2SeO_4, whereas in K_2ZnCl_2 the softening mode is evidenced well below T_i (13).

From these experimental results it seems that differences in the dynamics at the N-INC. transition would exist between potassium selenate and the other A_2MX_4 materials. The question arises: are the dynamics in K_2SeO_4 totally different from those of the others? As put by Axe: "in spite of their isomorphic crystallographic behavior a way of explaining these differences is a structural one." So far very few structural studies of incommensurate phases have been performed. Only recently in the frame of hyperspace theory proposed by de Wolf et al. [14], Yamada et al. [15] have given complete analyses of the structure of K_2SeO_4 and $(NH_4)_2BeF_4$.

In potassium selenate they regard each set of 2 SeO_4 groups and two K atoms as one unit--a carrier of structural change. Each unit has a dipole moment, depending on the phase of the modulation wave. The main modulation in the INC. state is given by the rotation of the SeO_4 groups around the c-axis and the associated translation of K atoms along the b-axis, in Pnma notation. The transition at T_i is connected with the librational modes of SeO_4 group regarded as rigid bodies. With this model the eigenvectors of the soft $\Sigma_3 - \Sigma_2$ mode are slightly different from those proposed by Axe and in good agreement with those calculated by Haque et al. [16]. For the tetrachlorozincates compounds it was suggested that the mechanism of the

133

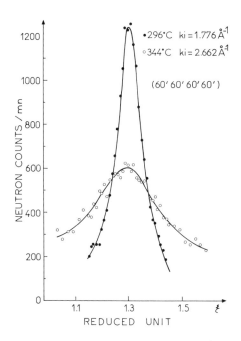

Fig. 5. Quasi elastic diffuse scattering at $(\xi,0,2)$ in K_2ZnCl_4.

transition at T_i would be of the order-disorder type. We have done neutron
diffraction measurements in the normal high temperature phase of K_2ZnCl_4.
Our experiment was done on a single crystal mounted in a furnace, which
allows the temperature fluctuations to be kept within ½ K at 588 K. It was
performed on the P 110 four-circle diffractometer at the ORPHEE Reactor at
Saclay. A total of 1841 reflections with $0 \leqslant h \leqslant 10$, $-8 \leqslant k \leqslant 0$,
$-16 \leqslant 1 \leqslant 15$ were measured by ω scans up to $\sin \theta/\lambda$ = .691 $Å^{-1}$. The raw
data were corrected for the Lorentz factor L and for absorption
(μ = 1.5 cm^{-1}). The internal R value of the reduced data was 0.012. We
observed that the intensities of the Bragg peaks are anomalously weak for
large values of Q, which we think is a signature of dynamical motions. The
lattice constants were determined by least squares refinement:
a = 9.052(1) $Å$, b = 7.370(8) $Å$, c = 12.676(7) $Å$. The refinements are based
on 310 unique unrejected reflections with $F > 3\sigma(F)$ out of 981 independent
ones, and calculations used the program system PROMETHEUS [17]. The
refinements of our data using anisotropic temperature factors for all atoms
evidence very large strongly anisotropic thermal motion for K atoms as well
as for Cl atoms (Fig. 6a). At this stage of refinement the question was
"does a state of static disorder exist which is correctly described by a
split-atom model or do the atoms show strong anharmonic temperature motions
not correctly described by standard harmonic temperature formalism (dyna-
mic-disorder)?"

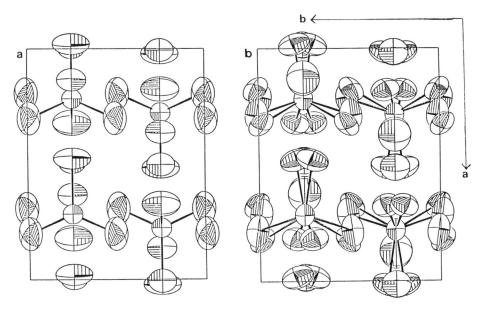

Fig. 6.   ORTEP projections on the (ab) plane of the structure unit cell.
(a) Harmonic model anisotropic temperature factors;
(b) Harmonic split model anisotropic temperature factors.

We have tested two models: a split-atom model and a so-called anharmonic model. In the first one, we succeeded in applying a model where all
the Cl atoms have two equally occupied positions (Fig. 6b). This split
model is the same as proposed for Rb_2ZnCl_4 [18-19]. It gives very large
mean-square displacement values for all atoms and especially for K atoms.
Although a reasonable R factor (R = .026) of the refinement is achieved,
this model cannot stand for a disordered state of the K_2ZnCl_4 normal phase,
explaining the proposed disorder-order behavior with a tripling of the unit
cell in the ferroelectric phase. Attempts with a split model following the
pseudo-hexagonal symmetry of the β-K_2SO_4 structure have not been successful
yet. Subsequently, a refinement including higher order cumulants has been
worked out. Starting with the non-split model we introduced higher order
cumulants up to the fourth order in the Gram-Charlier expansion for the
structure factors of the Cl atoms only. Keeping the significant parameters
we got a slightly improved R value (R = .019) with a total of 76 instead of
56 parameters. Attempts to introduce higher order cumulants also for the
description of the K atoms distribution failed. We have drawn sections of
the probability density function (p.d.f.) maps in the main crystallographic
planes for all the atoms. The p.d.f. map section (bc) of Cl_1 atom (Fig.
7a) seems to indicate a librational motion around the Cl_1-Zn bond. The
anharmonic contribution of this map is shown (Fig. 7b). The Zn atoms are
purely harmonic and almost isotropic (Fig. 8). At last from the p.d.f

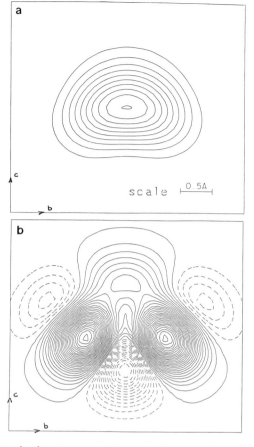

Fig. 7.   Section in (cb) plane of Cl_1 p.d.f. map obtained with:
a)   second, third, fourth order contribution;
b)   third and fourth order (anharmonic) contribution only.

section maps of the K atoms (Fig. 9) one sees that their position is not
well defined between two adjacent $ZnCl_4$ groups. It appears that this model
is better adapted to describe the highly dynamic situation of K_2ZnCl_4 in
the normal phase. It comes out also that there is a strong coupling
between the K position and the individual orientations of the two adjacent
equidistant $ZnCl_4$ groups which is not taken into account in our models.
So both structure models allow a good description of the observed density,
but they do not allow a definitive conclusion to be made on the mechanism
of the transition at T_i. As a temporary conclusion of what we have presented here, we think that one cannot ascertain that in K_2ZnCl_4, like Rb_2ZnCl_4
and Rb_2ZnBr_4, the relevant atomic displacements are in an effectively
double potential. We showed that a single anharmonic potential well cannot be rejected. Although it is appealing to look for a unique simple
model which could describe the mechanism of the N-INC transition, it seems
to us that the amonium salts for which the disordered model has been mainly
proposed have to be considered as a separate group within the whole

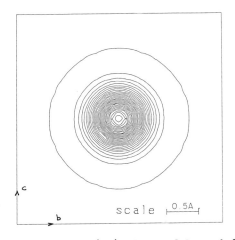

Fig. 8. Section in (cb) plane of Zn p.d.f. map.

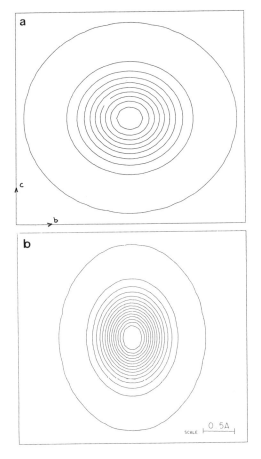

Fig. 9. Section in (cb) plane of:
a) K1 atom;
b) K2 atom.

A_1MX_4 family. In $(NH_4)_2SO_4$, $(NH_4)_2ZnCl_4$ and $(NH4)_2BeF_4$ the ammonium cat-ions are the disordered groups and the presence of hydrogen bonds makes the problem more complex than in the K_2SeO_4-like compounds. Moreover, in $(NH_4)_2SO_4$ the paraelectric-ferroelectric phase transition occurs without change of unit cell, while in $(NH_4)_2BeF_4$ where an INC. phase does exist, the lock-in wavevector is $q_\delta = 1/2\ a^*$; this may be the reason why the two-position split-atom model seems well adapted to describe the data.

4. ACKNOWLEDGMENTS

We would like to thank R. M. Pick with whom we had many fruitful dis-cussions, and C. Dugautier for his kind help.

5. REFERENCES

1. M. IIzumi, J. D. Axe, G. Shirane, K. Shimaoka, Physical Review B15, 4392 (1977).
2. H. Poulet, R. M. Pick, Journal of Physics C14, 2675 (1981).
3. M. Wada, A. Sawada, Y. Ishibashi and Y. Takagi, Journal of the Physical Society of Japan 42, 1229 (1977); 45, 1905 (1978).
4. H. G. Unruh, W. Eller, G. Kirf, Physica Status Solidi A55, 173 (1979).
5. M. Quilichini and R. Currat, Solid State Communications 48, 1011 (1983).
6. R. Zehyer and W. Finger, Physical Review Letter 49, 1933 (1982).
7. V. A. Golovko and A. P. Levanuyk, JETP 54, 1217 (1981).
8. V. Dvorak and J. Petzelt, Journal of Physics, C11, 4827 (1978).
9. K. Inoue and Y. Ishibashi, Journal of the Physical Society of Japan 48, 1785 (1980); 52, 556 (1983).
10. R. Currat, L. Bernard, and P. Delamoye, Incommensurate Phases in Dielectrics. 2 Materials, 161 (1986).
11. K. Gesi and M. Iizumi, Journal of the Physical Society of Japan 53, 4271 (1984).
12. S. R. Andrews and H. Mashiyama, Journal of Physics C16, 4985 (1983).
13. M. Quilichini, J. P. Matheiu, M. Le Postollec, and N. Toupry, Journal de Physique 43, 787 (1982).
14. P. M.de Wolff, Acta Crystallogr. A30, 777 (1974).
15. N. Yamada and T. Ikeda, Journal of the Physical Society of Japan 53, 2555 (1984).
16. M. S. Haque and J. R. Hardy, Physical Review B21, 245 (1980).
17. U. H. Zucker, E. Perenthaler, W. F. Kuhs, R. Bachmann, and H. Schulz, Prometheus, Program System for Structure Refinements, J. Applied Cryst. 16, 358 (1983).
18. K. Itoh, A. Hinasada, H. Matsunage and E. Nakamura, Journal of the Physical Society of Japan 52, 664 (1983).
19. M. Quilichini and J. Pannetier, Acta Crystallogr. B39, 657 (1983).

SOME ASPECTS OF RAMAN SCATTERING FROM A_2BX_4 COMPOUNDS

F. G. Ullman

Departments of Electrical Engineering and Physics

V. Katkanant, P. J. Edwardson and J. R. Hardy

Department of Physics
University of Nebraska
Lincoln, Nebraska 68588

1. INTRODUCTION

Since the development of ion lasers made Raman spectroscopy a routine
analytical tool, it has been used extensively in studies of structural
phase transitions in solids. Even though it is restricted in most cases
to zero wavevector excitations, the effects of non-zero wavevector insta-
bilities can also be evident in the spectra. It has also proved useful
for the study of the incommensurate phase transitions that were the subject
of this workshop. To illustrate how Raman scattering studies have contri-
buted to a better understanding of incommensurate phases in ferroelectric
insulators, we have selected for this article two quite different studies
in which Raman scattering has provided new information about the incommensurate ferroelectrics, K_2SeO_4 and Rb_2ZnCl_4.

First, we describe measurements of the temperature dependence of the
amplitude mode of K_2SeO_4 in its lock-in phase over a range from about
eighty degrees to just below its incommensurate transition, T_i, which
occurs at about 130 K. It will be shown that these data can be analyzed
by purely classical considerations with no need to invoke theories of critical behavior or to rely on partial fits to Curie-Weiss laws.

Second, we describe measurements of the internal mode spectra of
Rb_2ZnCl_4 which cannot be described by conventional factor group analysis.
The results support a model proposed earlier (1) to explain similar but
less obvious effects in the spectra of K_2SeO_4. The model attributes the
lack of agreement with conventional analysis to orientational disorder of
weakly coupled selenate ions. An explanation for the cause of the disorder

139

has been developed from an analysis of the dynamics of Rb_2ZnCl_4 and K_2ZnCl_4 [2].

2. TEMPERATURE DEPENDENCE OF THE AMPLITUDE MODE OF K_2SeO_4

Strictly speaking, the amplitude and phase modes exist only in the incommensurate phase but since they become normal Raman-active phonons in the lower-temperature, lock-in phase, the nomenclature of "amplitude mode" and "phase mode" has carried over to those phonon analogues. The observation of these modes by Raman-scattering was first reported by Wada et al. [3,4] and repeated somewhat more precisely by Unruh et al. [5]. The amplitude mode softens on approaching T_i from below and appears to extrapolate to zero frequency at T_i. According to the usual mean-field theory, the square of the mode peak-frequency should be linear with temperature but is found to be nonlinear although the plotted data can be fitted to a straight line over a small range close to the transition. Unruh et al. suggested that the data could be described by a temperature dependence with a critical exponent; Fleury et al. [6] proposed that their data fell on two intersecting lines of different slope. In our laboratory, we studied the dependence of the incommensurate transition temperature on uniaxial stress and used a linear extrapolation near the transition to determine T_i and its stress dependence [7]. In subsequent similar measurements by Wada et al. [8] the stress dependence was determined to be nearly a factor of two lower. (The value we reported in Reference 7 was in error because of an incorrect estimate of the piston area. However, even when corrected, it was about seventy percent larger than reported in Reference 8.) This led us to repeat the stress dependence studies by birefringence measurements [9] which gave close agreement with the results of Reference 8 and showed that measurements with stress along the c-axis (c < a < b) were poorly reproducible with the stress dependence increasing with each successive stress cycle. This also led us to question the extrapolation procedures used in our earlier Raman measurements [7]. Shortly after that work was completed, a paper by Kudo and Ikeda [10] appeared on the birefringence of K_2SeO_4 measured over the temperature range 30-140 K. They were able to fit their data to a function derived from a sixth-degree Landau free energy, depending on the order parameter for the incommensurate phase and a secondary order parameter (η_p) that is a polarization wave that becomes the macroscopic polarization in the lock-in phase.

Following Reference 10, the free energy, G, and the solution derived from it for the square of the order parameter, η , is given below in Eqs. 1 and 2.

$$G = G_0 + \alpha \eta^2/2 + \beta \eta^4/4 + \gamma \eta^6/6 + A\eta_p^2/2 - 2|B|\eta^3 \eta_p \tag{1}$$

$$\eta^2 = \frac{-\beta + [\beta^2 - 4(\gamma - 12|B|^2/A)\alpha]^{\frac{1}{2}}}{2(\gamma - 12|B|^2/A)} \tag{2}$$

where $\alpha = \alpha_0(T_i - T)$ and β, γ, and $|B|$ are constant coefficients.

Assuming that the squares of the amplitude mode peak frequency, ω, and the order parameter, η, have the same temperature dependence, and normalizing ω^2 to its value at the lowest measurement temperature, T_3, results in the expression given below in Eq. 3.

$$\frac{\omega^2(T)}{\omega^2(T_3)} = \frac{\beta \omega^2(T_3)}{\alpha_0(T_3 - T_1)} + \left\{ \frac{\beta^2 \omega^4(T_3)}{\alpha_0^2(T_3 - T_1)^2} + \left[4 \frac{1}{\alpha_0(T_3 - T_i)} \right. \right.$$

$$\left. \left. + \frac{\beta \omega^2(T_3)}{[\alpha_0(T_3 - T_i)]^2} \right] \alpha_0(T - T_i) \right\} \tag{3}$$

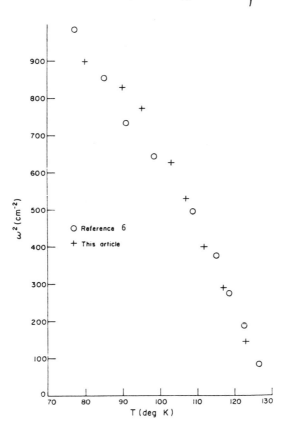

Fig. 1. The amplitude mode peak frequency vs. temperature as measured by Raman scattering [6,7].

141

In Fig. 1, zero-stress data measured by us [7] and by Fleury et al. [6] are shown. Since their data were presented in terms of a temperature difference relative to a transition temperature determined from the properties of the central peak, we have used their estimate of the transition temperature to plot their data on the same temperature scale as ours. Differences in the two sets of data can be attributed to the experimental error in reading the peak frequency which was estimated directly from the data with no attempt to obtain higher precision from a Lorentzian fit, and from a few degrees of laser beam heating in the scattering volume that was not accounted for in our measurements.

Both of these sets of data are fitted to Eq. 3 in Fig. 2 and the calculated parameters for both are given below in Table 1.

It can be seen from Fig. 2 that the data can be fitted with good precision by Eq. 3 over a relatively wide range of temperature (about fifty degrees in this case). Therefore, there is no need to invoke critical exponents or to assume different Curie-Weiss laws for different temperature regions to explain these data. It is only necessary not to make the mistake of assuming that the sixth degree term in the Landau expansion can be discarded because the incommensurate transition is of second order. Of course, at temperatures much closer to the transition, the system could exhibit critical behavior but so far, this has not been observed [10].

3. BEHAVIOR OF THE INTERNAL MODES OF K_2SeO_4 AND Rb_2ZnCl_4

Several years ago, when we first began work in our laboratory on incommensurate structures, Haque and Hardy [11] initiated a calculation of

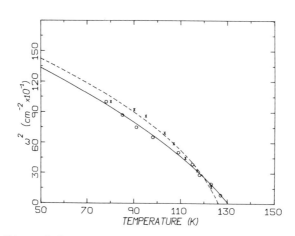

Fig. 2. Fits of data in Fig. 1 to Eq. 3. Reference 6 data, o;
Reference 7 data x.

Table 1. Calculated Parameters from Eq. 3 for Our Data [7],
and for Data from Reference 6.*

Reference 7	Reference 6
α_o = 0.2119	α_o = 0.1713
β = 0.4031 x 10^{-2}	β = 0.6313 x 10^{-2}
T_i = 126.2	T_i = 130.4

* Note the difference in the T_i values which indicates
the amount of laser beam heating unaccounted for in
our measurements.

the lattice dynamics of K_2SeO_4. For the purposes of comparison and to have
reasonable starting values for their calculations, they used the Raman frequencies then available in the literature [12]. They noted that in the
selenate bending and stretching mode regions, there were more lines than
predicted by the factor group analysis. At that time, we had just begun
our own Raman-scattering measurements on crystals grown in our laboratory,
and we found the same results. In most of the spectra, the observed internal mode lines were either strong, or weaker by one to two orders of magnitude. This led Fawcett et al. [12] to suggest that the weak lines were
artifacts arising from polarization leakage, i.e., from lines that are
strong in other scattering configurations leaking some of their intensity
into configurations in which they are symmetry-forbidden. Although this
explanation is reasonable, we questioned its validity in this case because
the same lines were observed in all of the samples studied. Consequently,
we began to seek other explanations. We considered several: the afore-
mentioned polarization leakage, birefringence depolarization, angular dispersion of polar phonons, a loss of the center of inversion through some
subtle orientational disorder of the selenate ions not apparent from the
X-ray data, and the effect of an orientational disordering combined with a
decoupling of the selenate sublattice.

The first, polarization leakage, is difficult to measure or even to
estimate for these birefringent systems. The incident laser beam is
slightly convergent in the crystal and the scattered light is collected
over a solid angle of about five degrees, so weak components of undesired
polarization can possibly be present. Further, the crystal is rarely
polished with faces exactly parallel to the crystal axes, permitting a
small amount of off-axis propagation with concomitant depolarization.
Probably, polishing errors as large as five percent are possible. In our

measurements, the errors introduced by laser beam convergence and finite
scattering angle were found to be small so the major leakage effects could
be attributed to imperfect face alignments with crystal axes. The
importance of this effect can be checked by comparing spectra with the
same incident and scattered <u>polarizations</u> but different incident and scattering <u>directions</u>; e.g., compare c(bc)a with c(bc)b or a(bc)b (this is
standard Raman notation with incident and scattered polarization respectively inside the parentheses). From these measurements, we could estimate
these leakage effects in one or two extreme cases to contribute as much as
fifty percent to the intensity of a given peak; in other cases, the effects
were much smaller. Leakage is therefore a contributing factor that needs
to be accounted for in any quantitative analysis of the data, but the presence of symmetry-forbidden lines in the K_2SeO_4 spectra cannot be attributed completely, or even largely, to leakage effects.

The effects of birefringence depolarization and angular dispersion
could be estimated and could not acccount for the observed multiplicity.

We considered next the effect of the loss of the center of inversion
symmetry which might occur through some subtle disorder whose effects might
not be noticeable in the X-ray diffraction patterns. Examination of the
correlation table given in Table 2 below for the D_{2h}^{16} point group shows that
18 Raman-active modes are expected but if the center of inversion is lost,
the infrared-active modes also would become Raman-active and then 36 modes
would be expected. Furthermore, the infrared-active lattice modes should
also become Raman-active. Neither of these effects was observed.
What was observed (1) was the correct number of lattice modes for the D_{2h}^{16}

Table 2. Correlation Table for the Internal Modes of the Selenate Ion in
K_2SeO_4 with D_{2h}^{16}

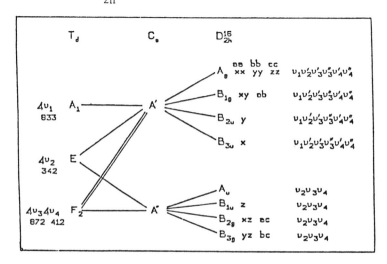

symmetry but only essentially the same nine internal modes in each of the four Raman-active factor group spectra. Thus, another explanation was sought which led to the model proposed in Reference 1, which attributes the observed effects to an orientationally disordered selenate sublattice with only weak coupling between the selenate ions. The expected internal modes for such a system would be those of the free selenate ion with the degeneracies removed by the crystal field of the potassium sublattice which would be nine in number.

Subsequently, with more precise measurements, at room temperature, of the B_{1g}, B_{2g}, and B_{3g} spectra (Unruh [13]) showed that the peak frequencies were more than nine in number with differences among nearly identical lines of at most three wavenumbers. We have recently confirmed those results with measurements (at 140 K) on a sample of better quality than we had previously, using neon lines for calibration, and with a numerical analysis of peak frequency interpolated from the digitized data, giving a precision of ± a few tenths of a wavenumber. We do not believe that these small differences in frequency negate the proposed disorder model because the calculated frequencies for the ordered structure [11] differ by as much as twenty wavenumbers, as shown below, along with the experimental frequencies, in Table 3. To be noted in the Table are the frequency differences between the B_{1g}, B_{2g}, and B_{3g} modes at the experimental frequencies of 342-345 cm^{-1}, 418-421 cm^{-1}, and 876-879 cm^{-1}, compared with calculated differences of 311-345 cm^{-1} 418-433 cm^{-1}, and 874-888 cm^{-1}. Since the model is qualitative and not the result of an exact calculation, such small differences among the same vibrations excited along different directions relative to the crystal axes are not unexpected.

Seeking further verification of the model, we turned to the isomorphs of K_2SeO_4, beginning with rubidium tetrachlorozincate, Rb_2ZnCl_4. This compound introduces two new problems in the measurements of the Raman spectra. First, the spectrum is compressed by a factor of three relative to K_2SeO_4 so that the bending mode region is superposed, or nearly so, on the lattice region and consequently may not be well-characterized by the internal mode approximation. Second, the incommensurate phase transition occurs at 302 K where line broadening interferes with resolution, even for K_2SeO_4, so for Rb_2ZnCl_4 the spectral compression makes the resolution even worse. To obtain better resolution, we measured the spectra over the temperature range 20-315 K, and concentrated on those in the incommensurate phase above 189 K, following their evolution with increasing temperature to see if any of the observed structure was introduced by the incommensurate transition. In Fig. 3, typical spectra for the stretching region are shown for the four factor groups.

Table 3. Comparison of Calculated Internal Mode Frequencies [11] with Measured Values.*

| A_g | | B_{1g} | | B_{2g} | | B_{3g} | |
Calc	Exp	Cal	Exp	Calc	Exp	Calc	Exp
292	332				332		332
	342-344	336	342	311	342	334	345
438	412-413				413		
	420	443	420	433	421	446	418
458	430-431	432	431		431		431
840	843-844	844	844		843		843
873	866-867				866		866
	876-878	888	877	876	879	874	876
910	905-906	906	905		905		905

*The calculated numbers do not all match well with experimental values because some of the measured values used for fitting were not correct at that time. The differences, however, between corresponding lines in different symmetries are reliable.

Note first that the line peaking at about 285 cm^{-1} seems to vanish between 200 and 298K. However, our numerical analysis of the four-peak c(bb)a spectrum showed [14] that a better fit is obtained at 298 K with that peak included, but at 315 K, the quality of the fit is just as good with or without it. However, as described in Reference 14, the integrated intensity of that peak does not have the same temperature dependence as the order parameter for the incommensurate phase so its apparent disappearance at the phase transition must have some cause other than the phase transition.

A second puzzling fact to note is that the strongest peak in the A_g spectra is not at the lowest frequency of the four, as in K_2SeO_4, and instead, the lowest peak corresponds to the strongest peak in the B_{2g} and B_{3g} spectra.

These two apparent discrepancies may be resolved by the results of our theoretical computations of the Raman frequencies (as yet unpublished) whose results can be cited here. In Table 4, the experimental and calculated frequencies for the stretching modes are given.

The theory (from examination of the atom displacements) shows the symmetric stretching mode, which should have the highest intensity of the four modes, to have the lowest frequency of the group, as for K_2SeO_4, in contrast to the already-cited experimental result. Further examination of

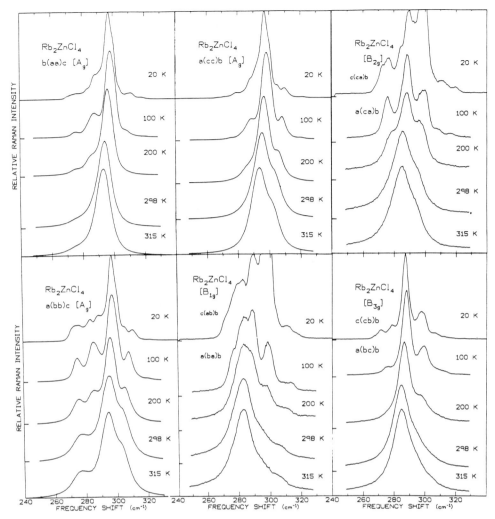

Fig. 3. Raman spectra of Rb_2ZnCl_4 at several temperatures from 20 K to 315 K.

the calculated lines, as can be seen in Table 4, shows that the symmetric stretching mode frequency and the three next-higher frequencies are nearly degenerate. If disorder exists as we have postulated, then these nearly degenerate lines can couple harmonically, which can produce the kind of intensity transfer between modes that was observed. Also, the disappearance of the 285 cm^{-1} mode above T_i becomes plausible when it is recognized that it is a coupled mode spectrum that is being observed. The increasing overlap with increasing temperature could produce increasing intensity transfer between the lines until one vanishes.

Table 4. Experimental and calculated (for Pnam) frequencies of the internal modes in the stretching region.

EXPERIMENT				THEORY			
A_g	B_{1g}	B_{2g}	B_{3g}	A_g	B_{1g}	B_{2g}	B_{3g}
275–277	278	278	278	276			
285–288	283	285	284		276	274	267
294–295	295	295	293	291	304		
302–306	306	304	303	320	333		

The arguments presented above are all based on the existence of the orientational disorder that was postulated previously [1]. For K_2SeO_4, the only supporting evidence for the disorder is the internal mode multiplicity in the Raman spectrum as discussed here and a slight anisotropy in the ORTEC plots of our X-ray structure analysis [1]. The effects in Rb_2ZnCl_4 are larger than in K_2SeO_4 and so the disorder in that compound has been seen in X-ray diffraction studies [15,16]. In fact, those authors state (Reference 16, p. 287) ". . . such a disordered structure should be widely seen in the normal phase of the beta-K_2SO_4 type compounds." That this is

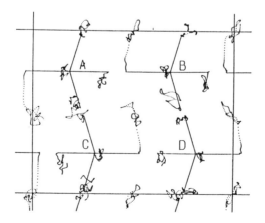

Fig. 4. Molecular Dynamics Simulation of the Pnam Rb_2ZnCl_4 Structure. The dotted lines show the evolution with time of the chlorine atom positions from their Pnam positions shown by the solid line termini. The zinc atoms are located at the center of the tetrahedra as indicated by the letters, A, B, C, and D.

so is supported by recent reports [17] of the Raman spectra of $CsZnI_4$, $CsZnBr_4$, and $CsZnCl_4$, in which can be seen the same type of internal mode multiplets as we have described here.

Finally, we have now confirmed the occurrence of the disorder in Rb_2ZnCl_4 from a molecular dynamics simulation [2] using Gordon-Kim potentials as in earlier work [18]. An example of the results is shown in Fig. 4 in which it can clearly be seen that while the Rb^+ and Zn^{+2} sublattices maintain their integrity, the $ZnCl^{2-}$ tetrahedra are highly disordered rotationally. Thus the Pnam starting points for the chlorine atoms have been shown to be unstable. The actual displacements are probably not as large as shown; forthcoming, more refined calculations will make these simulations more exact. It is not possible yet to extend these calculations to K_2SeO_4 in which we believe the same disorder occurs but on a much smaller scale. It remains to confirm this when similar calculations are possible.

4. CONCLUSIONS

Our purpose in this article was to give examples to demonstrate the usefulness of Raman spectroscopy for the study of incommensurate structures. The two examples selected, one concerned with low frequencies, the other with higher frequencies, were deliberately chosen to show that useful information is contained in the entire Raman spectrum. The importance of searching for low-frequency unstable modes associated with phase transitions with Raman spectroscopy is, of course, well-known but the importance of other parts of the spectrum should also be recognized. In this case, a detailed study of the internal modes has led to a new theory for incommensurate behavior [2].

5. REFERENCES

1. N. E. Massa, F. G. Ullman, and J. R. Hardy, Phys. Rev. B27, 1523 (1983).
2. V. Katkanant, P. J. Edwardson, J. R. Hardy, and L. L. Boyer, private communication, submitted for publication.
3. M. Wada, A. Sawada, Y. Ishibashi, Y. Takagi, and T. Sakudo, J. Phys. Soc. Jpn. 412, 1229 (1977).
4. M. Wada, H. Uwe, A. Sawads, Y. Ishibashi, Y. Takagi, and T. Sakudo, ibid. 43, 544 (1977).
5. H. G. Unruh, W. Eller, and G. Kirf, Phys. Stat. Sol. A55 (1979).
6. P. A. Fleury, S. Chiang, and K. B. Lyons, Solid State Commun. 31, 279 (1979).
7. N. E. Massa, F. G. Ullman, and J. R. Hardy, Solid State Commun. 42, 175 (1982).
8. M. Wada H. Shichi, A. Sawada, and Y. Ishibashi, J. Phys. Soc. Jpn. 51, 3681 (1981).
9. D. P. Billesbach, F. G. Ullman, and J. R. Hardy, Phys. Rev. B32, 1532 (1985).
10. S. Kudo and T. Ikeda, J. Phys. Soc. Jpn. 50, 3681 (1981).

11. M. S. Haque and J. R. Hardy, Phys. Rev. B21, 245 (1980).
12. V. Fawcett, R. J. B. Hall, D. A. Long, and V. N. Sankaranarayanan, J. Raman Spectroscopy 2, 629 (1974; 3, 299 (1975).
13. H. G. Unruh, "Raman Spectroscopy-Linear and Non-linear," J. Lascombe and P. Huong, eds. (John Wiley, New York, 1982).
14. V. Katkanant, F. G. Ullman, and J. R. Hardy, Japanese J. Applied Phys. 24, Suppl. 24.24, 775 (1986).
15. K. Itoh, A. Hinisada, H. Matsunaga, and E. Nakamura, J. Phys. Soc. Japn. 52, 664 (1983).
16. K. Itoh, A. Hinisada, M. Daiki, A. Ando, and E. Nakamura, Ferroelectrics 66, 287 (1986). (Supported by the U.S. Army Research Office).
17. O. P. Lamba and S. K. Sinha, Solid State Commun. 57, 365 (1986).
18. L. L. Boyer and J. R. Hardy, Phys. Rev. B24, 2577 (1981).

INTRINSIC DEFECTS IN INSULATING INCOMMENSURATE CRYSTALS

Pierre Saint-Grégoire

G.D.P.C. (L.A. cnrs n° 233)
Université des Sciences et Techniques du Languedoc
34060 Montpellier Cedex, France

1. INTRODUCTION

In spite of a considerable amount of studies in incommensurate systems, relatively little attention has been paid to the role that intrinsic defects can play. The aim of this paper is less to give a complete view on the theoretical and experimental situation than to present the fundamental ideas and some of their confirmations.

Generally, the incommensurate phase is stable between two normal crystalline phases (called N and C), and is predicted to be a modulation of the low temperature "C" phase ("lock-in phase") [1], with a wave vector which is temperature dependent. If the Lifschitz criterion is not satisfied between N and C phases, close to the lock-in transition at T_c, the incommensurate phase is predicted to be constituted of quasi-commensurate regions, regularly separated by walls (the discommensurations, "DC's") [1,2,3]. At the I-C transition, m, the density of discommensurations is going down to zero, and it is why some authors have proposed to consider this quantity as the O.P. [4]. In the phenomenological theory, the order of the transition is found to be the second one, but coupling with other degrees of freedom can explain the first order character [4,5] which is observed in practice; it implies that both phases (C and I) coexist in a small temperature range. We shall see that the interface is composed of linear defects. The exact mechanism for removing DC's is unknown in most of the cases. The particular interest of the invoked defects is that they provide for the wall density changes, another mechanism than the frontal motion of walls.

The next paragraph presents the problem of these intrinsic defects and a summary of theory; the third part is devoted to experimental evidences

for the presence of defects in some I-phases of insulators, and their role
is discussed.

2. THEORETICAL BACKGROUND

For the problem of interface structure, the situation is akin to that
in incommensurate C* smectic liquid crystals where the helicoidal structure
exists in the bulk, whereas because of anchoring, the molecules situated in
the neighborhood of the glass plates limiting the sample, are all parallel. The junction between both regions is made possible by linear defects
[6] called dechiralization lines (Fig. 1). As the chiral molecules are
polar, applying an electric field provokes the transition to the homogeneous structure which is observed to be governed by motion and nucleation of
dechiralization lines [7].

By the same way, in crystals with a modulation in a single direction,
linear defects can compose the interface I-C. For discussing this point,
it is convenient to use the concept of domain texture of the I phase. In
the case of the transition occurring in AFB, for example, the order parameter has two components p and q; the solution which holds close to T_i is
p=A sin(kx), q=A cos(kx), and can be represented on a circle of radius A,
described at a constant velocity (Fig. 2-a). The sense of rotation depends
on the sign of the coefficient d in front of the Lifschitz invariant in the
free energy expansion. Indeed, in polar coordinates, this term takes on
the form $A^2 d\theta/dx$ (θ is the phase of the modulation) and the stable state
corresponds to $d\theta/dx > 0$ if $d < 0$. Close to T_c, the circle is distorted and
the velocity no more constant: the system tends to create (commensurate)
regions with a constant θ (corresponding to domains of the C phase),

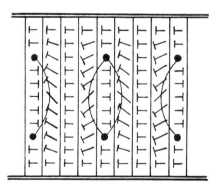

Fig. 1. Dechiralization lines in C* smectic liquid crystals. "Nails" are
         a perspective representation of the molecules. Chirality is con-
         served in the bulk, whereas anchoring prevents it close to the
         glass plates. Dots correspond to the section of singular lines
         (from ref. [6]).

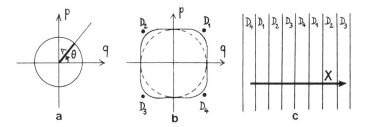

Fig. 2. 2-a Representation of the I phase for $T \sim T_i$ in the order parameter
space (case of ammonium fluoberyllate AFB).
2-b Representation of the I phase for $T \sim T_c$, and of domain states
D_i of the C phase (AFB).
2-c Domain texture of the I phase in AFB.

regularly disposed in the x direction (Fig. 2-b). The domain-texture
approximation (Fig. 2-c) consists in neglecting the size of transient regions: the I phase is then considered to be a succession of domains D_1, D_2,
$D_3, D_4, D_1, \ldots$. It can be easily seen that at the interface between I and
the C phase in e.g., the domain state D_1, the three domains D_i (i=2,3,4)
have to terminate in a line which is the defect in question (Fig. 3), called deperiodization line ("D-line").

The possibility of existence of defects in the DC's lattice, was first
considered for CDW systems by McMillan [8] and Walker [9]. A direct observation of 2H-TaSe$_2$ by electron microscopy [10] revealed actually their
presence in the I phase of this compound. Janovec studied the topology
[11] and their contribution to hysteresis phenomena [12]. D-lines can
induce changes in the wall density by their lateral motion, and they are
moreover apt to form loops, thus realizing nuclei of I or of C phases.

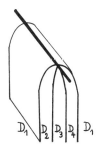

Fig. 3. Linear deperiodization defect in the case of AFB (N=4).

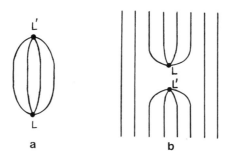

Fig. 4.   Elementary nuclei of I (a) and C (b) phases (AFB).

Such nuclei are represented in Fig. 4; L and L' lines annihilate mutually
when they meet together. Of course, other associations are possible, and
can be involved in particular in nucleation processes in samples in a poly-
domain state as it is illustrated for some cases in Fig. 5. In the general
case of a C phase with N domain states, N walls have to terminate at the
deperiodization line; N configurations are possible for the lines, and
$2N^2-N$ for the associations of lines, namely for elementary nuclei (with
(N-1) domains) plus the loops surrounding 0, 1,...,N-2 domains. The generalization to an incommensurate phase with a star (of modulation wave vector) having more than two arms, needs a special analysis in each case.

The homogeneous nucleation was investigated by Prelovsek for the case
of Rb_2ZnCl_4 (N=6) [13]. This author gave quantitative calculations of
nucleation energy and rate, and of the critical radius of nuclei. The
hysteresis was overestimated (20 K instead of few K according to experiments) perhaps due to the approximate determination of the energy or
because fluctuations of walls were neglected.

The mechanism for the wall density variation involving deperiodization
defects, differs considerably with a frontal motion of walls eliminated at
the surface of the sample. In systems in which the I-C transition is

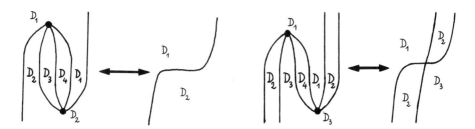

Fig. 5.   Examples of defects associations, and of configurations in
          the C phase which can give rise to them.

expected to be continuous, D-defects can exist only because of a pinning by extrinsic defects which cause the existence of metastable states. For the case of N-C transitions satisfying the Lifschitz condition, and where the intermediate I phase is attributed to a "quadratic-gradient" term $(\nabla \eta)^2$ in the free energy expansion [14], the I-C transition is predicted to be discontinuous [15] and D-defects can therefore be expected even in the ideal case when there is no extrinsic pinning. Moreover, it was demonstrated recently that in such systems, the domain walls can be intrinsically pinned relative to one another [16].

Let us now describe more specifically the consequences of these intrinsic defects on the thermal variation of the wall density. This problem was considered by Janovec et al. in Ref. [12]; pinning of walls was taken into account by introducing a minimal force f_{p_0} needed to shift them. On cooling, the density of walls is $m^c(T) = m^o(T + \Delta T)$ where m^o refers to the equilibrium density (without pinning) and ΔT is expected to increase on approaching the second order I-C transition temperature T_c. This favors the nucleation of C phase or of I phase with a m closer to equilibrium m^o. On heating, because of the pinning, the system is superheated up to a temperature T^h above T_c and the nucleation of I phase occurs. Again when T increases on, the system passes metastable states $m^h(T) = m^o(T - \Delta T)$ and eventually other D-lines are created to permit nucleation of I phase with a higher m value. If the nuclei are nearly free to expand, the transition is discontinuous; but if the deperiodization lines are in turn hindered in their motion or pinned by defects, the temperature range in which the phases coexist, is enhanced and the transition is blurred out.

3. EXPERIMENTAL EVIDENCES OF D-DEFECTS IN INSULATORS

Modulated Phases in the A_2BX_4 Crystalline Family

Most of the incommensurate phases encountered in this class of materials are due to the non-satisfaction of the Lifschitz criterion in the symmetry breaking $N = Pnma(D_{2h}^{16}) \to C$. The modulation occurs always in a single direction, but its wave vector q locks in at different values in the C phase: $q_c = 1/2$ in $(NH_4)_2BeF_4$, $1/3$ in Rb_2ZnCl_4, Rb_2ZnBr_4, $2/5$ in $(TMA)_2ZnCl_4$, etc. (see, e.g., Ref. [17]). The expected domain textures are therefore differing, as they do not involve the same number N of domains: one has $N=4$ ($q_c=1/2$), $N=6$ ($q_c=1/3$), $N=10$ ($q_c=2/5$).

Adjunction of impurities induces properties that are in good agreement with the analysis of Janovec et al. [12]: it is indeed a general feature that the hysteresis is then enhanced, I phases coexist in a wider temperature range, and the lock-in transition is blurred [18,19,20];

hysteresis phenomena appear in particular in the I phase, in the region where q changes [21]. The influence of extrinsic defects is well illustrated by the solid solution $(Rb_{1-x}K_x)_2ZnCl_4$: whereas both pure compounds (x=0 and x=1) have similar transition sequences N-I-C, the lock-in temperature decreases strongly for the mixture and finally the C phase vanishes for x between 0.1 and 0.9 [18]. The modulation wave vector is then constant, and the incommensurate satellites have a width which is clearly larger than the resolution. For the samples with a slight amount of impurities (for example with x=0.997), the thermal variation of q_δ is strongly modified, as q_δ varies by jumps between steps which can be horizontal or inclined [22]. This can be understood by the presence of D-lines which appear in the incommensurate phase when the discarding from equilibrium (due to pinning by defects) is important; horizontal plateaus would correspond to a pinned configuration, and the jumps to the nucleation of a phase with a q_δ value closer to equilibrium. Tilted plateaus can be explained by a hindered motion of D-lines which can be completely pinned in the samples with a high amount of impurities (0.1<x<0.9). The occurrence of such a discontinuous thermal variation of q_δ is also encountered in other systems: $(TMA)_2CoCl_4$ [23], $(TMA)_2FeCl_4$ [24], $(TMA)_2ZnCl_4$ [25], due to radiation damage of the samples. In Ref. [25], the study of $q_\delta(T)$ was performed with an X-ray beam of low intensity, whereas defects were introduced between two series of measurements by submitting the sample to a strong intensity radiation. The original sample, of very high quality, presents no hysteresis phenomena (except very close to T_c); after radiation damage, the curves $q_\delta(T)$ on cooling and on heating begin to differ already in the

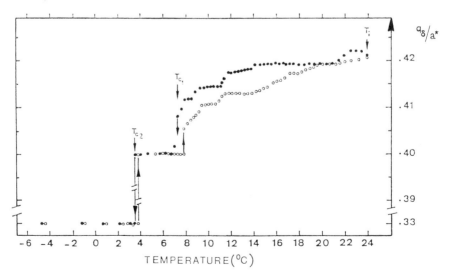

Fig. 6. Thermal variation of the satellite position in a $(TMA)_2ZnCl_4$ after X-radiation damage (from Ref. [25]).

I-phase, and plateaus are observed (Fig. 6). It is noteworthy that in small temperature ranges, there is a coexistence of q_δ values corresponding to the steps; the satellites have two components at say q_1 and q_2 and the transition from the plateau at q_1 to that at q_2 proceeds by the progressive vanishing of intensity at q_1. When the satellites are not completely resolved, an increase of the width is always observed between two consecutive steps. These results can be considered as an indirect evidence of D-defects in the crystal; the magnitude of the jumps between steps corresponds approximately to a "density" of one D-line (at which 10 walls terminate) for 200 walls. At the lock-in, the transition is somewhat smeared out, as the curve $q_\delta(T)$ is no more vertical; it can be due to a pinning of the D-lines. Coexistence of several q_δ values was also observed in "pure" samples (namely with an impurity concentration of approximately 10^{18} cm^{-3}), but in the close neighborhood of the lock-in [26], or within the I phase in relation with "memory effect" [27]. Indeed, after a temperature stabilization at T_1, the sample was cooled to T_2; the satellite has then two components, at $q(T_1)$ and at $q(T_2)$, and the intensity of the former vanishes progressively with time. The X-ray topography reveals that the mechanism involved is the expansion of the region modulated by q_2 to the detriment of that modulated by q_1. D-lines are unavoidable at the interface between both regions, and the observed relaxation expresses their slow motion.

In addition to these evidences for D-defects in these compounds, it is very difficult to get more direct information as in electron microscopy studies, the samples are quickly damaged by the beam. However, the study was exceptionally possible in the case of Rb$_2$ZnCl$_4$ [28], where it was observed that the formation of DC's occurred by nucleation and lateral propagation; unfortunately, it seems that the structure of the interface could not be clearly resolved.

Incommensurate Phase of Barium Sodium Niobate

The modulated phase of Ba$_2$NaNb$_5$O$_{15}$ (BSN) presents a lot of interesting properties which are described in detail in the excellent review by Tolédano et al. [29]. A large thermal hysteresis characterizes it, and relaxation processes and a strong memory effect were observed. The I phase exists between $T_c \sim 540K$ and $T_i \sim 570K$, but this range varies with the experimental conditions. In particular, an annealing within the I phase lowers T_c, whereas a temperature stabilization in the C phase induces an increase of T_c. It is noteworthy that the phase below T_c is not perfectly commensurate; the misfit δ varies from 0.01 in the C phase to 0.08 in the neighborhood of T_c, presents a thermal hysteresis, and varies continuously on cooling whereas the change is abrupt on heating. This transition is also marked by a broadening of the satellites in the modulation direction, which persists in the whole incommensurate phase. The satellite width decreases with time when T is stabilized within the modulated phase.

It seems that these properties can be understood by a mechanism of the transition at T_c, governed by the D-defects; being hindered in their lateral motion, these objects induce a continuous decrease of δ on cooling, and on heating the nucleation of D-lines implies a δ-jump. The broadening of satellites would be then attributed to a distribution of δ values; a slow motion of D-lines can induce (or contribute to) relaxation processes. Moreover, if the sample is annealed at a temperature with a given configuration of D-lines, it is natural to expect that diffusion processes of extrinsic defects will lead to a local modification of the impurities concentration around the lines. In addition to defect density waves ("DDW"), defect local centers ("DLC") should exist and contribute to memory effect. This can explain the T_c shifts if one supposes that the quasi-commensurate phase is due to a pinning of residual D-lines.

In BSN, 8 domain states can be expected by symmetry in the ideal (δ=0) C phase, and there are 4 vectors in the star of $\vec{q}_\delta$. However, the point symmetry reduction occurs already in the I phase where macroscopic domains were observed; they are modulated in a single direction, and within one domain, the domain texture of the I phase can be constructed with 4 walls, at which the phase of the modulation jumps by $\pi/2$. The observation of BSN by transmission electron microscopy (TEM) is possible [30,31] and the results are in agreement with theory. The TEM study revealed the presence of walls in the C phase, with fourfold nodes corresponding to D-lines as in Fig. 3, the configuration being unchanged at low temperatures. The change at T_c occurs by nucleation and multiplication of walls; in particular, objects as in Fig. 4 and Fig. 5 were observed to induce the transition [31]. The phenomena related to the existence of the modulated phase recently have been the object of a detailed TEM study by C. Roucau et al. in Toulouse (F).

The Intermediate Phases of Quartz and Berlinite

Between the α and β phases of quartz SiO_2, an incommensurate phase was discovered recently, the properties of which are described in detail by G. Dolino in this issue. Berlinite $AlPO_4$ is an isomorphous compound presenting closely related properties, including the I phase. Whereas in pure berlinite samples, the I phase is marked by only two anomalies (at T_i and at T_c) as in quartz, in samples containing impurities, a series of anomalies occurs. A detailed birefringence study [32] evidenced well-defined

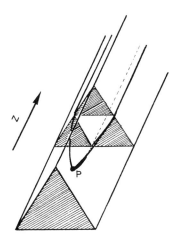

Fig. 7. Deperiodization "P-defect" (case of SiO_2 or $AlPO_4$). It limits
the extension of a prism in the Z direction, perpendicular to
the modulations (from Ref. [33]).

intermediate metastable states characterized by an absence of hysteresis,
and that seemed to correspond to constant q_δ values.

For the present modulated phase, the situation is not so simple as in
the preceding examples: three modulations at $120°$ are superimposed. The
domain texture consists of prisms having an equilateral base, disposed as
illustrated in Fig. 7. Two adjacent domains (white and hatched) are characterized by opposite values of the order parameter. It results from this
situation that the topology of D-defects is not the same as in the previous
cases. A first possibility for realizing a deperiodization is a change of
the wall density along Z; the D-defect limits the extension of the prism in
this direction, and corresponds to a coalescence of three edges (Fig. 7);
as the deperiodization defect is then a point, it is designed as P-defect.
Another simple intrinsic defect which can be imagined to change the
periodicity is an edge dislocation in the superstructure, with a Burger's
vector lying perpendicularly to Z (Fig. 8).

TEM studies can be performed in both compounds, and provide interesting information [33,34]. In impure $AlPO_4$, because of pinning, q_δ varies
discontinuously, and coexistence of different q_δ values occurs (Fig. 9).
P-defects can be invoked to permit the junction between both regions. It
happens that deperiodization defects are pinned by imperfections of the
crystalline lattice, thus accounting for the birefringence anomalies
observed in impure samples. In both compounds, the mechanism for the q_δ
variation is a motion of P-defects in the Z direction. The other type of
intrinsic defects could also be observed, and their core was resolved. It
is nevertheless not clear whether they participate to the q_δ thermal variation. At the lock-in to the α phase, an intermediate metastable state

$k + \delta k$

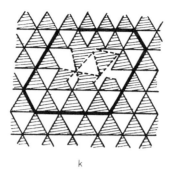

k

Fig. 8. Edge dislocation in the superperiod (I phase of $AlPO_4$ or quartz). The configuration in the core was observed in [33] (see the paper by G. Dolino in the same issue for the signification of full and broken lines).

takes place, likely by the motion along Z of "P'-defects" (which join also three edges of a prism, but in a different manner as P-defects).

The phenomena observed with the electron microscope are not static, but have a strong dynamic character. Thus, within the I phase, fluctuations of the incommensurate periodicity can create unstable pairs of P-defects (nuclei of α phase). Moreover, relaxation processes are directly observable; at a fixed temperature, the superstructure can change with time, due to the motion followed by a progressive stabilization of deperiodization point defects.

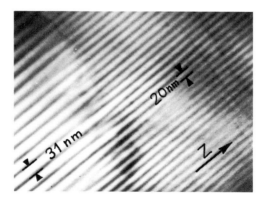

Fig. 9. Coexistence of different q_δ values in an impure berlinite sample (from ref. [33]).

4. CONCLUSION

The list of examples that were given is not exhaustive, and it is likely that other crystals exhibit behaviors related to the presence of such defects.

The presented experimental data are in good agreement with the ideas advanced in ref. [11] and [12]; when the incommensurate periodicity is not able to change by the frontal motion of walls, deperiodization defects can be created, either within the modulated phase itself, or at the lock-in transition. They are then expected to contribute to hysteresis phenomena. Direct or indirect evidence (according to the crystal which is studied) of their role in relaxation processes has been given, and this has to be stressed. Moreover, a memory effect can also be induced through them.

Janovec [11] mentioned also that deperiodization defects could induce the appearance of effects forbidden in regular I phases. They perhaps are responsible, too, for the enhancement of second harmonic generation of light, which is stronger than predicted by theory [35].

5. REFERENCES

1. A. P. Levanyuk and D. G. Sannikov, Sov. Phys. Sol. State 18, 245 (1976).
2. I. E. Dzyaloshinskii, Sov. Phys. JETP 19, 960 (1964).
3. W. L. McMillan, Phys. Rev. B14, 1496 (1976).
4. A. D. Bruce, R. A. Cowley, and A. F. Murray, J. Phys. C11, 3591 (1978).
5. H. Shiba and Y. Ishibashi, J. Phys. Soc. Jpn 44, 1592 (1978).
6. M. Brunet and C. William, Ann. Phys. 3, 237 (1978).
7. M. Glogarová, L. Lejček, Pavel, V. Janovec, and J. Fousek, Mol. Cryst. Liq. Cryst. 91, 309 (1983).
8. W. L. McMillan, Phys. Rev. B14, 1496 (1976).
9. M. B. Walker, Phys. Rev. B26, 6208 (1982).
10. C. H. Chen, J. M. Gibson, and R. M. Fleming, Phys. Rev. B26, 184 (1982).
11. V. Janovec, Physics Lett. 99A, 384 (1983).
12. V. Janovec, G. Godefroy, and L. R. Godefroy, Ferroelectrics 53, 333 (1984).
13. P. Prelovsek, Ferroelectrics 54, 29 (1984).
14. R. M. Horneich, M. Luban, and S. Shtrikman, Phys. Rev. Lett. 35, 1678 (1975).
15. A. D. Bruce, R. A. Cowley, and A. F. Murray, J. Phys. C11, 3591 (1978).
16. A. E. Jacobs, Phys. Rev. B33, 6340 (1986).
17. Y. Yamada, and N. Hamaya, J. Phys. Soc. Jap. 52, 3466 (1983).
18. K. Hamano, K. Ema, and S. Hirotsu, Ferroelectrics 36, 343 (1981).
19. P. S. Smirnov, Kh. Maak, and B. A. Strukov, Sov. Phys. Cryst. 30, 476 (1986).
20. E. Colla, P. Muralt, H. Arend, R. Perret, G. Godefroy, and C. Dumas, Solid State Com. 52, 1033 (1984).
21. K. Hamano, T. Hishinuma, and K. Ema, J. Phys. Soc. Jap. 50, 2666 (1981).
22. H. Mashiyama, S. Tanisaki, and K. Hamano, J. Phys. Soc. Jap. 51, 2538 (1982).
23. E. Fjaer, R. A. Cowley, and T. W. Ryan, J. Phys. C18, L41 (1985).
24. H. Mashiyama and S. Tanisaki, J. Phys. C15, L455 (1982).
25. M. Bziouet, R. Almairac, P. Saint-Gregoire, J. Phys. C20, 2635 (1987).

26. M. Ribet, Ferroelectrics 66 259 (1986).
27. M. Ribet, S. Gits-Leon, F. Lefaucheux, and M. C. Robert, J. Physique, 47, 1791 (1986).
28. H. Bestgen, Solid State Com. 58, 197 (1986).
29. J. C. Toledano, J. Schneck, and Errandonea, G., in "Incommensurate Phases in Dielectrics," ed. A. P. Levanyuk and R. Blinc (North Holland Publishing Co., 1985).
30. G. Van Tendeloo, S. Amelinckx, C. Manolikas, and Wen Shulin, Phys. Stat. Sol.(a)91, 483 (1985).
31. P. Xiao-Qing, H. Mei-shen, Y. Ming-Hui, and F. Duan, Phys. Stat. Sol. (a)92, 57 (1985).
32. P. Saint-Grégoire, F. J. Schäfer, W. Kleemann, J. Durand, and A. Goiffon, J. Phys. C17, 1375 (1984).
33. E. Snoeck, C. Roucau, P. Saint-Gregoire, J. Physique, 47, 2041 (1986) (and ref. therein).
34. E. Snoeck, Ph.D. thesis, University of Toulouse (1986).
35. V. A. Golovko and A. P. Levanyuk, Sov. Phys. JETP 50, 780 (1979).

THERMAL HYSTERESIS, SOLITONS AND DOMAIN WALLS

H.-G. Unruh and A. Levstik*

Fachbereich Physik der Universität des Saarlandes
D-6600 Saarbrücken
FRG

Abstract   The behavior of incommensurately modulated crystals is
essentially influenced by effects hindering the modulation from free
sliding and leading to metastable states. Thus nucleation processes
and pinning of discommensurations by defects have to be considered
when investigating and discussing static or dynamic properties of modulated phases. Low- and high-frequency dielectric measurements have
revealed a global thermal hysteresis and memory effects in the modulated phases of K_2SeO_4-type crystals, which are connected with the dynamics of the discommensurations.

1. INTRODUCTION

Macroscopic quantities of crystals with incommensurately modulated
structure (I-phases) are known by now to have often different values at
any temperature in the I-phase on heating and cooling runs. This phenomenon, sometimes designated as global thermal hysteresis, occurs when the
temperature of the sample is cycled in the I-phase without passing through
any phase transition. Quantitatively, the observed effects are more pro-
nounced in the vicinity of lock-in transitions to commensurate (C-) phases
than in the vicinity of the transitions to the parent or normal states
(N-phases). Furthermore, a thermal hysteresis is seen between the lock-
in temperatures on heating, T_c^h , and cooling, T_c^c. Experiments on various
systems show that thermal hysteresis is a common feature in I-phases as it
has been found as well in low dimensional conductors like TTF-TCNQ or

---

*Permanent address: J. Stefan Institute, Ljubljana, Jugoslavia.

2H-TaSe$_2$ as in insulators like K$_2$SeO$_4$ or SC(NH$_2$)$_2$. A summary on experimental results and corresponding references have been given by Toledano, recently [1]. Here we will concentrate on one type of dielectric materials but the phenomena and basic ideas of explanations discussed below are believed to be of rather general significance for incommensurate structures.

Some A$_2$BX$_4$-type crystals (K$_2$SeO$_4$, Rb$_2$ZnCl$_4$ and others) transform from an orthorhombic N-phase (space group Pcmn) at a temperature T$_i$ into an I-phase. The wavevector of the one-dimensional lattice modulation is given by $\vec{q}_i = (1 - \delta(T))\vec{c}^*/3$, where the reciprocal lattice vector $\vec{c}^*$ refers to the N-phase and $\delta(T) \ll 1$ denotes a parameter that characterizes the misfit between the wavelength of the modulation and an integer multiple of a lattice constant. At a temperature T$_c$ < T$_i$ the modulation locks into the commensurate value of c*/3. The C-phase is ferroelectric and the spontaneous polarization, $\vec{P}_s$, is directed along $\pm \vec{b}$. Further transitions which occur in some of these substances at temperatures above T$_i$ into a hexagonal structure (P6$_3$/mmc) and at temperatures below T$_c$ will not be considered.

Thermal hysteresis in A$_2$BX$_4$-type crystals has been observed in early experiments on K$_2$SeO$_4$ [2] and Rb$_2$ZnBr$_4$ [3] but Hamano and coworkers, [4,5] were the first to point out the global nature of this phenomenon in their investigations of the dielectric properties of Rb$_2$ZnCl$_4$ and Rb$_2$ZnBr$_4$. Soon it became evident that hysteresis of macroscopic quantities is intimately related to a corresponding behavior of the (average) modulation wavevector $\vec{q}_i$ or misfit-parameter $\delta(T)$, respectively [6,7]. Taking into account that higher harmonics of the modulation wave become important near T$_c$ the approximative description of the modulation by plane waves has to be replaced by the soliton picture, where regions of almost commensurate structure are separated by equally spaced discommensurations (DC's) [8]. These DC's, the widths of which are only weakly dependent on temperature, can be regarded as "walls" forming a nearly regular (one-dimensional) array with a lattice - constant L(T) at T $\gtrsim$ T$_c$ [9]. Approaching T$_c$ L increases and a chaotic behavior of DC's has to be expected eventually because of the exponentially decaying repulsive forces between DC's with increasing distances L, while various pinning forces hinder them from reaching equilibrium positions [10]. This will even lead to metastable quasi-commensurate states below T$_c$ where irregularly distributed DC's remain, acting as walls between antiparallel polarized ferroelectric domains in case of K$_2$SeO$_4$-type systems.

The interaction of DC's with lattice defects has been widely quoted as the cause of thermal hysteresis [4,5,7,11] although the mechanisms of the interaction are not understood in detail. Indeed, doping of Rb$_2$ZnCl$_4$ crystals by small amounts of potassium increases the thermal hysteresis strongly [4,12] and purification of the raw materials for crystal growth results in smaller hysteresis [13]. On the other hand, theoretical models often favor another explanation of this phenomenon in which nucleation and annihilation processes in combination with topological properties of DC's are considered [14-17]. Especially it has been shown by Nattermann [17] that metastable states will occur in the I-phase due to a finite nucleation energy for stripples of DC's, which lead to a hysteresis between heating and cooling, ΔT_c, independent on temperature as observed, e.g. in case of K_2SeO_4 and Rb_2ZnCl_4 [18]. Quantitative estimations of ΔT_c are in general by an order of magnitude too large compared to measured values [17,19].

Furthermore, a thermal hysteresis could be produced by intrinsic pinning of DC's through the potential of the underlying discrete lattice [20] or even through on oscillating potential of the DC's themselves [21]. However, any experimental evidence is missing up to now that these models are appropriate to explain the observations on A_2BX_4-type crystals.

The interpretation of experimental results is additionally complicated by special memory effects which appear when crystals are stored for some time at constant temperature in the I-state [18,22]. It suggests itself that interactions between defects and DC's or, more generally, between the lattice and/or defects on one side and the (spatially modulated) order parameter on the other side can be responsible for these phenomena too [23].

2. PHENOMENOLOGICAL DESCRIPTION OF INCOMMENSURATE PHASES IN K_2SeO_4-TYPE CRYSTALS

It seems obvious from the observations mentioned so far that the global thermal hysteresis is related to the appearance of DC's. Phenomenologically the N-I and I-C transitions are treated by considering a Landau-type functional of the free energy per unit length $f(z)$. In case of K_2SeO_4-type crystals $f(z)$ reads [24,25]:

$$f(z) = \frac{\alpha}{2} QQ^* + \frac{\beta}{4}(QQ^*)^2 + \frac{\gamma_1}{6}(QQ^*)^3 - \frac{\gamma_2}{12}(Q^6 + Q^{*6})$$

$$- i\frac{\sigma}{2}(Q\frac{dQ^*}{dz} - Q^*\frac{dQ}{dz}) + \frac{\kappa}{2}\frac{dQ}{dz}\frac{dQ^*}{dz} \tag{1}$$

$$+ \frac{1}{2\chi_o} P^2 - \zeta(Q^{*3} + Q^3)P + \frac{\eta}{2} QQ^*P^2 - PE \quad ,$$

where $Q = \rho(z) \exp(i\phi(z))$ is the complex order parameter and $\alpha = \alpha_o(T - T_o)$ is the only coefficient which is assumed to depend on temperature. Minimizing the free energy $F = \int f(z)dz$ with respect to the dielectric polarization P yields

$$P = \hat{\chi}_o(E + 2\zeta\rho^3 \cos 3\phi) \quad \text{and} \quad 1/\hat{\chi}_o = 1/\chi_o + \eta\rho^2 \quad . \tag{2}$$

Eliminating P from Eq. (1) and solving the Euler equations corresponding to
the variational problem of determining $\phi(z)$ that minimizes F in the
constant-amplitude approximation results in a periodic function $d\phi/dz$ with
the period

$$L(T) = \frac{kK(k)}{\rho^2(T)} \left(\frac{2\kappa}{3(\gamma_2 + 6\zeta^2\hat{\chi}_o)}\right)^{1/2} \qquad (3)$$

when E = 0. The temperature dependence of L(T) is mediated by the relation

$$k/E(k) = \rho^2(T)/\rho_c^2 \quad , \qquad (4)$$

where K(k) and E(k) are elliptic integrals of the first and second kind,
respectively, and ρ_c is the amplitude of the order parameter in the C-phase
[26,27].

By working out $\phi(z)$ it can be shown that the phase changes approximately stepwise in the z-direction when $T \gtrsim T_c$. The regions with almost
constant $\phi(z)$ are separated at distances L by discommensurations, where the
phase changes steeply by $\pi/3$ in case of K$_2$SeO$_4$-type crystals. From Eq. (2)
it follows that each DC separates regions with antiparallel polarizations
and the homogeneous polarization can be considered as a secondary order
parameter in these systems [24]. Because of the relation between primary
and secondary order parameter (see Eq. (2)) the dielectric susceptibility,
χ, is well suited for detailed investigations. It has been calculated within
various approximations and shows divergent behavior for T approaching T_c
from above, i.e., for $k \to 1$ [27-29]:

$$\chi \approx \hat{\chi}_o + \frac{6\zeta^2\hat{\chi}_o^2}{\gamma_2 + 6\zeta^2\hat{\chi}_o} \left(\frac{E(k)/K(k)}{1 - k^2} - 1\right). \qquad (5)$$

In a certain temperature range close to T_c, $\chi(T)$ obeys the Curie-Weiss
law in accordance with experimental results [2,27]. The dielectric properties of K$_2$SeO$_4$ and Rb$_2$ZnCl$_4$ crystals can be perfectly fitted to Eq. (5) as
long as hysteresis effects are disregarded. This also provides an appropriate set of those coefficients of the thermodynamic potential which are
included in Eq. (5) [27,29].

3. RESULTS OF DIELECTRIC INVESTIGATIONS

Let us now turn to the description of the thermal hystersis as seen by
measurements of the dielectric function, $\varepsilon(T,\omega)$. Figure 1 demonstrates a
low-frequency result for Rb$_2$ZnCl$_4$ in the vicinity of T_c. The real part of

166

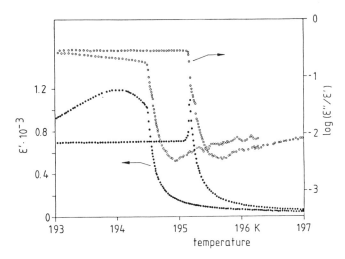

Fig. 1. Thermal hysteresis of the dielectric constant and loss tangent of Rb_2ZnCl_4 measured at 1 kHz and at a temperature rate of 0.04 K/min. The lock-in transitions occur at 194.5 K on cooling and at 195.2 K on heating, respectively.

the permittivity, ε', and the losses, in the form of $(\varepsilon''/\varepsilon')$, are shown for heating and cooling runs. Although the field strength of measurement was kept small (≈ 2 V/cm), the values of ε' and ε'' depend on field and frequency whenever $T < T_c$. This indicates a high mobility of the domain walls (DC's) in the ferroelectric C-phase. The observed behavior of $1/(\varepsilon' - \varepsilon_0)$ in the I-phase is depicted schematically in Fig. 2, where ε_0 is the temperature independent part of the dielectric constant [18]. Bold lines mean results on cooling and heating runs, which are extrapolated to T_c^c and T_c^h, respectively. A special situation occurs if the direction of temperature change is reversed: Doing this when the crystal is in a state marked on cooling or in ② on heating one follows the bold line ① → ② or ② → ①, respectively. Furthermore, state ③ can be reached at constant temperature from ① by applying an electric field high enough to induce the C-phase in the crystal temporarily.

Relations between states ① to ③ should be reflected by the number of DC's per unit length, $1/L(T)$, and, therefore, by the dielectric susceptibility through Eqs. (3) and (5). A sensitive measurement of ε during heating and cooling runs shows that $L(T)$ does not change continuously with temperature but--at least in real crystals--to some extent discontinuously [18]. The experimental set-up and results are shown in Fig. 3. A thin crystal plate of Rb_2ZnCl_4 cut perpendicular to the b-axis is covered on one side by a metal electrode and small stripe-like electrodes are applied to the other side. Then two of the capacitors C_1 to C_4 (see Fig. 3) are

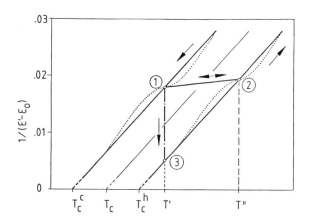

Fig. 2. Schematic representation of the thermal hysteresis in the I-phase
as seen in a Curie-Weiss plot of the real part of the permittivity.
See text for further explanations.

connected to a capacitance bridge, one as standard and the other as unknown.
The output of the balanced bridge, for example $\Delta C_{12} = C_1 - C_2$, is monitored
as a function of time while temperature changes slowly. As seen in Fig. 3
steps appear at the output which are negative or positive in case of ΔC_{12}
but negative only in case of ΔC_{24}. From the orientations of stripes relative to $\vec{q}_i$ it follows that a given change of L in the volume of one capacitor, C_i, will change the dielectric susceptibility in case of C_1 or C_2 by
1/b times the corresponding effect at C_3 or C_4, where l and b mean length and
width of the stripes. Hence, the sensitivity of this experiment resolves
steps from C_1 and C_2, whereas steps from C_3 and C_4 appear as noise. Consequently no definite steps are detected when measuring ΔC_{34}. Of course,
rearrangements of the pattern of DC's can be detected by this technique only
if the elementary processes leading to steps of the susceptibility are
restricted to the volumes of the sensed capacitors: A homogeneous change of
L would not unbalance the bridge except for effects of finite velocities of
the rearrangements. One must conclude, therefore, that sudden changes of
the DC-pattern occur at least in part in a volume comparable or smaller than
that of a capacitor C_i [18]. Further investigations of the density of DC's,
1/L(T), by this method using various sizes of electrodes and of their
distances seem to be promising for the determination of DC-kinetics.

With respect to the thermal hysteresis the experiment described above
clearly separates normal cooling or heating runs from those where the direction has been turned as, for example, at the transitions ① $\rightleftarrows$ ② in Fig. 2.
No step of any ΔC_{ik} could be detected during the latter changes of temperature so that states ① and ② may be characterized by identical
patterns of DC's. The temperature where this pattern represents the state of

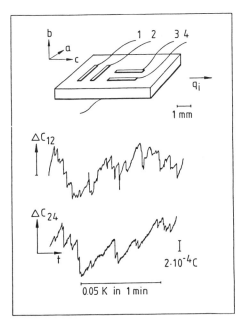

Fig. 3. Experimental arrangement of electrodes, 1 to 4, upon a plate of
Rb$_2$ZnCl$_4$ for the determination of local changes of the dielectric
susceptibility. Examples of a bridge output measuring the differences of capacitances, $\Delta C_{ik} = C_i - C_k$, when temperature changes
at a rate of 0.05 K/min are also given.

equilibrium can be assumed midway between ① and ② as indicated in
Fig. 2 by a thin line.

Metastable states as those on the cooling and heating runs are not
included in the phenomenological treatment of I-phases. Nevertheless, the
dielectric susceptibility can be calculated in the neighborhood of any state
of equilibrium as a function of temperature under the constraint of L being
constant [29]. Then Eq. (4) does not hold and the temperature parameter k
in Eq. (5) has to be evaluated from Eq. (3) for L = L$_{equilibrium}$ = const.
As shown in Ref. [29], the results account quantitatively for the measured
susceptibilities on turning branches like ① ⇄ ② of Fig. 2. It is
noteworthy that this holds true for Rb$_2$ZnCl$_4$ and K$_2$SeO$_4$ crystals so that
our interpretations seem to have a more general validity for I-phases.

Thermal hystersis is seen not only when measuring quasi-static behavior
as mentioned thus far but has been found also for dynamic properties. The
dispersion of the permittivity, which is known to obey a Debye-type relaxation for several dielectrics with I-phases [30], can be taken as an example.
Deguchi et al. [31] have demonstrated that the characteristic relaxation
frequency of Rb$_2$ZnCl$_4$, $f = 1/(2\pi\tau_\varepsilon)$, evaluated from

$$\varepsilon(\omega) = \varepsilon_\infty + (\varepsilon_s - \varepsilon_\infty)/(1 + (j\omega\tau_\varepsilon)^{1-h}) \tag{6}$$

169

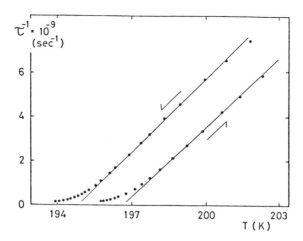

Fig. 4. Thermal hysteresis of the reciprocal relaxation time of Rb_2ZnCl_4
according to Deguchi et al. [3]

follows nearly a Curie-Weiss law, f = const$/(T - T_f)$, seperately for cooling
and for heating runs as shown in Fig. 4. Here T_f is equal to T_c^c or T_c^h,
respectively. This can be understood again as a consequence of different
metastable states which are characterized at a temperature $T \lesssim T_c$ by $L^c(T)$
or $L^h(T)$, respectively.

The critical slowing down of the characteristic frequency, f_ε, can be
brought into connection with the behavior of the (overdamped) phase mode or
phason [32]. Indeed, a homogeneous electric field parallel to the (polar)
b-axis of K_2SeO_4-type crystals induces a staggered sideways motion of DC's.
The same holds for the phason with a wavevector $\vec{K}(T) = \vec{c}^* - 3\vec{q}_i(T) = \delta(T)\vec{c}^*$
and in the plane-wave approximation it is $L(T) = \pi/K(T)$ [32]. Taking into
account the linear dispersion of the phason, $\omega_\phi \sim K$, and its damping, Γ, one
expects a relaxation time for an overdamped mode of the order

$$\tau_\varepsilon = \Gamma/\omega_\phi^2 \sim \Gamma/K^2(T) \sim \Gamma/\delta^2(T), \qquad (7)$$

which diverges as $\delta(T) \to 0$ for $T \to T_c$. Equation (7) relates the relaxation
time directly to $L(T)$.

Dispersion measurements demonstrate also a gradual transformation from
I- to C-phases. Figure 5 shows plots of $\varepsilon(\omega)$ in the complex plane for
K_2SeO_4 at some temperatures near $T_c \approx 93.4$ K. A few degrees above T_c the
dispersion can be well approximated by Debye's formula. A second relaxation
develops at the low frequency side with a nonzero distribution parameter
of relaxation times, h, when cooling through the lock-in transition. This
process can be followed to below T_c and is attributed to motions of domain
walls [33]. The coexistance of both dispersions proves that I-and C-regions

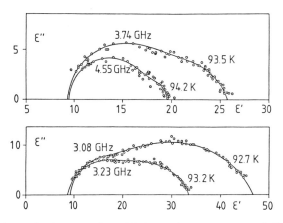

Fig. 5. Dispersion of the complex dielectric constant of K_2SeO_4 in the
vicinity of T_c [33]. The characteristic frequencies of the
high-frequency relaxation are indicated.

are present in the crystal simultaneously in a temperature range which seems
to be roughly of the order of the thermal hysteresis $T_c^h - T_c^c$.

Analogous results have been found in case of Rb_2ZnCl_4 by dielectric
[31] and by x-ray investigations [7,34], and corresponding conclusions must
be drawn also from low-frequency measurements. The loss tangent, $\varepsilon''/\varepsilon'$,
which is plotted in Fig. 1 for a measuring frequency of 1 kHz, begins to
increase above the transition temperature if this is identified by the peak
of ε or $d\varepsilon/dT$, respectively. The high-frequency dispersion due to the
soliton-lattice cannot account for these losses as $\varepsilon''/\varepsilon' = \omega\tau_\varepsilon \lesssim 10^{-4}$ for T
$\gtrsim T_c$ at 1 kHz [31].

4. MEMORY EFFECTS

The low-frequency dielectric measurements in the vicinity of T_c were
done at a constant temperature rate of 0.12 K/min or less, usually. If the
temperature is kept constant for several hours at $T = T' > T_c$ on a normal
cooling run, for example, then subsequent cooling and heating runs yield
values of ε which correspond to the dotted lines of Fig. 2 [18]. The same
is true if the crystal is stored in state ② (s. Fig. 2) on a heating run
provided that the temperature is not raised so much that the characteristic
time constant of the memory effect competes with the time of measurement.
As the slopes of the dotted lines at ① and ②, respectively, resemble
that of the turning run ① ⇄ ② one may conclude that the metastable
states ① and ② have been stabilized while leaving the DC-pattern
constant. It is apparent that the crystal acquires a memory of that

171

pattern. Because ε does not change appreciably at constant temperature above T_c there is no reason to assume that L changes with time. So an inter-action between the modulation and certain defects has to be considered.

The memory effect has been found in Rb_2ZnCl_4 not only near T_c, i.e., in the soliton limit, but also, though much weaker, near T_i, i.e., in the plane wave limit [35]. One may suppose, therefore, that a "defect modulation wave" is established in these crystals whenever they are in I-states. Concerning physical defects, the crystals used in our investigations were of high quality as proved by x-ray diffraction [34]. Point defects, which may have some internal complexity, are assumed to be responsible for the memory effects, therefore, and two basically different mechanisms can account for a defect modulation. Defects can either diffuse to preferred sites or reorient themselves according to the local lattice modulation. Assuming that the strength of the interaction in the constant–amplitude approximation is deter-mined by $\phi(z)$ or $d\phi/dz$ one expects a "volume-" or "wall-" effect, respec-tively, since DC's will interact with defects mainly in the latter case whereas the bulk between DC's in the former. If diffusion is the decisive mechanism, then a decoration of DC's will take place. Quantitative models have been worked out on this basis for memory effects of thiourea [23] and of barium sodium niobate, where mobile sodium vacancies are assumed to interact with the modulation [35].

The kind of defects which are responsible for the memory effects in Rb_2ZnCl_4 are not known, but further dielectric investigations may help to characterize them. Figure 6 shows dielectric hystersis loops of a typical sample at different conditions. The loop in the upper left is that of a sample just cooled from above T_i to $T = 193$ K $< T_c$. Depolarizing the sample to a state with zero overall polarization and waiting under zero field for eight hours at 193 K, one gets the loop shown in the upper right of Fig. 6. Repolarizing at this situation the sample partially to a somewhat positive polarization,the lower right loop was observed after 40 hours more. Leaving the freshly cooled sample in a fully plus-polarized state under zero field for two days resulted in the lower left loop, finally. Such a behavior is well known from classical ferroelectric materials and will not be considered here. The very similar development of bias fields in Rochelle salt, for example, has been shown to be caused by reorientable point defects interacting with the local order parameter [36]. Thus Fig. 6 demonstrates a "volume-" effect in Rb_2ZnCl_4 as well.

Another memory effect, which differs from the one mentioned above by the magnitude of its relaxation time, has been found when the coercive field, E_c, of Rb_2ZnCl_4 samples was measured. As Fig. 7 shows, E_c depends on whether the crystal (pure or doped with small amounts of potassium) has been stored for some hours at room temperature, i.e., 10 K below T_i, or for the same time at $T_i + 10$ K, i.e., in the N-phase.

Although these memory effects can be described at present only qualitatively, it becomes evident that strong interactions between the incommensurate modulation and lattice defects occur and it seems necessary to take them into account in further discussions of metastable states of I-phases.

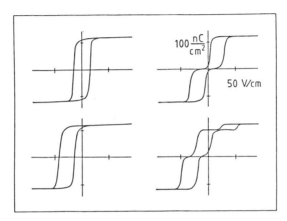

Fig. 6. Delectric hysteresis loops of a pure Rb_2ZnCl_4 cyrstal. Constrictions and shifts by biasing fields are due to aging effects at a constant temperature of T = 193 K.

5. CONCLUSIONS

Recently Bestgen succeeded in getting direct information about the "soliton" lattice in the I-phase and "domain walls" in the C-phase of Rb_2ZnCl_4 by electron microscopy [37]. These investigations show that DC's form a very regular pattern above T_c. Just below T_c the same objects appear no longer as straight lines in the microscope but are more or less wavy although the orientation stays the same on the average. At lower temperatures $T \lesssim 150$ K, domain walls seem to be planes again and their distances are somewhat larger than just below T_c. Occasionally six walls can be seen to meet at a line defect or dislocation [8] as shown in Fig. 8. Such structures have been proposed to take part in nucleation and

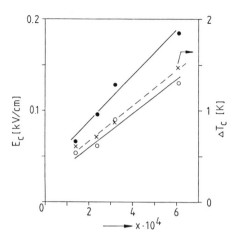

Fig. 7. Thermal hysteresis, ΔT_c, and coercive field, E_c, of nominally
pure ($x < 2.10^{-4}$) and potassium doped crystals of $Rb_{2-2x}K_{2x}ZnCl_4$.
E_c has been measured at 190 K by triangular fields with a slope
of 10^3 V/s after aging the sample for about twelve hours at 293 K
(dots) or at 313 K (circles), respectively.

annihilation processes of DC's [8,14,15,17,19] and have been observed in
Rb_2ZnCl_4 in between by other authors too [38].

As for the memory effects, the wavyness of the walls just below T_c may
prevent them from sensing the potential wells set up when a sample has been
aged in the I-state, as the corresponding distances of wells are small.
Indeed, constricted or biased loops as shown in Fig. 6 have not been
observed when samples treated in this way were cooled to below T_c. On the
other hand, the wells established below T_c correspond to the lower density
of DC's in the "chaotic regime" and will be seen by the walls when repolarizing the sample by an external field. Thus the memory effects described
for temperatures around T_c manifest themselves by biasing fields of polarized
volumes.

The global thermal hysteresis in the I-phases is due to the appearance
of metastable states which at a given temperature T differ by the density of
DC's. Certainly, a memory effect will hinder the crystal to reach equilibrium, but the main question is whether the width of the thermal hysteresis
ΔT_c, is determined crucially by nucleation (or annihilation) rates of objects
like in Fig. 8 or by pinning of the lattice of DC's. For experimental
reasons we mean by pinning all effects which may prevent DC's from free
sliding. First of all, a pinning due to discrete lattice effects [20] can
be excluded because of the extreme sensitivity of ΔT_c to crystal purity
[12,13]. Further, the increase of ΔT_c of Rb_2ZnCl_4 doped by small concentrations of potassium (see ref. [12] and Fig. 7) points to defect pinning

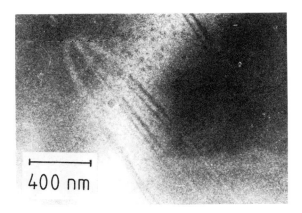

Fig. 8. Discommensurations of Rb_2ZnCl_4 at 127 K. Six walls are meeting in a line defect in the upper half of the figure.

as a homogeneous nucleation of DC's assisted by defects should lead to a decrease of ΔT_c at increasing impurity concentrations. On the other hand, the stepwise change of the DC-lattice as revealed by dielectric measurements shows that the distance L(T) of DC's is changed locally which is only possible by nucleation processes. However, each nucleus consists of six DC's because of a phase shift of $\pi/3$ at each DC in case of K_2SeO_4-type systems, and, therefore, a certain number of existing DC's must leave their positions (and potential wells) simultaneously if one single nucleus is growing. A quantitative estimation of the pinning potential would require to know the participating defect species and their interaction forces. As the increase of the coercive field of doped Rb_2ZnCl_4 crystals shows (cf. Fig.7), the mobility of DC's is strongly influenced by point defects which contribute here to the friction of moving walls. This again can be evaluated essentially in the same way as the memory effect [36,39], where the time constants are orders of magnitude shorter, however.

In conclusion, some examples of thermal hysteresis and memory effects in K_2SeO_4-type crystals have been discussed which are thought to have a general importance for modulated phases. From this follows that exhaustive investigations of the real structures of crystals with I-phases will be necessary in order to reveal details of their intrinsic properties.

6. ACKNOWLEDGMENTS

The authors are very much indebted to Dr. H. Bestgen for making Fig.8 available to them and to Mrs. H. Haas and Miss ·K. Kloss for some of the measurements. They also wish to thank N. Merl for the analyses of the

175

doped samples. This work has been supported by the Deutsche Forschungs-gemeinschaft-Sonderforschungsbereich 130.

7. REFERENCES

1. J. C. Toledano, Annales des Telecommunications (July-August 1984).
2. K. Aiki, K. Hukuda and O. Matumura, J. Phys. Soc. Japan, $\underline{26}$, 1064 (1969).
3. C. J. De Pater, Phys. Stat. Sol.(a) $\underline{48}$, 503 (1978).
4. K. Hamano, Y. Ikeda, T. Fujimoto, K. Ema and S. Hirotsu, J. Phys. Soc. Japan $\underline{49}$, 2278 (1980).
5. K. Hamano, T. Hishinuma and K. Ema, J. Phys. Soc. Japan $\underline{50}$, 2666 (1981).
6. H. Mashiyama, S. Tanisaki and K. Hamano, J. Phys. Soc. Japan $\underline{50}$, 2139 (1981).
7. H. Mashiyama, S. Tanisaki and K. Hamano, J. Phys. Soc. Japan $\underline{51}$, 2538 (1982).
8. W. L. McMillan, Phys. Rev. $\underline{B14}$, 1496 (1976).
9. H. Shiba and Y. Ishibashi, J. Phys. Soc. Japan $\underline{44}$, 1592 (1978).
10. R. Blinc, P. Prelovšek, A. Levstik and C. Filipič, Phys. Rev. $\underline{B29}$, 1508 (1984).
11. W. L. McMillan, Phys. Rev. $\underline{B12}$, 1187 (1975).
12. K. Hamano, K. Ema and S. Hirotsu, Ferroelectrics $\underline{36}$, 343 (1981).
13. K. Hamano, H. Sakata, H. Izumi, K. Yoneda and K. Ema, Jap. J. Appl. Phys. $\underline{24-2}$, 796 (1985).
14. P. Prelovšek and T. M. Rice, J. Phys. $\underline{C16}$, 6513 (1983).
15. K. Kawasaki, J. Phys. $\underline{C16}$, 6911 (1983).
16. V. Janovec, G. Godefroy and L. R. Godefroy, Ferroelectrics $\underline{53}$, 333 (1984).
17. T. Nattermann, J. Phys. $\underline{C18}$, 5683 (1985).
18. H.-G. Unruh, J. Phys. $\underline{C16}$ 3245 (1983).
19. P. Prelovšek, Ferroelectrics $\underline{54}$, 29 (1984).
20. D. A. Bruce, J. Phys. $\underline{C13}$, 4615 (1980).
21. A. E. Jacobs, Phys. Rev. $\underline{B33}$, 6340 (1986).
22. J. P. Jamet and P. Lederer, J. Physique Lettres $\underline{44}$, L-257 (1983).
23. P. Lederer, J. P. Jamet and G. Montamdaux, Ferroelectrics $\underline{66}$, 25 (1986).
24. M. Iizumi, J. D. Axe, G. Shirane and K. Shimaoka, Phys. Rev. $\underline{B15}$, 4392 (1977).
25. Y. Ishibashi, Ferroelectrics $\underline{20}$, 103 (1978).
26. D. G. Sannikov, Sov. Phys.-Solid State $\underline{23}$, 553 (1981).
27. A. Levstik, P. Prelovšek, C. Filipič and B. Zěkš, Phys. Rev. $\underline{B25}$, 3416 (1982).
28. D. G. Sannikov, J. Phys. Soc. Japan $\underline{49B}$, 75 (1980).
29. H. Mashiyama and H.-G Unruh, J. Phys. $\underline{C16}$, 5009 (1983).
30. M. Horioka and A. Sawada, Ferroelectrics $\underline{66}$, 303 (1986).
31. K. Deguchi, S. Sato, K. Hirano and E. Nakamura, J. Phys. Soc. Japan $\underline{53}$, 2790 (1984).
32. V. Dvorak and J. Petzelt, J. Phys. $\underline{C11}$, 4827 (1978).
33. W. Montnacher and H.-G. Unruh (to be published).
34. K. H. Ehses, Ferroelectrics $\underline{53}$, 241 (1984).
35. H.-G. Unruh, Ferroelectrics $\underline{53}$, 319 (1984).
36. H.-G. Unruh, Z. angew. Physik $\underline{16}$, 315 (1963).
37. H. Bestgen, Solid State Commun. $\underline{58}$, 197 (1986).
38. K. Tsuda, N. Yamamoto, K. Yagi and K. Hamano (1986), preprint.
39. P. Gurk, Phys. Stat. Sol. (a) $\underline{10}$, 407 (1972).

THERMAL MEMORY AND PHASE CONJUGATION EXPERIMENTS IN INCOMMENSURATE

BARIUM SODIUM NIOBATE

W. F. Oliver and J. F. Scott

Condensed Matter Laboratory
Department of Physics
University of Colorado
Boulder, Colorado 80309

Abstract   It is known that thermal memory in incommensurate (IC) insulators arises from the "decoration" of anti-phase boundaries of the IC structure by charged defects. In materials such as $BaMnF_4$ and $Ba_2NaNb_5O_{15}$ that are simultaneously IC and ferroelectric, it is possible to determine the sign and diffusivity of the charged defects by four-wave mixing experiments of the type used in phase conjugate optics. Our result in $Ba_2NaNb_5O_{15}$ is that the sign of the carriers is the same as that of the electro-optic coefficients r_{12}, r_{23}, r_{33}; these coefficients are known to have the same sign, but its absolute sense is not available in the literature.

1. INTRODUCTION

It is known from the initial work [1] by Hamano et al. (see the chapter in this book by Unruh [2]) that a strong thermal memory exists in incommensurate (IC) insulators, and that this effect arises from the decoration of the incommensurate modulation by charged defects. In Fig. 1a we illustrate this effect schematically, assuming that the charged defects are positive and accumulate at antiphase boundaries (APBs).

Feinberg [3] and others [4] have shown in commensurate ferroelectric insulators such as $BaTiO_3$ or $LiNbO_3$ that the presence of an external modulation, in particular, a standing wave in a crystal created by a laser beam, will also provide a spatial modulation of charged defects. In this case the defects will migrate over a period of time (normally a few s) to nodes in the standing wave, where their contribution to the energy of the system is minimum. It is possible to determine the sign, the defect

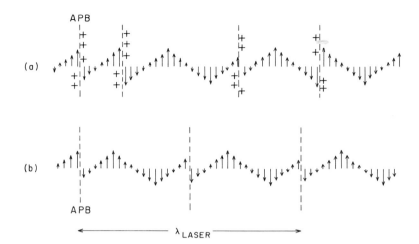

Fig. 1. Schematic diagram of charged defects trapped at APBs in an IC
         insulator.
         (a) In the dark; (b) after illumination with a standing wave
         generated by an external laser in a cavity configuration. It
         is assumed in the drawing that the laser wavelength is comparable
         to that of the initial APB spacing; in practice it might be much
         longer. It is also assumed that the diffusing charged defects
         will drag the APBs with them.

density, and diffusivity of the charged defects in this way, as outlined
in the section of this paper that follows. In this case the charged
defects form a diffraction grating that can be used for phase conjugation
experiments.

If we apply the "phase conjugation" technique to materials such as
$BaMnF_4$ and $Ba_2NaNb_5O_{15}$ that are simultaneously ferroelectric and incommensurate, we can rearrange the modulation of the charged defects, as shown
Fig. 1b: In the dark the average spacing of peaks in the density of charged
defects is that determined <u>thermally</u> by average APB separation; whereas
with the laser turned on this separation is $\lambda/2$ (laser). Fig. 1b illus-
trates the latter case and assumes that the charged defects will drag the
APBs with them when they diffuse.

Figure 1ab is rather schematic. More detailed information on the
structure of $Ba_2NaNb_5O_{15}$ in its upper IC phase from 270-300°C is given in
the chapter by Van Tendeloo in this book.

2. OTHER ANALYSES OF DIFFUSION IN IC $Ba_2NaNb_5O_{15}$

We have shown elsewhere [5] that there is very fast diffusion in the
upper IC phase of $Ba_2NaNb_5O_{15}$. A diffusivity of 1.3 cm^2/s has been deduced
very indirectly from optical data. This is about 65 times "faster" than
thermal diffusion at the same temperature and was tentatively assigned as

diffusion of APBs ("kinks"), a hypothesis compatible with theoretical models [6], which show that $D_{kinks} > D_{thermal}$. The process, however, probably does not involve charged defects or impurities, and it is much faster than the diffusion observed in phase conjugation experiments or in thermal memories. Hence, the attempt in the present experiment is to use phase conjugation experiments to determine information about such slow diffusion of charged defects as may be responsible for the thermal memory and hysteresis in $Ba_2NaNb_5O_{15}$.

3. STATISTICAL MECHANICAL MODEL FOR LIGHT-INDUCED CHARGE MIGRATION IN PHOTOREFRACTIVE MATERIALS

This treatment will basically follow that of Jack Feinberg who first proposed this model [7].

In the statistical mechanical model (also called the "Hopping Model") it is assumed that there exists within the crystal a certain number of migrant charges which can occupy any of a much larger number of sites. These charges remain fixed at a particular set of sites when in darkness, but in the presence of an optical intensity I, they tend to migrate or "hop" to other sites. According to the STM, the probability W_n that a migrant charge occupies the nth site at position x_n where the optical intensity is I_n satisfies

$$\frac{dW_n}{dt} = - \sum_m D_{mn} \left[W_n I_n e^{\frac{1}{2}\beta\phi_{mn}} - W_m I_m e^{\frac{1}{2}\beta\phi_{mn}} \right] \tag{1}$$

where the summation is over neighboring sites m. $D_{mn} e^{\frac{1}{2}\beta\phi mn}$ is a rate constant which measures the tendency for "hopping" from site m to site n. It is written in terms of a constant $D_{mn} = D_{nm}$, the static potential difference $\phi_{mn} = \phi_m - \phi_n$ between sites m and n, and a Boltzmann factor β, here defined to be $q/k\,T$, where q is the charge and T the lattice temperature. The static potential ϕ_n is equal to $\phi(x_n)$, and can be due to internal charge migration, externally applied fields and intrinsic chemical potentials. Note that in the steady state under weak uniform illumination the relative occupation probability is $W_m/W_n = e^{\beta\phi_{nm}}$ as expected by statistical mechanics.

Consider the case where the optical intensity I is spatially modulated as a result of interference between two optical beams of the same angular frequency ω. Let the complex electric field amplitudes of these two beams be given by

$$E^a(x) = \hat{e}^a_E E^a e^{ik^a \cdot x} \tag{2a}$$

$$E^b(x) = \hat{e}^b_E E^b e^{ik^b \cdot x} \tag{2b}$$

179

where $\hat{e}^a$ and $\hat{e}^b$ are the unit polarization vectors ($\hat{e^a} \cdot \hat{e^{a+}} = 1$). If these two optical beams are merged at an angle of 2ϑ within a photorefractive material, then the total optical field $E(x) = E^a(x) + E^b(x)$ will result in an optical intensity at position x_n of

$$I_n = I_0(1 + \text{Re}\left[me^{ik \cdot x_n}\right]) \tag{3a}$$

$$I_0 = |E^a|^2 + |E^b|^2 \tag{3b}$$

$$m = \frac{2E^aE^{b*}\hat{e}^a \cdot \hat{e}^{b*}}{I_0} \tag{3c}$$

where m is called the modulation index and satisfies $0 \le m \le 1$. The wave vector k for the modulated intensity I_n is equal to $k^a - k^b$, as shown in Fig.2.

For the case where $|m| \ll 1$, we assume that the solutions for W_n and ϕ_n can be approximated by keeping only the first spatial Fourier component

$$W_n = W_0 + \text{Re}W \, e^{ik \cdot x_n} \tag{4}$$

$$\phi_n = \phi_{0n} + \text{Re}\phi e^{ik \cdot x_n} \tag{5}$$

where ϕ_{0n} is due to external fields or to intrinsic chemical potentials. Poisson's equation $\nabla \cdot D = \rho_q$ relates the complex amplitudes ϕ and W,

$$\phi = \frac{W\rho q}{K^{dc}\varepsilon_0 k^2} \tag{6}$$

where ρ is the average number density of sites, $k = |k|$ and K^{dc} is the static (dc) dielectric constant, which for general anisotropic cases is $\hat{k}_i K_{ij} \hat{k}_j$, where $\hat{k} = k/k$.

Equations (3) - (5) are now substituted into equation (1) and the following simplifying assumptions are made:

1) Any uniform applied fields $-\nabla\phi_o$ are parallel to k;

2) We can thus assume that the sites are spaced by a rms distance ℓ in the $\hat{k}$ direction instead of being randomly spaced.

3) It is assumed that $k\ell \ll 1$, and thus that

4) The hopping rate ($D_{nm} = 0$ except for nearest neighbors where Dnm = Dmn = D. [8]

5) Finally, it is assumed that $1/2\beta\phi_{mn} \ll 1$. This last assumption is certainly reasonable for crystals where the sites involve defects. Defect densities are typically $\sim 10^{18}$ cm^{-3}, and thus, $\ell \sim 10^{-6}$ cm. Since k_BT 1/40 eV at room temperature, and so for $q \sim e$, the internal fields would have to be ~ 50 KV/cm for $1/2\beta\phi_{nm}$ to approach unity.

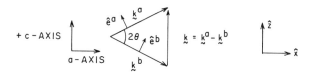

Fig. 2.   Wave vector relationship in a four-wave mixing experiment in which
          k is the wave vector of the "diffraction grating" produced by the
          density of charged defects whose diffusion is driven by incident
          beams at k^a and k^b. The two scattered beams are not shown in this
          four-wave configuration. This kind of configuration is referred
          to as a "phase conjugation" arrangement in the present paper
          because the "grating" so produced can be used for phase conjugate
          optics.

     With these approximations the coefficients of $e^{i\underset{\sim}{k}\cdot\underset{\sim}{x}_n}$ are equated in
equation (1) and the following differential equation results:

$$\frac{d\omega}{dt} = -\Gamma[(\omega + m)(\alpha^2 + i\alpha f) + \omega] \tag{7}$$

where $\omega = W/W_o$ is a normalized charge wave amplitude, $\Gamma = DI_o k_o^2 \ell^2$
is a characteristic "hopping" rate with D being the diffusivity, $\alpha = k/k_o$
is a normalized grating wave vector and $f \equiv -(\phi_{on} - \phi_{on-1})/\ell f_o$ is a uniform electric field normalized by a characteristic field f_o. The characteristic wave vector k_o and field f_o are defined as

$$f_o \equiv k_o k_B T/q \text{ and } k_o^2 \equiv \rho W_o q^2/(\epsilon_o K^{dc} k_B T).$$

     Integrating equation (7) yields for the occupation probability ampli-
tude

$$W(\underset{\sim}{x},t) = -W_o Re\left\{\frac{e^{-\Gamma(1+\alpha^2+i\alpha f)t}}{\Gamma(1 + \alpha^2 + i\alpha f)}\right.$$
$$\left.+ m\frac{\alpha^2 + i\alpha f}{1 + \alpha^2 + i\alpha f}\right\} e^{i\underset{\sim}{k}\cdot\underset{\sim}{x}} \tag{8}$$

which in the steady state and in the absence of applied fields (f=0)
reduces to

$$W(\underset{\sim}{x}) = -W_0\frac{m\alpha^2}{1 + \alpha^2}\cdot\cos \underset{\sim}{k}\cdot\underset{\sim}{x} \tag{9}$$

Note that information concerning Γ and D is not available from this steady-
state solution. We will therefore concentrate on extracting information
regarding the sign of the migrant charges in this paper [9]. The steady-

181

state space-charge density $\rho_q(x)$, static potential $\phi(x)$ and space-charge electric field $E^{dc}(x) = -\nabla\phi(x)$ are then given by

$$\rho_q(x) = -\rho q \ W_0 \ \frac{m\alpha^2}{1 + \alpha^2} \ \cos \ k \cdot x \tag{10}$$

$$\phi(x) = \frac{-\rho W_0 q}{\varepsilon_0 K^{dc} k^2} \cdot \frac{m\alpha^2}{1 + \alpha^2} \cdot \cos \ k \cdot x \tag{11}$$

$$E^{dc}(x) = \frac{-mkk_B T}{q} \cdot \frac{\sin \ k \cdot x}{1 + (k/k_0)^2} \ \hat{k} \tag{12}$$

When this static space-charge field is established in a photorefractive material, it locally alters the index of refraction via the linear electro-optic or Pockels effect. For a nonlinear dielectric medium, the electric displacement has the form

$$D_i = \varepsilon_{ij} E_j + \varepsilon_0 X^{NL}_{ijk} \ E_j E_k \tag{13}$$

where only the lowest-order nonlinear term $\varepsilon_0 X^{NL}_{ijk} E_i E_k$ has been kept. Here ε_{ij} is the usual linear dielectric permeability tensor $\varepsilon_{ij} = \varepsilon_0 (\delta_{ij} + X_{ij}) = \varepsilon_0 K_{ij}$ and X^{NL}_{ijk} is a third-rank nonlinear dielectric susceptibility tensor. We now write the real part of the total electric field in the crystal, assuming no applied field, as

$$\begin{aligned} E_i(x,t) = \frac{1}{2} \Big\{ &E^a \hat{e}^a_i e^{i(k^a \cdot x - \omega t)} \\ &+ E^b \hat{e}^b_i e^{i(k^b \cdot x - \omega t)} + c.c. \Big\} \\ &+ \frac{1}{2} \Big\{ E^{dc} \hat{k}_i e^{ik \cdot x} + c.c. \Big\} \end{aligned} \tag{14}$$

Substituting equation (14) into Maxwell's wave equation $\nabla^2 E - \mu_0 \ddot{D} = 0$ and phase matching for the $e^{i(k^a \cdot x - \omega t)}$ term yields

$$\nabla'^2 E^a \hat{e}^a_i + \frac{\omega^2}{c^2} (n^2)_{ij} \ E^a \hat{e}^a_j = -(\omega^2/c^2) \frac{[\Delta(n^2)]_{ij}}{2} E^b \hat{e}^b_j \ , \tag{15}$$

where

$$\nabla'^2 = \nabla^2 + ik^a \cdot \nabla - k^a \cdot k^a \tag{16}$$

In equation (15), $(n^2)_{ij}$ is equal to K_{ij} and $[\Delta(n^2)]_{ij}$ is the change in the dielectric tensor (or index of refraction squared) defined by

$$\Big[\Delta(n^2)\Big]_{ij} = X^{NL}_{ijk} E^{dc} \hat{k}_k \tag{17}$$

The nonlinear susceptibility X^{NL}_{ijk} for the case where one field is static and one is at optical frequencies can be written in terms of the linear electro-optic coefficients r_{ijk} as [10]

$$X^{NL}_{ijk} = -K_{i\alpha} \ r_{\alpha\beta k} \ K_{\beta j} \tag{18}$$

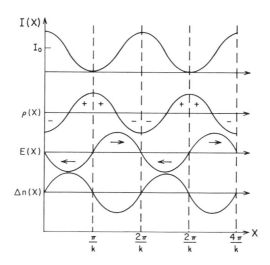

Fig. 3. Optical intensity I(x), charge density $\rho(x)$, electric field E(x), and refractive index change $\Delta n(x)$ versus real-space coordinate x in a ferroelectric undergoing four-wave mixing in a "phase conjugation" configuration in which a standing wave is cavity-generated in the specimen. q is assumed positive.

Since $\Delta(n^2) = 2n\Delta n$, an expression for Δn can be obtained from equations (17) and (18), which in terms of principal coordinates reduces to the familiar form of the linear electro-optic effect

$$\Delta n_i = -\frac{1}{2} n_i^3 \, r_{ik} \, E^{dc} \hat{k}_k \tag{19}$$

where the usual reduced subscript notation is being used for the third-rank tensor r_{ijk}. Using equation (12), equation (19) takes the form

$$\Delta n_z = \frac{1}{2} \frac{mkk_B T}{q} \cdot \frac{n_z^3 \, r_{33}}{1 + (k/k_0)^2} \sin \underset{\sim}{k} \cdot \underset{\sim}{x} \tag{20}$$

Figure 3 shows a plot of the quantities I(z), $\rho_q(z)$, $E^{dc}(z)$ and $n_z(z)$ for the case q>0 (see equations (3), (10), (12) and (20)). Note that for q<0, the quantities ρ_q, E^{dc} and Δn will be π out of phase with their values in the case of positive q. In particular, note that the index grating Δn is shifted by $\pi/2$ with respect to the intensity interference pattern for one sign of the migrant charge carriers, and by $-\pi/2$ for the opposite sign.

Thus far we have seen how the intersection of two optical beams in a photorefractive medium can produce a quasistationary spatially modulated optical susceptibility given by $\mathrm{Re}\ \{X_{ijk}^{NL}\ \varepsilon^{dc}\hat{k}_k\ e^{i\underset{\sim}{k}\cdot\underset{\sim}{x}}\}$. One of the effects of this grating is that it can "scatter" the two optical beams which create it. To see the nature of this steady-state two-beam coupling,

we note that the amplitude of the nonlinear optical polarization density
$P_i^{NL} = \varepsilon_o(X_{ijk}^{NL} E^{dc}\hat{k}_k) E^b e_j^b$ (see the right-hand side of equation (15)) with
wave vector $\underset{\sim}{k}^a$ is linearly proportional to ε^a. This is true since P_i^{NL} is
proportional to the modulation index through ε^{dc}, which is proportional
to ε^a. P^{NL} thus alters the dispersion relation $\underset{\sim}{k}^a(\omega)$ resulting in a new
wave vector $\underset{\sim}{k}^a(\omega) + \delta k^a$. If we assume that both the real and imaginary
parts of δk^a are small compared to those of $\underset{\sim}{k}^a$, then the following relation
can be derived from equations (8) and (14)-(18):

$$\underset{\sim}{k}^a \cdot \delta \underset{\sim}{k}^a = -\frac{i\omega^2 f_0}{2c^2}\left(\frac{\alpha + if}{1 + \alpha^2 + i\alpha f}\right) \cdot \frac{2|E^b|^2}{I_o} \cdot \left(e_i^{a*} K_{i\alpha}\left(r_{\alpha\beta k}\hat{k}_k\right)K_{\beta j}\hat{e}_j^b\right)\left(\hat{e}^a \cdot \hat{e}^{b*}\right)$$

(21)

From a similar derivation we find that $\underset{\sim}{k}^b \cdot \delta \underset{\sim}{k}^b = (\underset{\sim}{k}^a \cdot \delta \underset{\sim}{k}^a)*$ to within a
constant real factor.

Immediately one can see from equation (21) and the result for $\underset{\sim}{k}^b \cdot \delta \underset{\sim}{k}^b$
that the imaginary parts of δk^a and δk^b will give rise to an exponential
growth in one of the optical beams and a similar decay in the other. The
determination of which beam grows and which decays depends on the signs of
the appropriate electro-optic coefficients and on the sign of f_0 which is
equal to the sign of q.

4. EXPERIMENT

We have performed a simple two-beam coupling experiment in a poled
sample of $Ba_2NaNb_5O_{15}$ BSN. The predictions of the statistical mechanical
model of exponential growth in one beam and decay in the other have been
experimentally verified. In our experiment there were no applied fields
(f=0),the two incident beams were set to have nearly equal intensities
($2|\varepsilon^a|^2 = 2|\varepsilon^b|^2 = I_o$) of about 20 mW/mm^2, and they were merged at an angle
of about 2θ equal to $35°$ such that $\hat{k} = \hat{z}$, i.e., the grating pointed along
the polar c-axis. Since $Ba_2NaNb_5O_{15}$ has the orthorhombic point group symmetry mm2 at room temperature, the electro-optic coefficients appropriate
to our case which are allowed by symmetry are r_{13}, r_{23} and r_{33} (Nye [11]).
These coefficients were measured in BSN by Singh et al. [12] to be
$r_{13} = 15 \pm 1$, $r_{23} = 13 \pm 1$ and $r_{33} = 48 \pm 2$ all in units of 10^{-12} m/V. They
did not, however, measure the sign of these coefficients relative to the
positive c-axis, other than the fact that they all have the same sign.
Singh et al. also measured the room temperature indices of refraction at
the wavelength of our experiment (5145 Å Ar$^+$ radiation) to be $n_a = 2.3767$
and $n_c = 2.2583$.

Using the geometry of Fig. 2, equation (21) expressed in principal
coordinates becomes

$$\underset{\sim}{k}^a \cdot \delta \underset{\sim}{k}^a = -i\ \frac{\omega^2 f_0}{2c^2} \cdot \frac{\alpha}{1 + \alpha^2} \cdot \left[\hat{e}_i^{a*} \left(n_{ii} \right)^2 r_{ij3} \right.$$

$$\left. \times \left(n_{jj}^2 \right) \hat{e}_j^b \right] \left(\hat{e}^a \cdot \hat{e}^{b*} \right) \tag{22}$$

Also from Fig. 2, we may write $\hat{e}^a = -\sin\theta\ \hat{x} + \cos\theta\ \hat{z}$ and $\hat{e}^b = -\sin\theta\ \hat{x}$ $+ \cos\theta\ \hat{z}$. If we assume the direction of δk^a to be nearly parallel to that of $\underset{\sim}{k}^a$, then we may write

$$\delta k^a \cong \frac{-iC}{q} \cos 2\theta\ (n_c^4\ r_{33}\ \cos^2\theta - n_a^4\ r_{13}\ \sin^2\theta) \tag{23}$$

where

$$C = \frac{\omega^2}{2c^2}\ \frac{kk_B T}{1 + (k/k_0)^2} > 0$$

Thus,

$$\delta k^a \cong -i\ 895C/q$$
$$\delta k^b \cong +i\ 895C/q \tag{24}$$

In our experiment we observed the beam incident on the crystal at an acute angle with respect to the positive c-axis to always emerge with the weaker intensity. This is beam k^b in Fig. 2. Its intensity decreased by 12% after passing through the crystal, while beam ka in Fig. 2 had a 12% increase. Since the exponential growth or decay has the form $\exp\{-2\,\text{Im}\ (\delta kL)\}$, where L is the interaction length of the two beams in the crystal, we see from equation (24) that q is positive in our experiment. Note, however, that if the signs of the r_{ij3} are negative with respect to the positive c-axis, then the sign of the charge carriers would be negative. Unfortunately, even this conclusion is different for different models [13].

5. CONCLUSION

If α_{33} is negative, then the charge of the defects responsible for thermal memory is also negative. This would be in accord with the hypothesis of J. C. Toledano [14] that they are sodium vacancies. J. Schneck [15] has shown that the concentration of sodium vacancies in BSN can be as high as 10% of the sodium sites. Furthermore, it's quite possible that

these ions could be fairly mobile along the c-axis of the crystal due to
the low packing of atoms in the BSN structure. We hope to gain more information on the defect density by studying how the fraction of the energy
coupled between beams k^a and k^b depends on the grating wave vector k, i.e.,
on the intersection angle θ. These measurements will yield a value of k_o,
and thus, a value of the density of filled trap sites ρW, and will be
reported on in a future paper. Finally, from a study of the transient (or
non-steady-state) properties of the grating formation and decay, it might
be possible to obtain information on the diffusivity of the charged
defects.

6. ACKNOWLEDGMENTS

These experiments were performed in the laboratory of Prof. D. Z.
Anderson with the help of Marie Erie. Thanks are also due to Ragini
Saxena for helpful discussions.

7. REFERENCES

1. K. Hamano et al., J. Phys. Soc. Japan 49, 2278 (1980).
2. H. G. Unruh, this book.
3. J. Feinberg, "Optical Phase Conjugates," Chap. II (Academic Press,
    New York, 1983).
4. D. L. Staebler and J. J. Amodei, J. Appl. Phys. 43, 1042 (1972).
5. J. F. Scott, Ferroelectrics 66, 11 (1986); 52, 35 (1983); and
    Proceedings Int. Conf. Nonlinear Physics (Plenum, 1986).
6. G. H. Hassold, J. Dreitlein, P. D. Beale, and J. F. Scott, Phys.
    Rev. B. 33, 3581 (1986).
7. J. Feinberg, D. Heiman, A. R. Tanguay, Jr., and R. W. Hellwarth,
    J. Appl. Phys. 51, 1297 (1980).
8. In the more general case where assumptions 3 and 4 do not hold,
    and where charge transport can also be due to photoexcitation of
    electrons to the conduction band, see: N. V. Kukhtarev, V. B. Mar-
    kov, S. G. Odulov, M. S. Soskin, and V. L. Vinetskii, Ferroelectrics,
    22, 949 (1979).
9. The sign of the charge carriers is also discussed by F. S. Chen,
    J. Appl. Phys. 40, 3389 (1969); see also ref. 4.
10. Amnon Yariv and Pochni Yeh, "Optical Waves in Crystals" (Wiley-
    Interscience, 1984), Problem 12.1. If this problem is restated for
    the case where the total electric field is that of equation (14)
    in this paper, then $P_i^{NL} = \varepsilon_0 X_{ij} E_j^a + 2d_{ijk} E_k^{dc} E_j^b$, where the d_{ijk} are
    related to the X_{ijk}^{NL} as $d_{ijk} = \tfrac{1}{2} \varepsilon_0 X_{ijk}^{NL}$.
11. J. F. Nye, "Physical Properties of Crystals" (Clarendon Press,
    Oxford, 1985).
12. S. Singh, D. A. Draegert, and J. E. Geusic, Phys. Rev. B2, 2709
    (1970).
13. N. V. Kurkhtarev, V. B. Markov, S. G. Odulov, M. S. Soskin, and
    V. L. Vinetskii, Ferroelectrics 22, 949 (1979).
14. J. C. Toledano, J. Schneck and G. Errandonea, "Incommensurate Phases
    in Dielectrics - Materials," Vol. 14.2 in the series "Modern Problems
    in Condensed Matter Sciences" (North-Holland, 1986), Chapter 17.
15. J. Schneck, 1982, Ph.D. Thesis (Universite Paris VI), unpublished.

PHASONS IN QUASI-CRYSTALS AND INCOMMENSURATE LIQUID CRYSTALS

T. C. Lubensky

Department of Physics
University of Pennsylvania
Philadelphia, PA 19104

Abstract   Phason elasticity and hydrodynamics for incommensurate
liquid crystals and icosahedral quasi-crystals are derived under the
assumption that the free energy is analytic in the gradient of the
relative phase, w, of incommensurate density waves. The phason mode
is always diffusive in the hydrodynamic limit, but the diffusion constant for quasi-crystals is small. In strongly coupled incommensurate
systems, the free energy can be non-analytic in $\vec{\nabla}w$, leading to a
breakdown of elasticity and hydrodynamics. The phason mode is pinned. For both types of phason dynamics non-uniform spatial distortions of w can easily be quenched into icosahedral quasicrystals.
Evidence for quenched phason strains in these materials will be discussed.

1. INTRODUCTION

    Quasi-crystals [1,2] are materials with a point group symmetry that is
incompatible with the existence of a regular periodic lattice. Experimentally observed [1,3,4] quasi-crystals all have either icosahedral symmetry
or five-fold symmetry [5] in a plane. Quasi-crystals with other forbidden,
such as seven- or eight-fold rotation [6], symmetries are, however, theoretically conceivable. The diffraction pattern from ideal quasi-crystals
has a dense set of Bragg peaks in reciprocal space and co-linear peaks,
the ratio of whose wavevectors is irrational (incommensurate). Quasi-crystals are, thus, fundamentally incommensurate structures. The energy of
incommensurate systems is invariant with respect to uniform translations
of the relative phase (phason variable) of two incommensurate density waves
[7]. The dynamical mode associated with these translations is called a

phason. The properties of phasons [8-11] and phason variables in icosahedral and pentagonal quasi-crystals will be the subject of this paper.

Many of the static and dynamic properties associated with non-uniform
distortions of phason variables w are common to all incommensurate systems. One of the simplest imaginable incommensurate systems is the recently reported [12,13] incommensurate smectic-A phase of liquid crystals.
This phase is characterized by two co-existing one-dimensional density
waves (arising from mass density and electric polarization) and their harmonics. Phason elasticity and hydrodynamics [14] will be developed for
this system and generalized to quasi-crystals. The important conclusion
to this analysis is that the phason mode is diffusive when there exists an
analytic expansion of the free energy in gradient of w. The phason diffusion constant in icosahedral quasi-crystals is, however, very small [10]
so that non-equilibrium phason strains can easily be quenched into a
sample.

Even though the free energy F of incommensurate systems is invariant
with respect to spatially uniform increments of w, it is possible for w to
be pinned so that F is not analytic in $\vec{\nabla}w$. The properties of the one-
dimensional Frenkel-Kontorowa [15] model where the breaking of analyticity
was first identified [16] will be reviewed briefly in Sec. 4. It will
then be shown how the unit cell or tiling construction of quasi-crystals
leads naturally to a non-analytic interpretation of phason strains [11,
17]. If phason dynamics is correctly described by the pinned or nonanalyt-
ic rather than the hydrodynamic model, phason strains will again be easily
quenched into non-equilibrium configurations.

Theory predicts that quenched phason strains are likely, and it is of
some interest to know if they have been observed experimentally. Section 5
will discuss what effects [18,19] quenched phason-strains will have on the
scattering intensities and the experimental evidence for their existence
from electron and x-ray diffraction [20,21] and from high resolution electron micrographs [22].

2. INCOMMENSURATE SMECTICS

The existence of a truly incommensurate liquid crystal smectic-A phase
(S_{AI}) has recently been reported [12]. The x-ray scattering intensity from
this phase shows two strong co-linear quasi-Bragg peaks, the ratio of whose
wavenumbers k_1 and k_2 is irrational (Fig. 1). It can be useful to think of
these peaks as arising from co-existing one-dimensional mass and polarization density waves with fundamental periods $d_1 = 2\pi/k_1$ and $d_2 = 2\pi/k_2$ which
can be represented as layers as shown in Fig. 2. The experiments also
show a peak of $2k_1$. In general one would expect peaks to occur at all

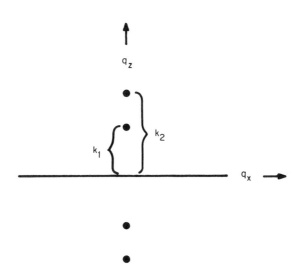

Fig. 1. X-ray scattering intensity from an incommensurate smectic-A liquid
crystal showing the two fundamental Bragg peaks at incommensurate
wavevectors k_1 and k_2. The z-axis is normal to the smectic
layers. In general, there should be Bragg peaks at all wavevectors of the form $G = pk_1 + qk_2$ for p and q integers.

integer combinations of k_1 and k_2 though their intensity might be unobserv-
ably small. Thus the equilibrium mass density of the S_{AI} phase can be
expanded in a Fourier series in k_1 and k_2:

$$<\rho(\vec{x})> = \sum_{p,q} \rho_{pq} e^{i(pk_1z+qk_2z)} \quad , \tag{2.1}$$

where p and q are integers, $\vec{x} = (x,y,z)$ with z the co-ordinate normal to
the smectic layers, and

$$\rho_{pq} = |\rho_{pq}| e^{i\phi_{pq}} \tag{2.2}$$

is a complex amplitude.

The phase ϕ_{pq} can be written as

$$\phi_{pq} = pk_1u_1 + qk_2u_2 + \phi_{pq}^0 \quad . \tag{2.3}$$

Spatially uniform increments of u_1 and u_2 correspond to translations of the
fundamental layers associated wave numbers k_1 and k_2. Because k_1/k_2 is
irrational, the free energy of the S_{AI} is invariant with respect to
spatially uniform translations of either of these layers. This means that
there will be an elastic free energy F proportional to gradients of u_1 and
u_2. In addition to being invariant with respect to uniform translations of

189

Fig. 2. Schematic representation of the two coexisting colinear density
waves with periods of irrational ratio in an incommensurate
smectic-A liquid crystal.

u_1 and u_2, F must be invariant with respect to uniform rotations of the
entire system but not with respect to rotation of one set of layers relative to the other. The elastic free energy incorporating these constraints
is conveniently expressed as follows [14]:

$$F = \int d^3x[\tfrac{1}{2}B_1(\nabla_\| u_1)^2 + \tfrac{1}{2}K_1(\nabla_\perp^2 u_1)^2]$$

$$+ \int d^3x[\tfrac{1}{2}B_2(\nabla_\| u_2)^2 + \tfrac{1}{2}K_2(\nabla_\perp^2 u_2)^2] \qquad (2.4)$$

$$+ \int d^3x[B_{12}(\nabla_\| u_1)(\nabla_\| u_2) + K_{12}(\nabla_\perp^2 u_1)\,\nabla_\perp^2 u_2) + \tfrac{1}{2}C(\vec{\nabla}_\perp u_1 - \vec{\nabla}_\perp u_2)^2].$$

where $\|$ refers to the z-direction (perpendicular to the layers) and $\perp$ to
the xy plane (parallel to the layers). The first two terms of this equation describe the free energy of the two sets of independent layers. They
are identical in form to the Landau-Peierls elastic energy [23] for a smec-
tic-A liquid crystal, and contain no $(\vec{\nabla}_\perp u_1)^2$ or $(\vec{\nabla}_\perp u_2)^2$ terms. The third
term describes the coupling between the two sets of layers. The term pro-
portional to $(\vec{\nabla}_\perp u_1 - \vec{\nabla}_\perp u_2)^2$ describes the energy cost of rotating the two
layers relative to each other.

It is useful to describe incommensurate systems not in terms of the
phases u_1 and u_2 of the constituent incommensurate waves but in terms of
variables u and w describing, respectively, translations of the entire
system and relative translations of the two waves:

$$u_1 = u + sw, \qquad u_2 = u - (1-s)w, \qquad (2.5)$$

or

$$u = \tfrac{1}{2}[(1-s)u_1 + su_2], \qquad w = u_1 - u_2 \qquad (2.6)$$

It is clear that u displaces both layers by the same amount. u couples to external forces and its gradients are the physical strains of the system. w is the phason variable unique to incommensurate systems. s depends on the detailed dynamics of the system. Its precise value is relevant to the present discussion. In terms of the variables u and w, the phase ϕ_{pg} becomes

$$\phi_{pq} = Gu + G_\perp w , \tag{2.7}$$

where

$$G = pk_1 + qk_2 \tag{2.8}$$

is the wavevector of the quasi-Bragg peak associated with pq.

$$G_\perp = spk_1 - (1 - s)k_2 \tag{2.9}$$

is a "perpendicular" wavevector that has a natural interpretation in the two-dimensional lattice from which the incommensurate one dimensionally modulated mass density of the smectic phase may be obtained by projection [24]. Note that since both G and $G_\perp$ are uniquely specified by p and q, there is a unique $G_\perp$ associated with each G. The elastic free energy is now

$$F = \int d^3x \frac{1}{2}[B(\nabla_\parallel u)^2 + K(\nabla_\perp^2 u)^2 + B_w(\nabla_\parallel w)^2 + C_w(\vec{\nabla}_\perp w)^2]$$

$$+ \int d^3x[B_{uw}(\nabla_\parallel u)(\nabla_\parallel w) + K_{uw}(\nabla_\perp^2 u)(\nabla_\perp^2 w)]. \tag{2.10}$$

Because layers cannot be rotated relative to each other, there is a term in F proportional to $(\vec{\nabla}_\perp w)^2$. This term implies that w is more rigid than u. Its presence has important ramifications for the properties of dislocations in incommensurate smectics.

An interesting and apparently general consequence of Eq. (2.10) is that the effective elastic constants describing the response of u to external forces are non-analytic in wave number. The phason field w will relax to its local equilibrium value in the presence of a non-uniform u created by external forces. This leads to an effective elastic energy,

$$F^{eff} = \int d^3x \frac{1}{2}[B(\theta)(\nabla_\parallel u)^2 + K(\theta)(\nabla_\perp^2 u)^2]. \tag{2.11}$$

which has the same form as for a normal smectic but with elastic constants,

$$B(\theta) = B - B_{uw}^{2}[B_w + C_w \cot^2\theta]^{-1}$$

$$K(\theta) = K - 2K_{uw} B_{uw}[B_w + C_w \cot^2\theta]^{-1} , \tag{2.12}$$

depending on the angle θ between the gradient of u and the z-axis. Equations (2.10) and (2.11) imply that the response of u to external forces imposed at time t = 0 will be determined by B and K for times short compared to the relaxation time of w and by $B(\theta)$ and $K(\theta)$ for long times [10].

Since there is no free energy increase resulting from uniform spatial increments of w, the frequencies of modes associated with long wavelength distortions of w will go to zero with wavenumber q. w is a broken symmetry hydrodynamic variable [25] that must be included along with the other conserved and broken symmetry variables to describe properly the hydrodynamics of the system. The full hydrodynamic equations for the S_{AI} phase can be derived using standard procedures and will be presented elsewhere [14]. The important result for the present purposes is that w always relaxes diffusively at the longest wavelengths. Let v_w be the relative velocity along the z-direction of the incommensurate layers. The equation of motion for w is then,

$$\partial_t w - v_w = -\Gamma' \frac{\delta F}{\delta w} , \tag{2.13}$$

where Γ' is a transport coefficient. Note the right hand side of this equation is zero if w is spatially uniform. Thus Eq. (2.13) alone would say that a steady state in which the two incommensurate waves move relative to each other at constant velocity is possible. Such a situation is almost realizable if the coupling between the two waves is small as would be the case for mercury chain salts [26] for example. In real systems, however, there is always friction between the two waves, and v_w must be zero in steady state. This can be described by looking at the momentum $g_w = \rho p_w$ conjugate to w. Its force equation is

$$\partial_t g_w + \gamma g_w = -\frac{\delta F}{\delta w} \tag{2.14}$$

where γ is a friction coefficient and $\delta F/\delta w$ is a generalized force density. At low frequency $\partial_t g_w$ is small compared to γg_w and $g_w = -\gamma^{-1}(\delta F/\delta w)$ leading to the purely dissipative hydrodynamic equation,

$$\partial_t w = -\Gamma \frac{\delta F}{\delta w} , \tag{2.15}$$

for w where $\Gamma = \Gamma' + (\rho\gamma)^{-1}$. $\delta F/\delta w$ is proportional to $\nabla^2 w$ and $\nabla_{\parallel}^2 u$, i.e., w relaxes diffusively, though the detailed mode structure requires a complete treatment of the dynamics of u. Note. that if $\gamma = 0$, g_w becomes a conserved variable, and there is a propagating rather than a diffusive mode associated with w. When $\gamma \neq 0$, the phason mode becomes propagating for $\omega \gg \gamma$.

3. PENTAGONAL AND ICOSAHEDRAL QUASI-CRYSTALS

Quasi-crystals are structures with a point group symmetry that is incompatible with regular periodic translational symmetry. The most common experimental quasi-crystal [1] is an Al-Mn alloy with icosahedral point group symmetry. There are, however, a number of materials exhibiting icosahedral symmetry [3,4] and phases without full icosahedral symmetry but with a five-fold rotation axis [5]. In addition to experimentally observed quasi-crystals, there are an essentially infinite number of mathematical tilings [27] of two and three space with arbitrary point group symmetries [6]. The forbidden point group symmetry of quasi-crystals leads to diffraction patterns with a dense set of Bragg peaks at wavevectors which include colinear wavevectors, the ratio of whose magnitudes is irrational. Quasi-crystals are fundamentally incommensurate systems with irrational ratios determined by point group symmetry.

The reciprocal lattice L_R of a quasi-crystal can be formed by integer linear combinations of a basis set of vectors $\{\vec{G}_n\}$ on some fundamental star. For pentagonal quasi-crystals, $\{\vec{G}_n\}$ is the set of five vectors pointing to the vertices of a pentagon whereas for icosahedral quasi-crystals, $\{\vec{G}_n\}$ can conveniently be taken as six of the twelve vectors pointing to the vertices of an icosahedron. The fundamental star of vectors and four subsequent stars formed from linear combinations of $\{\vec{G}_n\}$ for pentagonal symmetry is shown in Fig. 3. Note that $\vec{G}_1 + \vec{G}_4 = \tau^{-1}\vec{G}_0$ where $\tau = (1 + \sqrt{5})/2$ is the Golden mean.

As in the case of incommensurate smectics, the mass density can be expanded in a Fourier series:

$$<\rho(\vec{x})> = \sum_{\vec{G}\epsilon L_R} |\rho_{\vec{G}}| e^{i\vec{G}\cdot\vec{x}} e^{i\phi_{\vec{G}}} , \qquad (3.1)$$

where $\phi_{\vec{G}}$ is the phase of the mass density wave at wavevector $\vec{G}$. It can be expressed in terms of displacement and phason variables

$$\phi_{\vec{G}} = \vec{G}\cdot\vec{u} + \vec{G}_{\perp}\cdot\vec{w} + \phi_{\vec{G}}^o \qquad (3.2)$$

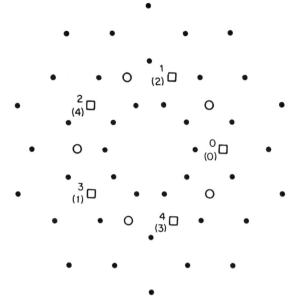

Fig. 3. The basis set $\{\vec{G}_n\}$ for a pentagonal quasi-crystal and the next generation of reciprocal lattice vectors obtained by adding all pairs in the basis set. The open squares are the five vectors $\vec{G}_n$, and the open circles are the five vectors $-\vec{G}_n$. The dark circles are the second generation of vectors obtained by summing all pairs of vectors in the basis set. The numbers in parentheses next to the open squares indicate the set of perpendicular vectors $\{G_{\perp n}\}$.

where $\vec{G}_\perp$ is the vector analog of $G_\perp$ in the incommensurate smectic liquid crystal. For pentagonal quasi-crystals the vectors $\vec{G}_{\perp n}$ corresponding to the fundamental set $\{\vec{G}_n\}$ are $\vec{G}_{\perp n} = \vec{G}_{[2n \ (\mathrm{mod}5)]}$. $\vec{G}_{\perp n}$ is defined for icosahedral systems in Ref. 10. Again there is a unique $\vec{G}_\perp$ associated with each $\vec{G}$. The intensity of scattering peaks at $\vec{G}$ decreases with increasing $\vec{G}_\perp$ [2,24].

The elastic energy for quasi-crystals [8-10] can be expanded in gradients of $\vec{u}$ and $\vec{w}$:

$$F = \int d^3x [K_{ijkl} u_{ij} u_{kl} + K^w_{ijkl} \nabla_i w_j \nabla_k w_l + C^{uw}_{ijkl} u_{ij} \nabla_k w_l] \qquad (3.3)$$

where $u_{ij} = \frac{1}{2}(\nabla_i u_j + \nabla_j u_i)$ is the symmetrized strain. For both icosahedral and pentagonal quasi-crystals, there are two independent elastic constants in K_{ijkl} and K^w_{ijkl}. For icosahedral and pentagonal systems with D5 symmetry, there is one elastic constant in C^{uw}_{ijkl}. For pentagonal systems with C5 symmetry, there are two constants in C^{uw}_{ijkl}, one of which can be

integrated to the surface. As for incommensurate smectics, the effective
strain elastic energy is non-analytic.

The hydrodynamic mode associated with $\vec{w}$ is diffusive [9,10], obeying
an equation exactly analogous to Eq. (2.15). The diffusion time is thus of
order

$$\tau_w^{-1} = \Gamma_w K^w q^2 \equiv D_w q^2 \tag{3.4}$$

where Γ_w is a mobility, K^w a phason elastic constant, and q the wavenumber. It is clear that phasons in quasi-crystals involve local rearrangements of atoms. Thus a reasonable estimate for Γ_w is that it is of the
same order of magnitude as the mobility for vacancy or impurity diffusion.
Furthermore K^w should be of the same order of magnitude as the shear elastic constant. Thus D_w is of order the vacancy diffusion constant of a
metal. This number is typically very small[28] ($<10^{-10}$ cm^2/sec) so that the
relaxation time of hydrodynamic phason strains in quasi-crystals should be
very long, probably so long that it would be difficult to observe in typical laboratory times.

4. PINNED DYNAMICS

In the previous section, it was argued that there should be a continuum elasticity and hydrodynamics associated with the phason variable
because its spatially uniform changes do not alter the free energy. It is
possible, however, as shown by Aubry [16], for the phason variable to be
pinned so that continuum elasticity and hydrodynamics do not describe the
energetics and dynamics of its spatially non-uniform distortions. The
simplest model where the breakdown of hydrodynamics can be seen is the
one-dimensional discrete Frenkel-Kontorowa model [15] in which harmonically
coupled atoms at positions x_n interact with a periodic external potential.
The Frenkel-Kontorowa Hamiltonian is

$$H_{FK} = \tfrac{1}{2}K \sum_n (x_{n+1}-x_n-\ell_0)^2 - V \sum_n \cos(2\pi x_n/\ell_2). \tag{4.1}$$

The separation between adjacent atoms is ℓ_0 when V = 0, and the period of
the external potential is ℓ_2. The positions of the atoms at extrema of H_{FK}
satisfy

$$x_n = n\ell_1 + u_1 + f(n\ell_1 + u_1 - u_2) \tag{4.2}$$

where ℓ_1 is the average separation of the atoms when V $\neq$ 0 and f(y) is a
function of period ℓ_2. u_1 and u_2 are displacement variables introduced in

the discussion of incommensurate smectics in Section 2. In the FK model, the phase of the periodic potential is rigidly fixed so that u_2 is a constant which can be set to zero. In more general interacting systems (e.g., two interacting harmonic chains with different preferred interatomic separations), u_2 could vary.

For small values of V/K, the function f(y) is continuous, and the frequency of the phason mode (described in this case by u_1) goes to zero with wavevector $\vec{q}$, as in the hydrodynamic description of the previous section. Above a critical value of V/K, however, f(y) becomes non-analytic, and the phason frequency tends to a finite value as $\vec{q} \to 0$. As a result, the free energy is not analytic in u_1, as it is when the usual elastic expansions apply. Similarly, the fundamental assumption of continuum hydrodynamics that an analytic gradient expansion exist is no longer valid.

To see the effects of the breaking of analyticity in higher dimensions, it is instructive to consider a toy model in which there are rows of atoms parallel to the y-axis with x coordinates $x_n(y)$. Now assume that the equilibrium atomic positions are determined by the simple analytic function,

$$x_n(y) = n + \alpha(y) + \rho[\sigma n + \beta(y)] \tag{4.3a}$$

$$= nT + u_1 - \rho \left[\frac{nT + u_1 - u_2}{T/\sigma} \right] \tag{4.3b}$$

where $T = (1 + \rho\sigma)$, [x] is the greatest integer less than or equal to x, and $\{x\} = x - [x]$ is the fractional part of x. Eq. (4.3b) is in the same form as Eq. (4.) with $\ell_1 = T$, $u_1 = T(\alpha + \rho\beta)$, and $u_2 = T(\alpha - \sigma^{-1}\beta)$. $\alpha = T^{-1}(\sigma u_1 + \rho u_2) = T^{-1}u$ and $\beta = \sigma T^{-2}(u_1-u_2)$ are, respectively, linearly proportional to the strain and phason variables u and $w = u_1 - u_2$. Eq. (4.3) implies that the distance between adjacent atoms can take on only two values: 1 and (1 + ρ). When $\rho = \sigma = \tau^{-1}$ where τ is the golden mean, the resulting sequence is the Fibonacci sequence of long and short intervals shown in Fig. 4. (Fibonacci sequences can be generated by the inflation rules $L \to LS$ and $S \to L$.) Each value β gives a Fibonacci sequence. A change in β will give rise to a new sequence obtained from the old by interchanging long and short intervals. Thus if $\beta(y)$ increases from zero at y = 0 to 0.5 at y = L, there will be exactly one change in length for each interval as shown in Fig. 4. The changes in length give rise to discontinuities or jags at discrete points.

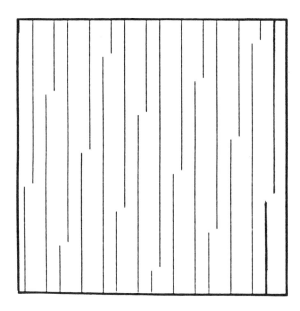

Fig. 4. A strained Fibonacci sequence described by Eq. (4.3a) with
$\rho = \sigma = \tau^{-1}$ with $\beta(y)$ = const. x y. The x-axis is horizontal
and the y-axis is vertical. at each value of successive lines
are either a unit distance or $1 + \tau^{-1} = \tau$ times a unit
distance apart, and the sequence of long and short intervals
is an acceptable Fibonacci sequence (i.e., one that obeys
the Fibonacci inflation rules). As $\beta(y)$ changes, long and
short intervals interchange at discrete values of y leading
to jags in the linear grid.

It is natural to assign an energy ε to each jag shown in Fig. 4.
Since the number of jags per unit length in the x-direction is proportional
to the change, $\Delta\beta$ in β along y, the total energy is

$$E = N_T\varepsilon \sim \int dx\Delta\beta = \int dxdy|\nabla\beta| \tag{4.4}$$

where N_T is the total number of jags. In higher dimensions, this result
generalizes to $E \sim \int d^2x \, |\nabla\beta|$. Note that $E \to 0$ as $\nabla\beta \to 0$, but unlike the
elastic case E is not analytic in β.

Penrose tilings for pentagonal (icosahedral) quasi-crystals can be
constructed by taking the dual lattice to Amman grids [30] consisting of
parallel lines (planes) normal to the vectors to the vertices of a pentagon
(icosahedron) with positions given by the Fibonacci sequence of Eq. (4.2)
with $\rho = \sigma = \tau^{-1}$ (see Fig. 5). The tiles resulting from this construction
obey certain matching rules which can be encoded by decorating each tile as
shown in Fig. 6. Phason strains give rise to jags in the Amman grid which
lead to tiles which violate matching rules with their neighbors as shown in

Fig. 6. As in the single grids, it is natural to associate an energy with each matching rule violation. This leads again to an energy proportional to $\int d^d x\, |\vec{\nabla w}|$. It is not yet entirely clear whether the continuum or pinned energetics correctly describes icosahedral quasi-crystals, there are theoretical indications [17] that the pinned picture is more likely to be correct in systems such as AlMn where individual atoms are well localized in space.

Fig. 5. Construction of the Penrose tiling of a two-dimensional plane
        from the Amman pentagrid consisting of five grids of parallel
        lines with separations obeying the Fibonacci sequence and
        oriented normally to the five vectors pointing to the vertices
        of a pentagon. The "fat" and "skinny" Penrose tiles and the
        intersection of the lines of the Amman pentagrid with them are
        shown at the bottom of the figure. The Penrose tiling is the
        dual of the Amman pentagrid (see Ref. 30).

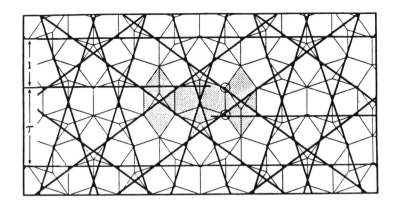

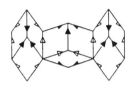 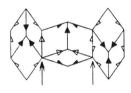

Fig. 6. This figure shows the effect of a jag in the amman grid on the
Penrose tiling. The Penrose tiles in an undefected lattice
obey matching rules depicted in the set of tiles below the
figure on the left. Each tile has two sides with a black half-
arrow and two sides with a white half-arrow. When adjacent tiles
join, the half-arrows on their common side must form a full
arrow of a single color. When there is a jag in the Amman grid
(produced by a phason strain), tiles near the jag violate the
the matching rules. The set of tiles below the figure on the
left represent the shaded tiles in the figure. There is a
matching rule violation at the edge marked with the single
pointed arrow: the edge has two white half-arrows that do not
form a full arrow. If the position of the jag were moved to the
left, the matching rule violation would move to the left, say to
the edge marked by the double-pointed arrow. Note that the
matching rule violations occur at only discrete places, and an
energy can be associated with these violations as in Eq. (4.4).

5. QUENCHED PHASON STRAINS

It was just argued that phason strains are likely to be present in
almost any laboratory sample of an icosahedral quasi-crystal because of
their long relaxation time. These strains can easily be detected in scattering experiments. The scattering intensity of electrons and x-rays is
proportional to the Fourier transform of the density correlation function

$$I(\vec{q}) = \int d^3x d^3x' <\rho(\vec{x})\rho(\vec{x'})> e^{-i\vec{q}\cdot(\vec{x}-\vec{x'})} \; , \tag{5.1}$$

which when there is long-range positional order is dominated by the coherent scattering part,

$$I(\vec{q}) = \int d^3x\, d^3x' <\rho(\vec{x})><\rho(\vec{x}')> e^{-i\vec{q}\cdot(\vec{x}-\vec{x}')} \quad , \tag{5.2}$$

where $<\rho(\vec{x})>$ is given by Eq. (3.1). When there are quenched-in phason strains, the phason field $\underline{w}(\vec{x})$ has some fixed positionally dependent value $\underline{\tilde{w}}(\vec{x})$. The displacement field will then relax to some local equilibrium value $\underline{u}(\vec{x})$ in the presence of $\underline{\tilde{w}}(\vec{x})$. The complex amplitude of the mass density at wave vector $\vec{G}$ is then

$$<\rho_{\vec{G}}(\vec{x})>_w = |\rho_{\vec{G}}| e^{i\vec{G}\cdot\underline{u} + i\vec{G}_\perp\cdot\underline{w}} \tag{5.3}$$

If $\underline{\tilde{w}} = \tilde{M}\vec{x}$, where $\tilde{M}$ is a 3x3 matrix, is approximately linear over the scattering volume V and $\underline{u}$ is constant,

$$I(\vec{q}) = \int d^3x\, d^3x' \sum_{\vec{G}} |<\rho_{\vec{G}}>|^2 e^{i\vec{G}_\perp\cdot[\underline{w}(\vec{x})-\underline{w}(\vec{x}')]} e^{i\vec{G}\cdot(\vec{x}-\vec{x}')} \tag{5.4a}$$

$$= V^2 \sum_{\vec{G}} |<\rho_{\vec{G}}>|^2 \delta_{\vec{q},\vec{G}+\Delta\vec{G}} \quad , \tag{5.4b}$$

where $\Delta\vec{G} = \vec{G}_\perp \tilde{M}$. Thus a uniform phason strain anisotropically shifts Bragg peaks by an amount proportional to $|\vec{G}_\perp|$. Since the peaks with the smallest intensity have the largest $|\vec{G}_\perp|$, the dimmest Bragg peaks are shifted the most, as shown in Fig. 7. High resolution electron diffraction patterns [18,20] from the small regions of icosahedral patterns do show spots whose deviation from ideal icosahedral positions scale with $|\vec{G}_\perp|$ in a way which is consistent with a uniform phason strain.

X-ray powder diffraction probes a much larger volume than does electron scattering and effectively averages over the possible realizations of the quenched phason variable $\underline{\tilde{w}}(\vec{x})$, i.e., $I(\vec{q})$ is an ensemble average over the probability distribution for $\underline{\tilde{w}}(\vec{x})$. If components of $\underline{\tilde{w}}(\vec{x})$ that grow at large $\vec{x}$ are sufficiently weighted, the peaks $I(\vec{q})$ will be broadened, as is clearly the case if a powder average (i.e., average over directions) of a single linear strain is taken. If averages with respect to $\underline{\tilde{w}}(\vec{x})$ are denoted by square brackets, the scattering intensity becomes

$$I(\vec{q}) = \int d^3x\, d^3x' [<\rho(\vec{x})>_w <\rho(\vec{x}')>_w] e^{-i\vec{q}(\vec{x}-\vec{x}')} \quad . \tag{5.6}$$

The exact form of $I(\vec{q})$ depends on the probability distribution for $\underline{\tilde{w}}$, which is not known in any detail. A reasonable model is that $\underline{\tilde{w}}(\vec{q})$, the Fourier transform of $\underline{\tilde{w}}(\vec{x})$, is an independent random variable with zero mean and variance (in three dimensions) proportional to $q^{-(\alpha+3)}$

$$[w_i(\vec{q})w_j(\vec{q}')] \sim \delta_{ij}\delta_{\vec{q},\vec{q}'}\, q^{-(\alpha+3)} \quad . \tag{5.7}$$

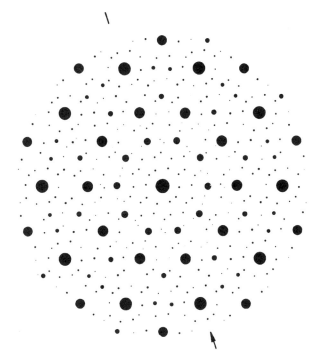

Fig. 7.   Theoretically calculated electron diffraction pattern along a
          five-fold icosahedral symmetry direction in the presence of a
          uniform phason strain. The dark spots have a higher intensity
          and smaller $\vec{G}_\perp$ than do the light spots. The light spots are
          shifted more than the dark spots as can be seen by sighting
          along the line marked by the arrow. In the unstrained pattern
          dark and bright spots would lie along common straight lines.
          Experimental electron diffraction patterns from small
          scattering volumes look similar to this.

In this case,

$$[\underline{\vec{w}}^2(\vec{x})] \sim \begin{cases} \text{const.} & \alpha < 0 \\ \ln L & \alpha = 0 \\ L^\alpha & \alpha > 0 \end{cases}, \tag{5.8}$$

in a sample of size L . When $\alpha > 0$,

$$[(\underline{\vec{w}}(\vec{x}) - \underline{\vec{w}}(0))^2] = a|\vec{x}|^\alpha \tag{5.9}$$

for large $\vec{x}$ where a is a constant. Since $w(\vec{x})$ is a Gaussian random variable, the evaluation of $\langle \rho_{\vec{G}}(\vec{x}) \rangle_w$ is straightforward. If $\alpha < 0$, there
will still be Bragg peaks in $I(\vec{q})$ with a non-thermal Debye-Waller factor

proportional to $\exp\{\frac{1}{2}G_\perp^2[\underline{w}^2(\vec{x})]\}$. There will also be additional diffuse scattering associated with the randomness of $\vec{\underline{w}}$. If $\alpha > 0$,

$$I(\vec{q}) = \int d^3x \sum_{\vec{G}} |<\rho_{\vec{G}}>|^2 e^{-i(\vec{q}-\vec{G})\cdot\vec{x}} e^{-\frac{1}{2}G_\perp^2 a|\vec{x}|^\alpha}, \qquad (5.10)$$

and the peaks of $I(\vec{q})$ are broadened. If $\alpha = 1$ the peaks have a Lorentzian-squared line shape, and if $\alpha = 2$, a Gaussian lineshape. If $\alpha = 1$, and there is an additional term in $[(\vec{\underline{w}}(\vec{x}) - \vec{\underline{w}}(0))^2]$ proportional to $\ln|\vec{x}|$, the line-shape can be Lorentzian. Since $\vec{\underline{u}}$ is driven by $\vec{\underline{w}}$, it will make similar contributions to $I(\vec{q})$ depending on the magnitude of the $\vec{\underline{u}} - \vec{\underline{w}}$ coupling energy. Dislocations always carry both $\vec{u}$ and $\vec{w}$ strains [8]. Like simple phason strains, they can be quenched into non-equilibrium and can contribute to broadening of x-ray linewidths [19].

High resolution electron micrographs obtained by refocusing electrons scattered from a thin strip of material onto a photographic plate yield information about the electron charge density integrated over the coordinate normal to the strip. Thus electron images obtained from a five-fold electron diffraction pattern contain information about the electron density integrated along a coordinated parallel to a five-fold icosahedral axis.

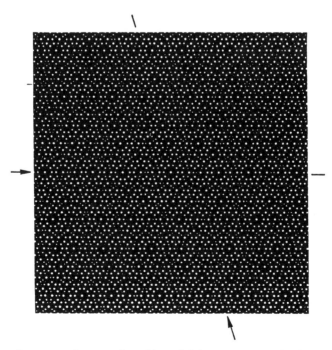

Fig. 8.   Density wave image of a five-fold pattern with phason strain produced as explained in the text (Eg. (5.10)).

These images [22] show regular grids of long and short intervals following a Fibonacci sequence like the Amman grids used to generate the Penrose tiles. The grids are interrupted by jags similar to those produced by phason strains (Fig. 6). Figure 8 shows a theoretical picture that closely approximates the density probed by electron imaging. It was obtained by placing a black dot at every point $\vec{x}$ in an array where the function,

$$\rho(\vec{x}) = \sum_{n=0}^{4} \cos(\vec{G}_n \cdot \vec{x} + \vec{G}_{\perp n} \cdot \vec{w}(\vec{x})) , \tag{5.11}$$

was greater than 2/5 its maximum value and a white dot at other points on in the array. $\vec{G}_n$ are the pentagonal basis vectors shown in Fig. 3, and $\vec{w}(\vec{x})$ was given a uniform linear strain. This picture is very similar to the electron images of Ref. 22.

Electron [18,21] and x-ray [21] scattering and high resolution micrographs]22] from icosahedral AlMn all show deviations from ideal icosahedral order. All of these deviations are consistent with the existence of quenched phason strains in an otherwise ideal icosahedral structure. They also appear to be consistently described by a model of an icosahedral glass based on random packing of icosahedral units [31]. Since this model has built-in disorder that cannot be expressed solely in terms of phason strain, it is different from the ideal icosahedral state with quenched phason strains discussed in this article. Further refinements in theory and experiments will be needed to determine if icosahedral AlMn should be viewed as an icosahedral glass or a defected quasi-crystal.

6. ACKNOWLEDGMENTS

The work reported here is really the result of extended collaborations with a number of colleagues including Peter Bancel, Paul Heiney, Dov Levine, Stellan Ostlund, Sriram Ramaswamy, Josh Socolar, Paul Steinhardt, and John Toner. It was supported in part by the National Science foundation under grant No. DMR 8540332 and MRL program grant No. DMR 8519059.

7. REFERENCES

1. D. S. Schechtman, I. Blech, D. Gratias, and J. W. Cahn, Phys. Rev. Lett. 53, 1951 (1984).
2. D. Levine and P. J. Steinhardt, Phys. Rev. Lett. 53, 2477 (1984); Phys. Rev. B (1986) (to be published).
3. Peter A. Bancel and Paul A. Heiney, Phys. Rev. B33, 7917 (1986).
4. S. J. Poon, A. J. Drehman, and K. R. Lawless, Phys. Rev. Lett. 55, 2324 (1985).
5. L. Bendersky, Phys. Rev. Lett. 55, 1461 (1985).

6. J. E. S. Socolar, P. J. Steinhardt, and D. Levine, Phys. Rev. B32, 5547 (1985); F. Gahler and J. Rhyner, J. Math Phys. A19, 267 (1986).
7. See, for example, P. Bak, Rep. Prog. Phys. 45, 587 (1981); V. L. Pokrovsky and A. L. Talopov, "Theory of Incommensurate Crystals," Sov. Science Review (Horwood Acad. Pub., Switzerland, 1983).
8. D. Levine, T. C. Lubensky, S. Ostlund, S. Ramaswamy, P. J. Steinhardt, and J. Toner, Phys. Rev. Lett. 54, 1520 (1985); P. Bak, Phys. Rev. Lett. 54, 1517 (1985); Phys. Rev. B32, 5764 (1985).
9. P. A. Kalugin, A.Yu Kitayev, and L. S. Levitov, Pis'ma Zh. Eksp. Teor. Fiz. 41, 119 (1985) [Soviet Phys. JETP 41, 119 (1985)]; J. Phys. Lett. (Paris) 46, L-601 (1985).
10. T. C. Lubensky, Sriram Ramaswamy, and J. Toner, Phys. Rev. B32, 7444 (1985).
11. Joshua E. S. Socolar, T. C. Lubensky, and Paul J. Steinhardt, to be published in Physl. Rev. B.
12. R. Ratna, S. Shashidhar, and V. N. Raja, Phys. Rev. Lett. 55, 1476 (1985); R. Ratna and S. Shashidhar, this volume.
13. P. Barois, Phys. Rev. A33, 3632 (1986); Jiang Wang (unpublished thesis, University of Pennsylvania, 1985).
14. T. C. Lubensky, S. Ramaswamy, and J. Toner (unpublished).
15. Y. I. Frenkel and T. Kontorowa, Zh. Eksp. Teor. Fiz. 8, 1340 (1938).
16. S. Aubry, in "Solitons in Condensed Matter Physics," Vol. 8 of Springer Series in Solid State Physics, edited by A. Bishop and T. Schneider (Springer, Berlin, 1978), p. 264; M. Peyrard and S. Aubry, J. Phys. C7, 1593 (1983).
17. D. Frenkel, C. L. Henley, and E. D. Siggia, Phys. Rev B34, 3649 (1986).
18. T. C. Lubensky, Joshua E. S. Socolar, Paul. J. Steinhardt, Peter A. Bancel, and Paul A. Heiney, Phys. Rev. Le H. 57, 1440 (1986).
19. P. M. Horn, W. Malzfeldt, D. P. DiVincenzo, J. Toner, and R. Gambino, Phys. Rev. Lett. 57, 1444 (1986).
20. M. Tanaka, M. Terauchi, K. Hiraga, and M. Hirabayashi, Ultramicroscopy 17, 279 (1985).
21. P. Bancel, P. A. Heiney, P. W. Stephens, A. I. Goldman, and P. M. Horn, Phys. Rev. Lett. 54, 2422 (1985).
22. K. Hiraga, M. Hirabayashi, A. Inoue, and T. Masumoto, Sci. Rep. RITU, A32, No. 2, 209 (1985).
23. See, for example, P. G. De Gennes, "The Physics of Liquid Crystals" (Clarendon, Oxford, 1974).
24. V. Elser, Acta Cryst. A42, 36 (1986); M. Duneau and A. Katz, Phys. Rev. Lett. 54, 2688 (1985); A. Katz, this volume.
25. P. C. Martin, O. Parodi, and P. S. Pershan, Phys. Rev. A6, 2401 (1972); D. Forster, "Hydrodynamic Fluctuations, Broken Symmetry, and Correlation Functions" (Benjamin, Reading, MA, 1975).
26. J. P. Pouget, G. Shirane, J. M. Hastings, A. J. Heeger, N. D. Miro, and A. G. MacDiarmid, Phys. Rev. B18, 3465 (1978); See also references in Ref. 10.
27. R. Penrose, Bull. Inst. Math. and Its Appl. 10, 266 (1974); M. Gardner, Sci. Am. 236, 110 (1977).
28. S. Pantelides, private communications; see also Ref. 10.
29. See for example Ref. 1 or B. Grunbaum and G. C. Shepard, "Tilings and Patterns" (Freeman, San Francisco), to be published.

THE FRUSTRATED SPIN-GAS THEORY OF MULTIPLY REENTRANT LIQUID CRYSTALS

A. Nihat Berker

Department of Physics, Massachusetts Institute of Technology
Cambridge, Massachusetts 02139, USA

J. O. Indekeu

Laboratorium voor Technische Natuurkunde
Technische Universiteit Delft
2600 GA Delft, The Netherlands
        and
Department of Physics, Katholieke Universiteit Leuven*
Celestijnenlaan 200 D, B-3030 Leuven, Belgium

Abstract   Quadruply reentrant phase diagrams (nematic ↔ smectic A_d ↔ nematic ↔ smectic A_d ↔ nematic ↔ smectic A_1), smectic layer thicknesses, and transition enthalpies have been obtained from the frustrated spin-gas model, in satisfactory agreement with experiments. The stringent requirement on the molecular tail length for quadruple reentrance, seen in experiments, also occurs in this theory. The microscopic mechanism underlying the two smectic A_d peninsulae is found to be relief of dipolar frustration by permeation either on atomic length scale, or on librational length scale, respectively, at higher or lower temperatures. The distinctive transition enthalpies can be related to the molecular dimers, which form a nematic background in the smectic A_d phase, but which break up at the onset of the smectic A_1 phase.

The spin-gas theory [1-5] addresses the microscopics of (multiply) reentrant liquid crystals [6-12], by incorporating the dipolar frustration of the molecules, as well as their positional fluctuations. To wit, a system that is composed of dipolar molecules under the close-packing conditions of a liquid [13] is unavoidably subject to frustration effects, in

---

*Permanent address

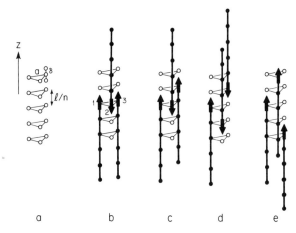

Fig. 1.   Examples of configurations of a triplet of molecules.   (a) The
atomic permeation positions of the dipole heads.   Librational
permeation positions are illustrated only in the upper right-
hand corner.   (b) A frustrated configuration: A zero net force
is felt by either of dipoles 1 and 3.   (c) Another frustrated
configuration: A zero net force is felt by dipole 3. Configurations
(b) and (c) are thus not conducive to layering.   On the other
hand, (d) and (e) are configurations in which frustration is
relieved by permeation, respectively conducive to interdigitated
partial-bilayer (A_d) and monolayer (A_1) smectic layering.

the sense that many molecules can be in a position of mutually cancelling
interactions from neighboring molecules (Fig. 1).

The theory considers triplets of molecules, the basic unit in which
frustration can be seen, evaluating whether frustration is relieved by positional fluctuations [14].   In the latter case, the molecules associate
in a "polymer" which extends into the xy plane (normal to the molecular
axes) and which underpins a smectic layer.   Calculations show that positional fluctuations in the molecular axes direction, namely permeation
fluctuations, rather than positional fluctuations within the xy plane,
affect the phase diagram [1,15].

The molecular tails have many important functions in these
systems[16].   Firstly, tail-tail van der Waals attraction or steric hindrance modify the dipolar intermolecular interaction.   A pure dipole-dipole
interaction has the form

$$V_d(\vec{r}_1,\hat{s}_1,\vec{r}_2,\hat{s}_2) = B\,[\hat{s}_1 \cdot \hat{s}_2 - 3\,(\hat{s}_1 \cdot \hat{r}_{12})(\hat{s}_2 \cdot \hat{r}_{12})]\,/\,|\vec{r}_{12}|^3 , \qquad (1)$$

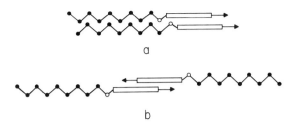

a

b

Fig. 2.   Schematic representation of molecules of the compound DB9ONO2.
The molecular pairs have dominant tail-tail attraction (a) or
steric hindrance (b).

where $\vec{r}_i$ is the position of the dipole of molecule i, $\hat{s}_i$ is the unit vector describing the dipolar orientation, and $\vec{r}_{12} = \vec{r}_1 - \vec{r}_2$ and $\hat{r}_{12} =$
$\vec{r}_{12}/|\vec{r}_{12}|$. The simplified potential

$$V_t(\vec{r}_1,\hat{s}_1,\vec{r}_2,\hat{s}_2) = J\ \hat{s}_1{\cdot}\hat{s}_2/|\vec{r}_{12}|^3 \qquad\qquad (2)$$

incorporates the alternate possibilities (Fig. 2) of dominant tail-tail
attraction (J<0) or steric hindrance (J>0). Combining the two potentials,

$$V(\vec{r}_1,\hat{s}_1,\vec{r}_2,\hat{s}_2) = V_d + V_t = [A\ \hat{s}_1{\cdot}\hat{s}_2 - 3B\ (\hat{s}_1{\cdot}\hat{r}_{12})(\hat{s}_2{\cdot}\hat{r}_{12})]/|\vec{r}_{12}|^3,\ (3)$$

where A = B + J.

Secondly, the corrugation of the molecular tails causes preferred
mutual permeation positions between neighboring molecules (e.g., Fig. 2a).
A mismatch occurs between the length scales of the tail corrugation and of
the dipole potential. Molecular displacements respecting the permeation
minima will be referred to as "atomic permeation." Smaller permeation
displacements away from the minima will be referred to as "librational
permeation" (Fig. 1).

The approximate treatment of the microscopic model has been in terms
of a prefacing transformation [1] that is designed to project, as usual,
the strength of the couplings, but also their variance which may lift frustration. The positional degrees of freedom are summed over in the expression for the partition function, and thus temperature-dependent effective
interactions are obtained between the orientational degrees of freedom of
the molecules. In a triplet of molecules, if the intermediate-strength
bond is closer to the strongest bond than to the weakest bond, frustration
is lifted and long-range correlations in the xy plane can occur.

Figure 3 shows two phase diagram topologies that are calculated readily [1]. Figure 3a corresponds to tail-tail interactions that are

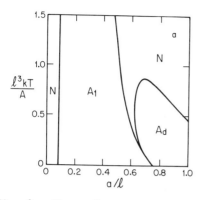
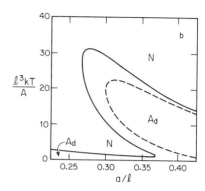

Fig. 3. Phase diagram topologies readily obtained (Ref. 1) with the frustrated spin-gas model. The ratio a/ℓ of the lateral separation to the length of the molecules serves as a pressure and/or composition scale. In (a) and (b), the number n of atomic permeation minima is respectively 7 and 5. In (a), the ratio of interaction constants B/A is 2, reflecting dominant tail-tail attraction. Librational permeation has not been included, but does not affect the topology. In (b), the ratio B/A is 0 and 0.1 for the solid and dashed lines, respectively, reflecting dominant tail-tail repulsion. Librational permeation over a length δ = 0.01 ℓ/n is included for the solid line. No librational permeation is included in the calculation of the dashed line.

predominantly attractive (J<0). This phase diagram topology reproduces the experimental results [8] for the compound "T8". In the underlying theory, frustration is relieved, in the partial bilayer (A_d) and monolayer (A_1) smectic phases, by mutually permeated triplets of molecules with antiferroelectric (e.g., Fig. 1d) and ferroelectric (Fig. 1e) orientational interactions, respectively. The reentrant nematic phase is due to the competition of these two tendencies. The two smectic phases meet at a bicritical point [1] located at zero temperature. The zero-temperature location of this bicritical point is dictated by the approximation of our calculations. A more sophisticated calculation could very possibly lift this bicritical point to finite temperature, pulling a first-order A_d-A_1 transition into the phase diagram, as seen in experiment [10] and phenomenological theory [17].

Figure 3b corresponds to tail-tail interactions that are predominantly affected by steric hindrance. The upper and lower smectic A_d phases are due to triplets with frustration relieved by atomic and librational permeation, respectively. The full topology of this phase diagram, to our knowledge, has not been experimentally observed to date.

It should be commented that our currently "bare-essentials" microscopic theory does not include tilt fluctuations of the molecular axes out of the z axis. We do obtain important permeation fluctuations, as shown in

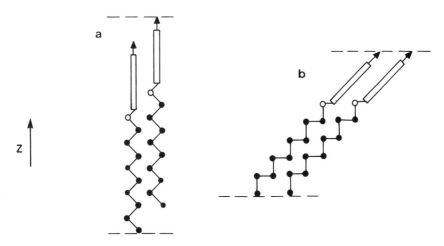

Fig. 4. (a) Permeation alone leaves vacant spaces. (b) Permeation
and local tilting optimizes packing. The dashes delineate the
smectic layer.

Fig. 4a. It is likely that in reality, while molecules still relieve frustration by the mutual displacements along their axes, simultaneous tilting
as in Fig. 4b will achieve better packing. Thus, the dipolar frustration
mechanism of reentrance would be associated with the presence of smectic C
fluctuations. This has been recently seen experimentally [18].

Another recent experimental development has been the discovery [9,10]
of quadruple reentrance (nematic $\leftrightarrow$ smectic A_d $\leftrightarrow$ nematic $\leftrightarrow$ smectic A_d $\leftrightarrow$
nematic $\leftrightarrow$ smectic A_1) with the compound DB_9ONO_2 and its mixtures, as shown
in Fig. 5a reproduced here from Ref. 10. We therefore searched the frustrated spin-gas model, and found [2] this quadruple reentrance, as shown in
Fig. 5b, but for very narrow ranges of microscopic parameters:

$$1.856 < B/A < 1.896 \quad \text{for} \quad n=4,$$
$$1.445 < B/A < 1.466 \quad \text{for} \quad n=5,$$
$$1.468 < B/A < 1.479 \quad \text{for} \quad n=6, \tag{4}$$

where n is the number of permeation minima, as illustrated in Fig. 2a. We
have not found quadruple reentrance for n = 3,7,8,9. It is thus satisfactory to note that, experimentally in the considerable number of reentrant
liquid crystal systems, quadruple reentrance is seen seldom. Correspond-
ingly, in the frustrated spin-gas theory, only a very narrow range of microscopic parameters yields quadruple reentrance. Moreover, the number of
permeation minima n being equal to 5 is in fact fully consistent, as shown
in Fig. 2a, with the existence of 9 carbon atoms in the molecular tail of
DB_9ONO_2. The upper and lower smectic A_d phases in Fig. 5b are due to triplets with frustration relieved by atomic and librational permeation, respectively [5].

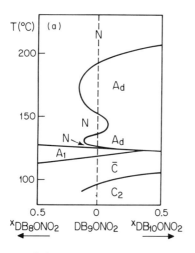
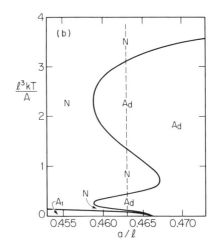

Fig. 5.   (a) Experimentally observed quadruple reentrance, from Ref. 10, of DB₉ONO₂ and its mixtures with homologs of concentration x. (b) Quadruply reentrant phase diagram calculated (Ref. 2) with the frustrated spin-gas model, with n = 5, B/A = 1.451, and δ = 0.015ℓ/n. The dashed line in (b) marks where layer thicknesses were calculated, given in Table 1 and compared with experiment.

Using the spin-gas theory, relative thicknesses of smectic layers have been calculated [2] along the dashed line drawn in Fig. 5b, as

$$d = \ell + <(\Delta h)^2>^{\frac{1}{2}}, \qquad\qquad (5)$$

Table 1.   Calculated Layer Thickness Ratios and Nearest-Neighbor Dipolar Correlations $<s_1 s_2>$ in the Smectic Phases, at Different Temperatures.*

			d/d$(A_1$,T=0)	d/d(A_1)
Phase	$\ell^3 kT/A$	$<s_1 s_2>$	Theory	Expt.[a]
A_d	2.75	−0.22	1.30	1.28
A_d	2.25	−0.22	1.30	1.29
A_d	1.75	−0.23	1.30	1.30
A_d (reent.)	0.30	−0.12	1.28	1.42
A_1	~0	1.00	1	1

[a] Reference 9.

* The phase diagram of Fig. 5b is scanned at constant a/ℓ = 0.463 (i.e., along the dashed line in Fig. 5b). For comparison, experimental thickness ratios from Ref. 9 are also given.

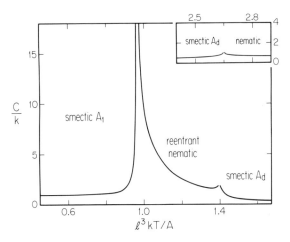

Fig. 6. Specific heat per molecule versus temperature, calculated along
the path a/ℓ = 0.59 for n = 9, B/A = 2, and δ = 0. The inset,
on the same scale, shows the result for the high temperature NA$_d$
transition.

using the standard deviation of Δh, the difference between the z coordinates of the centers of two nearest-neighbor molecules. In Eq. (5), d is
the layer thickness and ℓ is the molecular length. Our results are shown
in Table 1. It is seen that good agreement with experiment is obtained in
the upper A$_d$ phase. The current calculation fails to reproduce the trend
of increasing layer thickness into the lower A$_d$ phase. We speculate that
this trend is due to smectic A$_2$ (fully bilayer) correlations which occur in
the experimental systems, but which are not included in the current calculation.

Further quantitative properties can be calculated with the frustrated
spin-gas theory. Fig. 6 shows a specific heat curve calculated [4] for a
temperature scan across a (doubly reentrant) phase diagram of the type
shown in Fig. 3a. Two characteristics of the transition enthalpies, i.e,
of the areas under the phase transition peaks, emerge from such calculations: (i) The transition enthalphies are of comparable magnitude on each
side of the smectic A$_d$ phase. (ii) The transition enthalpy for the onset
of the A$_1$ phase is much larger than the transition enthalpies into the A$_d$
phase. These results emerge independently of model microscopic parameters.
They also are in agreement with experiments [4,19].

The qualitative difference between the transition enthalpies into the
smectic A$_1$ and A$_d$ phases can be understood within context of the microscopics of our model: We have calculated the molecular dimer concentrations
[5] as a function of temperature. We find an essentially steady dimer
concentration, consistently with dielectric measurements [20], throughout
the nematic and smectic A$_d$ phases. In the latter case, the dimers form
a nematic background to the smectic modulation caused by the infinite

"polymer" of molecules. By contrast, upon lowering the temperature, the dimers break up around the onset of the smectic A_1 phase, making a large contribution to the transition enthalpy. These calculational results are in drastic contrast to an alternatively proposed scenario [21].

ACKNOWLEDGMENTS

We are grateful to Professors N. A. Clark and J. F. Scott for organizing a very instructive and enjoyable NATO Workshop on Incommensurate and Liquid Crystals, where this paper was presented.

We thank Professors R. J. Birgeneau and C. W. Garland for many useful discussions and an enjoyable collaboration (Ref. 4). One of us (J.O.I.) was a Postdoctoral Research Fellow of the Dutch "Stichting F.O.M.". Research at M.I.T. was supported by the National Science Foundation under Grant No. DMR84-18718.

REFERENCES

1. A. N. Berker and J. S. Walker, Phys. Rev. Lett. 47, 1469 (1981).
2. J. O. Indekeu and A. N. Berker, Phys. Rev. A33, 1158 (1986).
3. J. O. Indekeu and A. N. Berker, Physica (Utrecht) 140A, 368 (1986).
4. J. O. Indekeu, A. N. Berker, C. Chiang, and C. W. Garland, Phys. Rev. A35, 1371 (1987).
5. J. O. Indekeu and A. N. Berker, M.I.T. preprint (1987)
6. P. E. Cladis, Phys. Rev. Lett. 35, 48 (1975).
7. P. E. Cladis, R. K. Bogardus, and D. Aadsen, Phys. Rev. A18, 2292 (1978).
8. F. Hardouin and A. M. Levelut, J. Phys. (Paris) 41, 41 (1980).
9. N. H. Tinh, F. Hardouin, and C. Destrade, J. Phys. (Paris) 43, 1127 (1982).
10. F. Hardouin, A. M. Levelut, M. F. Achard, and G. Sigaud, J. Chim. Phys. 80, 53 (1983).
11. A. R. Kortan, H. von Känel, R. J. Birgeneau, and J. D. Litster, J. Phys. (Paris) 45, 529 (1984).
12. R. Shashidhar, B. R. Ratna, V. Surendranath, V. J. Raja, S. Krishna Prasad, and C. Nagabhushan, J. Phys. (Paris) 46, L445 (1985).
13. In retrospect, this microscopic model could be more aptly referred to as the "frustrated spin-liquid"! We shall continue the spin-gas appellation for historic reason (Ref. 1).
14. For an alternate approach and mechanism, see F. Dowell, Phys. Rev A28, 3526 (1983); 31, 2464 (1985); 31, 3214 (1985).
15. This theoretical prediction was subsequently confirmed experimentally by Ref. 11.
16. Most basically, as presumed quite generally, molecular tails delay crystallization upon cooling and make liquid crystal phases possible.
17. J. Prost and P. Barois, J. Chim. Phys. 80, 65 (1983). For a general review, see J. Prost, Adv. Phys. 33, 1 (1984).
18. K. W. Evans-Lutterodt, J. W. Chung, R. J. Birgeneau, J. W. Goodby, and C. W. Garland (to be published).
19. N. H. Tinh, G. Sigaud, M. F. Achard, H. Gasparoux, and F. Hardouin, in "Advances in Liquid Crystal Research and Applications," ed. L. Bata (Pergamon, Oxford, 1980), p. 147.

20. B. R. Ratna, R. Shashidhar, and K. V. Rao, in "Proceedings of the International Liquid Crystal Conference," Bangalore (1979). These results are also exhibited in S. Chandrasekhar, Mol. Cryst. Liq. Cryst. $\underline{124}$, 1 (1985).

21. L. Longa and W. H. de Jeu, Phys. Rev. A $\underline{26}$, 1632 (1982).

MACROSCOPIC DESCRIPTION OF FERROELECTRIC CHIRAL SMECTIC C* LIQUID CRYSTALS

B. Zeks[*],[**], T. Carlsson[***], C. Filipic[*] and A. Levstik[*]

[*] Institute of Biophysics, Medical Faculty
University of Ljubljana, Lipiceva 2
61105 Ljubljana, Yugoslavia

[**] J. Stefan Institute,
University of Ljubljana, Jamova 39
61111 Ljubljana, Yugoslavia

[***] Institute of Theoretical Physics
Chalmers University of Technology
S-41296 Goteborg, Sweden

Abstract. The generalized Landau model is reviewed, which describes thermodynamic behavior of ferrolectric chiral smectic C* liquid crystals. Polarization and tilt temperature dependences are studied. For small biquadratic tilt-polarization coupling as compared to the bilinear one, the dependences deviate only slightly from the classic square root behavior, while for large biquadratic coupling both tilt and polarization develop S-shape behavior. The most convenient way to analyze the data is to study the tilt-polarization relation. In dimensionless form this nonlinear dependence is determined by a single parameter. The model is extended to allow for the polarization sign reversal, which has been observed for some systems recently.

I. INTRODUCTION

Primary order parameter for the transition between the smectic A phase and the smectic C phase is the two-component tilt vector $\vec{\xi} = (\xi_1, \xi_2)$, which describes the magnitude and the direction of the tilt of molecules from the normal to smectic layers. The secondary order parameter is the two-component in-plane polarization $\vec{P} = (P_x, P_y)$, which is related to the rotational order of molecules around their long axes and corresponds to the averaged transverse molecular dipole moment. For non-chiral molecules there is no linear coupling between the tilt and the polarization and only the biquadratic coupling exists. At the smectic A to smectic C transition the tilt therefore does not induce the spontaneous polarization and only the quadrupolar transverse ordering appears. For chiral systems, on the other hand, the bilinear coupling exists and the tilt locally induces the in-plane polarization, which is perpendicular to the tilt. Because of the chirality of molecules, the tilt as well as polarization precess around the normal to smectic layers as one goes from one smectic layer to another. One obtains a helicoidal structure which is locally polar (helicoidal smectic C* phase). Such "ferroelectric" liquid crystalline systems [1], where the helicoidal polarization is induced by the tilt, should actually be called improper helielectrics [2].

It has been shown recently [3,4,5], that the thermodynamic properties of smectic C^* ferroelectric liquid crystals can be described by a Landau free energy expansion in the tilt, which couples bilinearly and biquadratically to the polarization, in contrast to previous models which included only bilinear coupling [6] and led to a qualitative disagreement with experiments [7]. In smectic A phase and in smectic C^* phase close to the transition where the tilt is small, the biquadratic coupling has no effect. At lower temperatures, when the tilt becomes large, the biquadratic coupling terms become relevant. They induce a transverse quadrupole ordering and indirectly amplify the effect of bilinear coupling, thus increasing the transverse polarization. As the bilinear coupling is of chiral character and therefore small, while the biquadratic coupling is nonchiral and large, the cross-over temperature, below which the biquadratic coupling becomes important, can be relatively close to the transition temperature T_c.

The free energy density describing the smectic A to the smectic C^* transition is expressed as [3,4],

$$
g(z) = \frac{1}{2}a|\vec{\xi}|^2 + \frac{1}{4}b|\vec{\xi}|^4 + \frac{1}{6}c|\vec{\xi}|^6 - \Lambda(\xi_1\frac{d\xi_2}{dz} - \xi_2\frac{d\xi_1}{dz})
$$
$$
+ \frac{1}{2}K_{33}[(\frac{d\xi_1}{dz})^2 + (\frac{d\xi_2}{dz})^2] - d|\vec{\xi}|^2 (\xi_1\frac{d\xi_2}{dz} - \xi_2\frac{d\xi_1}{dz})
$$
$$
+ \frac{1}{2\varepsilon}|\vec{P}|^2 + \frac{1}{4}\eta|\vec{P}|^4 - \mu(P_x\frac{d\xi_1}{dz} + P_y\frac{d\xi_2}{dz})
$$
$$
+ C(P_x\xi_2 - P_y\xi_1) - \frac{1}{2}\Omega(P_x\xi_2 - P_y\xi_1)^2 \tag{1}
$$

where the z-axis is normal to the smectic layers. Only the term quadratic in tilt is explicitly temperature dependent: $a = \alpha(T - T_0)$. K_{33} is the elastic modulus, Λ the coefficient of the Lifshitz term responsible for the modulation and μ and C are the coefficients of the flexo- and piezo-electric bilinear coupling. Ω is the coefficient of the biquadratic coupling term and the η-term has been added to stabilize the system. The d-term is describing the observed monotonous increase of the pitch of the helix with temperature at low temperatures. The sixth order term in tilt (the c-term) has been added to account for the measured specific heat temperature dependences [8-10], which indicate that the phase transitions in these systems are of second order but close to the tricritical point.

Predictions of this model qualitatively and also quantitatively agree with experimental data [5] (mostly obtained for DOBAMBC):
- the spontaneous polarization is not proportional to the tilt [11-13]
- the pitch of the helix has a maximum below T_c [14,15]
- the dielectric constant has a broad maximum below T_c [16,17]
- the specific heat displays a relatively sharp anomaly at T_c in contrast to a simple mean field discontinuity [8-10]
- the critical magnetic field for unwinding the helix can be shown to be inversely proportional to the pitch and therefore displays a minimum below T_c [18].

2. THEORY

In this contribution we shall analyze in more detail the temperature dependences of the spontaneous polarization and of the tilt as predicted by the model (Eq. 1). These two quantities are usually experimentally determined for unwound systems. Perpendicular to the helical axis an electric field is applied, which is large enough to destroy the helicoidal structure, and therefore transforms the helicoidal smectic C^* phase into homogeneous smectic C phase. The results for the polarization and the tilt are then extrapolated to zero electric fields. We are therefore interested in homogeneously ordered structures with polarization $\vec{P}$ being perpendicular to tilt $\vec{\xi}$

$$\vec{\xi} = (\theta, 0), \qquad \vec{P} = (0, P) \qquad (2)$$

As θ and P are constants, all the derivative terms in Eq. 1 equal zero and the free energy equals

$$g = \frac{1}{2}a\theta^2 + \frac{1}{4}b\theta^4 + \frac{1}{6}c\theta^6 + \frac{1}{2\varepsilon}P^2 + \frac{1}{4}\eta P^4$$
$$-CP\theta - \frac{1}{2}\Omega P^2\theta^2 \qquad (3)$$

The values of the tilt and of the polarization are obtained by minimizing this free energy. The requirement that the partial derivatives of g with respect to P and θ equal zero leads to two coupled equations for P(T) and θ(T)

$$\alpha(T - T_0)\theta + b\theta^3 + c\theta^5 - CP - \Omega P^2 \theta = 0 \tag{4a}$$

$$\frac{1}{\varepsilon}P + \eta P^3 - C\theta - \Omega P\theta^2 = 0 \tag{4b}$$

It can be seen from Eqs. 3 and 4 that the sign of the induced polarization P is given by the sign of the piezoelectric coupling coefficient C. Here we shall limit ourselves to the case of positive C and therefore positive P. For negative C only the sign of P should be taken as negative. It is convenient to introduce dimensionless quantities $\tilde{P}$, $\tilde{\theta}$ and τ expressing the spontaneous polarization, the tilt and the temperature difference $T_c - T$ as

$$\tilde{P} = \frac{P}{P*}, \quad \tilde{\theta} = \frac{\theta}{\theta*}, \quad \tau = \frac{T_c - T}{T*} \tag{5}$$

where

$$P* = \frac{1}{\sqrt{\varepsilon\eta}}, \quad \theta* = \frac{1}{\sqrt{\varepsilon\Omega}}, \quad T* = \frac{b}{\alpha\varepsilon\Omega} \tag{6}$$

and T_c is the transition temperature, where the disordered phase with $\theta = P = 0$ becomes unstable with respect to homogeneous tilt and polarization fluctuations. The free energy density g can be expressed as a function of τ, $\tilde{\theta}$ and $\tilde{P}$ as

$$g = \frac{1}{\varepsilon^2\eta}\{\frac{1}{2}\gamma(-\tau + \beta)\tilde{\theta}^2 + \frac{1}{4}\gamma\tilde{\theta}^4 + \frac{1}{6}\rho\tilde{\theta}^6 + \frac{1}{2}\tilde{P}^2$$
$$+ \frac{1}{4}\tilde{P}^4 - \frac{1}{2}\tilde{P}^2\tilde{\theta}^2 - (\gamma\beta)^{1/2}\tilde{P}\tilde{\theta}\} \tag{7}$$

where the three parameters β, γ and ρ are defined as

$$\beta = \frac{\varepsilon^2\Omega C^2}{b}, \quad \gamma = \frac{b\eta}{\Omega^2}, \quad \rho = \frac{c\eta}{\varepsilon\Omega^3} \tag{8}$$

218

The parameter β is related to the piezo-electric bilinear coupling coefficient C. It therefore reflects the chiral character of the system and is
expected to be small. The parameter ρ is related to c, i.e., the sixth
order coefficient in the expansion in $\vec{\xi}$, and has only small saturation
effects on $\tilde{P}$ and $\tilde{\theta}$.

The equations for $\tilde{P}(\tau)$ and $\tilde{\theta}(\tau)$ now read

$$\gamma(-\tau + \beta)\tilde{\theta} + \gamma\tilde{\theta}^3 + \rho\tilde{\theta}^5 - \tilde{P}^2\tilde{\theta} - (\gamma\beta)^{1/2}\tilde{P} = 0 \tag{9a}$$

$$\tilde{P} + \tilde{P}^3 - \tilde{P}\tilde{\theta}^2 - (\gamma\beta)^{1/2}\tilde{\theta} = 0 \tag{9b}$$

These equations can be solved numerically and the results for $\tilde{P}(\tau)$ and
$\tilde{\theta}(\tau)$ are shown in Figs. 1 and 2 for ρ = 0, γ = 1.5 and for various values of
β.

3. DISCUSSION

It is easy to see that Eq. 9 reproduces for γβ = 1 the usual square
root temperature dependences for both the polarization and the tilt, i.e.,
$\tilde{P} = \theta \propto \tau^{1/2}$. When β is decreasing one observes a deviation from the square
root behavior, which is small and therefore hard to see for large β, while
for smaller β one obtains an S-shape temperature dependence for polarization and for even smaller β also for the tilt. The deviation from the
classical behavior can be more clearly observed in the temperature

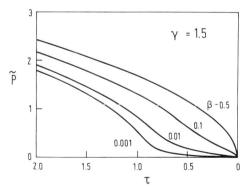

Fig. 1. Temperature dependence of spontaneous polarization in dimensionless units for ρ = 0, γ = 1.5 and four different values for β.

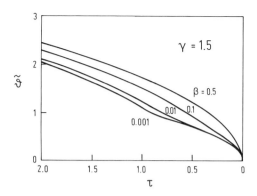

Fig. 2. Temperature dependence of the tilt in dimensionless units for
$\rho = 0$, $\gamma = 1.5$ and four different values of β.

dependences of the ratio $\tilde{P}/\tilde{\theta}$ (Fig. 3), where one sees for large β a cross-over behavior from one value close to T_c to another one far below T_c.

According to Eq. 9, the temperature dependences of the polarization and of the tilt depend on six parameters: the three dimensionless parameters β, γ and ρ, and the three scale factors P^*, θ^* and T^* for the polarization, for the tilt and for the temperature, respectively. Such a large number of parameters makes the analysis of experimental data complicated, especially when the deviation from the classical behavior is small. In such cases it is more convenient to use the Eq. 9b and analyze the dependence of the spontaneous polarization on the tilt. The dependence of the ratio $\tilde{P}/\tilde{\theta}$ on $\tilde{\theta}^2$, which is shown in Fig. 4, is similar to the temperature dependence of $\tilde{P}/\tilde{\theta}$ (Fig. 3), but depends only on a single combination of the parameters, that is on the product $\beta\gamma$. The ratio $\tilde{P}/\tilde{\theta}$ is increasing monotonically from a finite value at T_c to its low temperature value.

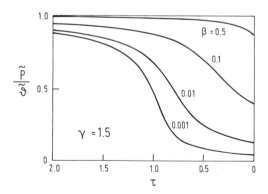

Fig. 3. Temperature dependence of the ratio P/θ in dimensionless units
for $\rho = 0$, $\gamma = 1.5$ and four different values of β.

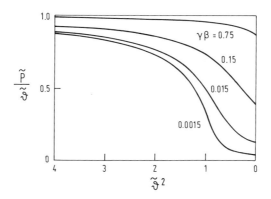

Fig. 4.   Dependence of the ratio $\tilde{P}/\tilde{\theta}$ on $\tilde{\theta}^2$ for four different values of
the product $\beta\gamma$.

The present model does not seem to be able to account for a non-monotonic
temperature dependence of this ratio, which has been observed for some
ferroelectric liquid crystalline systems [19].

Recently the systems were reported where the spontaneous polarization
changes its sign as a function of temperature [20].   Such a behavior can be
explained qualitatively by adding to the free energy density (Eq. 1) an
additional term with the structure

$$f|\vec{\xi}|^2(P_x\xi_2 - P_y\xi_1) \qquad (10)$$

Effectively this makes the bilinear piezoelectric coupling coefficient tilt
dependent

$$C \to C + f\theta^2 \qquad (11)$$

and allows for the polarization reversal at large tilts.   The detailed
analysis of such systems will be published elsewhere.

REFERENCES

1.   R. B. Meyer, L. Liebert, L. Strzelecki, and P. J. Keller, J. Phys.
     (Paris) lettres 36, L69 (1975).
2.   P. E. Cladis, and H. R. Brand, this meeting.
3.   B. Zeks, Ferroelectrics 53, 33 (1984).
4.   B. Zeks, Mol. Cryst. Liq. Cryst. 114, 259 (1984).
5.   T. Carlsson, B. Zeks, A. Levstik, C. Filipic, I. Levstik and R. Blinc,
     Phys. Rev. A36, (1987) (in press).
6.   S. A. Pikin and V. L. Indenbom, Uspekhi Fiz. Nauk 125, 251 (1978).
7.   R. Blinc, B. Zeks, I. Musevic, and A. Levstik, Mol. Cryst. Liq.
     Cryst., 114, 189 (1984).
8.   C. C. Huang and J. M. Viner, Phys. Rev. A25, 3385 (1982).

9. T. Carlsson and I. Dahl, Mol. Cryst. Liq. Cryst. 95, 373 (1983).
10. S. C. Lien, C. C. Huang, T. Carlsson, I. Dahl, and S. T. Lagerwall, Mol. Cryst. Liq. Cryst. 108, 149 (1984).
11. B. I. Ostrovskii, A. Z. Rabinovich, A. S. Sonin, B. A. Strukov, and S. A. Taraskin, Ferroelectrics 20, 189 (1978).
12. S. Dumrongrattana and C. C. Huang, Phys. Rev. Lett. 56, 464 (1986).
13. C. Filipic, A. Levstik, I. Levstik, R. Blinc, B. Zeks, M. Glogarova, and T. Carlsson, Ferroelectrics (in press).
14. I. Musevic, B. Zeks, R. Blinc, L. Jansen, A. Seppen, and P. Wyder, Ferroelectrics 58, 71 (1984).
15. K. Yoshino, M. Ozaki, H. Agawa, and Y. Shigeno, Ferrolectrics 58, 283 (1984).
16. A. Levstik, T. Carlsson, C. Filipic, and B. Zeks, to be published.
17. A. Levstik, T. Carlsson, C. Filipic, I. Levstik, and B. Zeks, Phys. Rev. A35, 3527 (1987).
18. I. Musevic, B. Zeks, R. Blinc, Th. Rasing, and P. Wyder, Phys. Rev. Lett. 48, 192 (1982).
19. D. S. Parmar, M. A. Handschy, and N. A. Clark, poster presented at the 11th International Liquid Crystal Conference, Berkeley, 1986.
20. J. W. Goodby, E. Chin, J. M. Geary and J. S. Patel, poster presented at the 11th International Liquid Crystal Conference, Berkeley, 1986.

ON THE MOLECULAR THEORY OF SMECTIC-A LIQUID CRYSTALS

W. H. de Jeu

FOM-Institute for Atomic and Molecular Physics
Kruislaan 407, 1098 SJ Amsterdam, Netherlan

(Also, The Open University, P.O. Box 2960
6401 DL Heerlen, Netherlands)

Abstract   The intermolecular interactions leading to the various
types of smectic-A phase are considered both for effectively symmetric
and for asymmetric molecules. At least a qualitative understanding
of the molecular organization in the various phases can be obtained
by combining the tendency of aromatic and aliphatic parts to segregate
with optimal packing and minimal dipolar repulsions. In particular,
it is shown that probably two rather different ways of molecular
ordering may lead to a smectic-A phase with a period approximately
equal to the molecular length.

1.   INTRODUCTION TO SMECTIC-A PHASES

In this paper I propose to review what I believe to be the present
state of our understanding at the molecular level of the various types of
smectic-A phases that have been observed. I will do so in an informal
way, trying to emphasize the basic ingredients and the results of the various models, rather than going into the formal details.

Figure 1 shows the textbook picture of a nematic (N) and a smectic-A
(S_A) phase of rodlike molecules. The molecules are, on the average, aligned with their long axis parallel to a certain direction in space (taken as
the z-direction). In the following this orientational order will be considered to be a constant background and occasionally assumed to be perfect. The smectic phase is distinguished from the nematic one by a density
modulation along the z-axis. This leads to what has loosely been called:

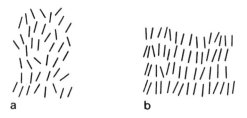

Fig. 1. Nematic (a) and smectic-A phase (b).

a structure of stacked liquid layers. However, it should be emphasized that the density modulation is very weak, contrary to the suggestion implied by the word layer. Indicating the repeat distance by d, the density wave can be described by a Fourier-series [1,2]:

$$\rho = \Sigma \; \rho_n \; \cos(2\pi nz/d) \qquad (1)$$

where the ρ_n = $\langle \cos(2\pi nz/d) \rangle$ are the smectic order parameters, the brackets indicating the average over the distribution function. Series (1) can be truncated after one term, leading to

$$\tau = \rho_1 = \langle \cos(2\pi z/d) \rangle, \qquad (2)$$

which is called the smectic order parameter. In the simplest case d is taken equal to ℓ, the length of the molecules.

Essential to simple molcular models of the smectic-A phase is the combination of Eq. (2), which contains the layer-spacing $d \cong \ell$, with a second length. The most important aspect is the division of the molecule into an aromatic and an aliphatic part (Fig. 2). Without paying attention to the actual interaction mechanisms I mention the well-known fact that these parts have a tendency to avoid each other and prefer like ones as neighbors. Thus it can be expected that the interaction energy of two molecules of the type pictured in Fig. 2 is minimal in a smectic-like configuration with the aromatic centers in equidistant planes. Working-out such models the smectic phase is, as expected, found to be stabilized for small enough values of r_0/ℓ. Of course, to make these ideas quantitative, the details of the particular model must be considered [3]. The resulting behavior is qualitatively in reasonable agreement with the phase-behavior of many homologous series. In these series r_0 can be taken as approximately constant and ℓ as variable.

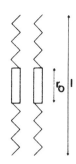

Fig. 2.   Schematic representation of two molecules in a smectic-like configuration.

For the present purpose it is important to note that for many systems d as measured by X-ray diffraction is indeed approximately equal to ℓ. Furthermore, the absence or weakness of higher-order reflections indicates that Eq. 2 is a good approximation.   Centro-symmetric molecules as pictured in Fig. 2 are rare, but the same type of argument can be applied to other series, provided there is on a local level an equal distribution of molecules pointing "up" (positive z-direction) or "down."   This also implies that in a first approximation permanent dipole moments play, at most, a minor role.   These extensions of the original argument are justified because the phase behavior of many homologous series of non-symmetric molecules (which thus necessarily have dipole moments) is very similar to that of the few series with symmetric molecules [4].

2.   ASYMMETRIC MOLECULES, THE S_{Ad}-PHASE, AND REENTRANT NEMATIC BEHAVIOR

Now I will consider the extreme case of an asymmetric molecule, one with an aromatic head and just one aliphatic tail.   Evidently the type of argument of the previous section cannot be extended directly.   The combination of a smectic density modulation with period $d \cong \ell$ and a random up-down distribution will certainly not promote an optimum neighborhood for the aromatic and aliphatic parts.   Assuming that these parts still prefer to segregate, clearly two solutions are possible, shown in Fig. 3.   These two configurations each have their own specific problems:

- Situation I provides an unfavorable packing with crowded aromatic regions and open space around the aliphatic parts.   From the packing point of view case II will be favored.

- In general, dipoles will be present in asymmetric molecules.   As far as the component along the long molecular axis is considered, in case II dipolar repulsions will occur. From the dipolar point of view situation I will be favored.

From these observations one is tempted to conclude that under the appropriate conditions each of the two situations might be observed.   This is

225

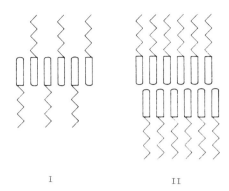

Fig. 3. Idealized sketch of the two possibilities for segregation of aromatic and aliphatic parts of asymmetric molecules.

indeed the case. In the remaining part of this section I will discuss case I, which in principle leads to the S_{Ad} phase. A discussion of the so-called S_{A2} phase that results from situation II will be postponed until the next section.

Any molecular model of the S_{Ad} phase should account for the following observations:

(1) For the layer spacing d one finds $\ell < d < 2\ell$. If the aromatic and aliphatic parts have length a and b, respectively, one finds for several compounds approximately [5]:

$$d \cong a + 2b \tag{3}$$

There is no universal trend for the temperature dependence of d. In many cases temperature variations are minor, but both increases and decreases have also been observed.

(2) Apart from the phase-transition NS_{Ad} often also a transition $S_{Ad}N_{reenter}$ (reentrant nematic) occurs.

(3) In high-resolution studies of the phase transitions no difference is found between the critical behavior of the "classical" NS_A transition and the NS_{Ad} transition [6]. Here "classical" stands for the effectively symmetric case discussed in section 1.

(4) The S_{Ad} phase usually occurs in compounds with a strongly polar end group at the end of the aromatic part, such as -CN or -NO$_2$. The average dielectric permittivity is given by

$$\overline{\varepsilon} \sim \frac{\mu_{eff}^2}{3k_BT} \approx \frac{1}{2}\frac{\mu_{free}^2}{3k_BT} \tag{4}$$

indicating a considerable amount of anti-parallel dipole correlation. Most importantly this behavior is not only found in the S_{Ad} phase, but also in the nematic and the isotropic phase occurring at higher temperatures [7].

Some years ago Longa and I formulated a dimer model for the S_{Ad} phase
[8]. Guided by Eq. 3 and similar observations by Cladis [9], it was
assumed that the dipolar and other interactions between the compounds
involved would lead to a dynamic monomer-dimer equilibrium

$$2M \rightleftarrows D \tag{5}$$

For the purpose of a model calculation the monomers and dimers were represented by the bodies pictured in Fig. 4. In this way the basic mechanism
discussed in section 1 can still be effective. The monomer has no inten-
tion of supporting a smectic density wave, but the dimers are symmetric
again, and will do so for small enough values of r_0/d' (where now d' is the
length of the dimer). In addition to this requirement there must be, of
course, a sufficient number of dimers, which makes the phase transiton
NS_{Ad} percolation like. Hence it is again the tendency of aromatic and
aliphatic parts to segregate (but this is only possible for the dimers)
which causes the smectic density wave. In the model of Fig. 4 the free
energy describing the system contains in addition to the rotational entropy
S_r, an extra packing entropy S_p:

$$F = U - TS_r - TS_p \tag{6}$$

Once the S_{Ad} phase has been established, with decreasing temperature or
increasing pressure (increasing dimer concentration and/or increasing
smectic order parameter) this term becomes increasingly important due to
the central slightly bulkier part of the dimers. Because of the unfavorable packing entropy, dimers then have to move out of the smectic planes.
Finally this destablizes the smectic phase, leading to reentrant nematic
behavior.
   Several comments can be made concerning this type of model:
- Once Eq. 5 is accepted, in practice all the experimental observations
  follow in a natural way, in particular also the equivalence of the
  "classical" NS_A and the NS_{Ad} transition.
- The driving force for reentrant nematic behavior is the packing entropy.
  Hence, one can imagine that other mechanisms leading to such a term could
  give similar effects. Dowell has emphasized the importance of the temperature dependence of the conformation of the alkyl chains [10]. Provided the right packing entropy evolves, thus also reentrant nematic
  behavior is possible with non-polar molecules.

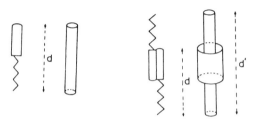

Fig. 4.   Monomer and dimer with their model representation.

- The definition of the dimer and the underlying forces is not very precise. It is quite feasible that several types of dimers are possible
with a somewhat different overlap, in particular if this promotes a favorable conformation and packing of the alkyl chain. Incorporation of
such a possibility into the model would probably lead to multiple reentrant behavior with various nematic regions separated by S_{Ad} phases with
different periodicities.
- The position of the dipole moment within the aromatic part of the molecule is rather unimportant, the most favorable situation being obtained
with strongly conjugated dipoles. This contrasts with the situation to
be discussed in the next section.
- One might wonder whether n-mers with n > 2 should be taken into account
as well. In particular, Berker ascribed the stability of the S_{Ad} phase
to the existence of a two-dimensional network of antiparallel molecules
in a background of n-mers of which many with n = 2 [11]. Though one might
question whether such a structure is compatible with the liquid nature of
the S_{Ad} layers, it is interesting to note that this mechanism and the
dimer-model are not mutually exclusive. Whether one of the two is dominant could probably be inferred from a more full calculation of the interaction energy of the molecules. I conclude that in spite of the ob-
jections that can and have been made against the dimer-model, it still
seems to provide the only rather complete microscopic account of the
present experimental situation.

3.   ANTIFERRO-ELECTRIC SMETIC-A PHASES

In this section I will discuss possibility II for aromatic and aliphatic molecular parts to segregate as mentioned at the beginning of the
previous section (see Fig. 3). As this leads to ferro-electric layers,
the succession of neighboring layers must be considered. This succession
can evidently be parallel (leading to macroscopic ferro-electricity, S_{Af}
phase) or anti-parallel (leading to longrange antiferro-electricity and

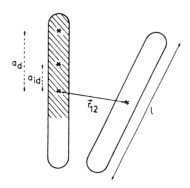

Fig. 5. Molecular model for calculating the interactions in asymmetric
molecules; a dipole moment and a polarizable center are situated
at a_d and a_{id} from the moleculed center, respectively.

bilayer periodicity, S_{A2} phase). The important molecular interactions
have been calculated by Longa and De Jeu using the molecular model of
Fig. 5 [12]. The results for the dipolar repulsions are displayed in
Fig. 6. As is to be expected, the S_{Af} phase is strongly destabilized, in
agreement with the fact that it has not been observed experimentally.
More interestingly, the dipolar repulsions become quite small in the antiferro-electric S_{A2} phase, provided the dipoles are localized at the end of
the molecules. This is due to the favorable contribution of the interaction between the dipoles in the neighboring layers. In combination with
the optimal packing this causes the S_{A2} phase to be the lowest temperature smectic-A phase, as observed experimentally. In a truncated Fourier-

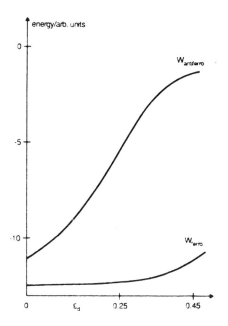

Fig. 6. Dipole repulsions for the S_{Af} and S_{A2} phase as a function of the
dipole position $\varepsilon_d = a_d/\ell$ along the axis.

expansion analogous to Eqs. 1 and 2 the S_{A2} phase can be described by an order parameter

$$\xi = < s \cos(\pi z/d) >, \tag{7}$$

where $s = +1$ or -1 indicates whether the polar head is respectively up or down. In agreement with this model, the dielectric permittivity in the S_{A2} phase is observed to be strongly reduced compared with the nematic phase at higher temperature [13]. This provides direct evidence for the long-range antiferro-electricity.

The situation is experimentally more complex than described so far. In the case of asymmetric molecules with strongly polar end groups such as -CN and NO_2, several more types of S_A phase have been observed. It will be clear from the discussion so far that a competition can exist between the tendency to condense as S_{Ad} or as S_{A2}. As the two wavelengths involved have no direct relation, this could in principle lead to an incommensurate smetic-A phase as predicted by Prost et al. [14]. Such a phase, characterized by two sharp X-ray reflections at incommensurate values, has recently been observed [15]. It seems that this type of phase is rare. A probable explanation is that the (in)commensurability problem can be solved in a different way, that being by imposing a modulation upon the S_{A2} structure as pictured in Fig. 7. X-ray results for this so-called $S_{\tilde{A}}$ phase indicate that the modulation varies from "walls" at lower temperatures to a more sinusoidal modulation at higher temperatures. It seems that this additional modulation is one-dimensional, thus introducing a unique direction within the smectic planes [16]. This makes the $S_{\tilde{A}}$ phase biaxial, quite different from the phases discussed so far, and the name Smetic A not very proper.

I cannot infer any mechanism for the asymmetric molecules discussed so far to condense as a "classical" S_A phase with $d \cong \ell$ and a random up-down distribution of the molecules. Such a situation would completely counteract a segregation of aliphatic and aromatic parts, and probably even enhance the free energy compared with the nematic phase. Nevertheless, S_A phases with $d \cong \ell$ have been observed with asymmetric molecules with terminal -CN or NO_2 groups, often below a reentrant nematic phase. They have sometimes been named S_{A1}, to emphasize $d \cong \ell$, and to distinguish them from the $S_{\tilde{A}}$ and S_{A2} phases that usually occur in the same region of the phase diagram at lower temperaturs. The molecular theory of Longa and de Jeu

230

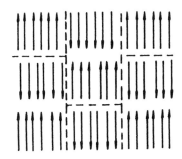

Fig. 7.   Schematic structure of the $S_{\tilde{A}}$ phase.

allows for such a phase, characterized by an order parameter

$$\zeta_1 = \langle\, s\, \cos(2\pi z/d)\, \rangle \tag{8a}$$

with in addition

$$\tau = \zeta = 0 \tag{8b}$$

The only way to accommodate this combination of order parameters is a superposition of two ferro-electric stacks of layers that are antiparallel and shifted over $d/2$. There is no evidence that this corresponds to the real situation in the S_{A1} phase.

An alternative model for an S_{A1} phase from polar anisotropic molecules is a variation on the $S_{\tilde{A}}$ structure. It consists of S_{A2} regions separated by "walls" as in the $S_{\tilde{A}}$ phase, in which the double layer structure shifts over a single molecular length, but with a random distribution of walls. The phase transition $S_{\tilde{A}}S_{A1}$ then corresponds to an order-disorder transition of the walls. Such a model predicts the S_{A1} phase to be antiferro-electric, though weaker than the S_{A2} phase. As far as experimental evidence is available, this seems indeed to be the case [12].

Recent high-resolution surface X-ray reflectivity measurements show that for a particular S_{A1} phase, two or three S_{A2}-like double layers exist at the surface, in spite of the single-layer X-ray reflection in the bulk [17]. The reconstructed electron density is shown in Fig. 8. The correlation length varies around the NS_{A1} phase transition from 4 to 6 molecular lengths, in quantitative agreement with the behavior of the diffuse scattering from S_{A2}-like fluctuations in the bulk. In the above-mentioned model this probably indicates that the walls are expelled from the surface, which thus reveals the underlying short-range structure. If this

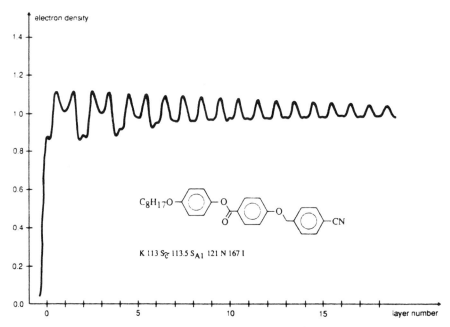

Fig. 8. Reconstructed electron density at 0.8°C above the NS_{A1} phase transition.

interpretation is correct, two microscopically rather different S_A phases exist with both d ≅ ℓ, and as far as known the same macroscopic symmetry. One is the "classical" S_A phase discussed in section 1 with molecules that, if not symmmetric by themselves, have locally a random up-down distribution, and secondly an S_{A1} phase from asymmetric polar molecules that has locally a double layer structure. Interestingly, there are recent indications that the critical behavior of the latter NS_{A1} phase transition differs from the former NS_A and the NS_{Ad} phase transitions [18].

4. CONCLUSIONS

     The tendency of aromatic and aliphatic parts of the molecules to segregate is taken as the basic mechanism for the formation of the smectic density modulation. In combination with optimal packing and minimal dipolar repulsions this provides at least a qualitative understanding of the molecular organization of the various types of smectic-A phase, both for effectively symmetric and for asymmetric molecules. In particular, it is shown that probably two rather different ways of molecular ordering can lead to a smectic-A phase with d ≅ ℓ. The implications of this result in terms of miscibility and of critical behavior around the nematic-smectic-A phase transition have not fully been investigated yet.

5. REFERENCES

1. P. G. de Gennes, Solid State Commun. $\underline{10}$, 753 (1972).
2. R. B. Meyer and T. G. Lubensky, Phys. Rev. $\underline{A14}$, 2307 (1975).
3. W. L. McMillan, Phys. Rev. $\underline{A4}$, 1238 (1971).
4. D. Demus, H. Demus, and H. Zaschke, "Flussige Kristalle in Tabellen," 2nd ed. (VEB Verlag, Leipzig, 1983).
5. G. J. Brownsey and A. J. Leadbetter, Phys. Rev. Lett. $\underline{44}$, 1608 (1980).
6. B. M. Ocko, R. J. Birgeneau, and J. D. Litster, Z. Phys. $\underline{B62}$, 487 (1986).
7. L. G. P. Dalmolen, S. J. Picken, A. F. de Jong, and W. H. de Jeu, J. Phys. $\underline{46}$, 1443 (1985).
8. L. Longa and W. H. de Jeu, Phys. Rev. $\underline{A26}$, 1632 (1982).
9. P. E. Cladis, R. K. Bogardus, and D. Aadsen, Phys. Rev. $\underline{A18}$, 2292 (1978).
10. F. Dowell, Phys. Rev. $\underline{A31}$, 2464 (1985).
11. A. N. Berker and J. S. Walker, Phys. Rev. Lett. $\underline{47}$, 1469 (1981); J. O. Indekeu and A. N. Berker, Phys. Rev. $\underline{A33}$, 1158 (1986).
12. L. Longa and W. H. de Jeu, Phys. Rev. $\underline{A28}$, 2380 (1983).
13. C. Druon, J. M. Wacrenier, F. Hardouin, Nguyen Huu Tinh, and H. Gasparoux, J. Phys. $\underline{44}$, 1195 (1983).
14. J. Prost and P. Barois, J. Chim. Phys. $\underline{80}$, 65 (1983).
15. B. R. Ratna, R. Shashidar, and V. N. Raja, Phys. Rev. Lett. $\underline{55}$, 1476 (1985).
16. A. M. Levelut, J. Phys. Lett. $\underline{45}$, L-603 (1984); P. E. Cladis, private communication.
17. E. F. Gramsbergen, W. H. de Jeu, and J. Als-Nielsen, J. Phys. $\underline{47}$, 711 (1986).
18. R. J. Birgeneau, 11th International Liquid Crystal Conference (Berkeley, 1986).

ORDER ELECTRICITY AT SMECTIC LIQUID CRYSTAL INTERFACES

G. Durand

Laboratoire de Physique des Solides, Bat.510
Université de Paris-Sud-91 405 Orsay-Cedex

Abstract   The spontaneous tilt of molecules inside thin smectic films
and the observation of antiferroelectric order SA_2 at a free surface of
a SA_1 material can be explained by the onset of the electric polarization associated with the gradient of orientational order.

1. INTRODUCTION

Order-electricity has been introduced recently [1,2] as a generalization of flexoelectricity in nematic liquid crystals. An electric
polarization must be associated with a gradient of the nematic order
parameter modulus S. This order polarization could couple directly with
an external field giving rise, for instance, to polar surface instabili-
ties, as recently observed [3]. When the gradient of the order parameter
appears on a scale shorter than the Debye screening length d for the
material, one must include in the free energy of the system the dielectric
self energy of the order polarization. It has been shown for instance [1]
that this dielectric energy can explain the director tilt at a nematic-isotropic interface. In this paper we try to show that the order polarization can also explain some behavior of smectic liquid crystal-air interfaces previously not clearly understood.

Let us recall what is the order-electric polarization. Most nematic
liquid crystal molecules possess permanent dipoles. However, a uniform
texture of director $\vec{n} = \vec{n}_0$ at constant order parameter S is not ferroelectric, because rotational diffusion around $\vec{n}_0$ and flip-flop $\vec{n}_0 \rightarrow -\vec{n}_0$
cancels on the average the macroscopic resultant polarization $\vec{P}$. Meyer
has shown [4] that the presence of a texture curvature $\vec{\nabla}\vec{n}$, at constant S,

Fig. 1. a. The curvature of n breaks the inversion symmetry and gives
            rise to the macroscopic flexoelectric polarization P_f, at
            constant S.

         b. The gradient of S at uniform $\vec{n}$ breaks the inversion symmetry
            and gives rise to the macroscopic order electric polarization
            P_o.

breaks this centrosymmetry and allows for a "flexo" polarization $\vec{P}_f \sim e\ \vec{n}\vec{\nabla}\vec{n}$.
(Fig. 1a). The Bordeaux group has demonstrated later [5] that $\vec{P}_f$ was
mostly of quadrupolar origin and could be written as $\vec{P}_f = -\vec{\nabla}.\overline{\overline{Q}} = e\vec{\nabla}.(\vec{n}\vec{n}$
$- \overline{\overline{I}}\ 3)$, where $|\overline{\overline{Q}}| \sim -e$ is the quadrupole density of the nematic liquid
crystal, which must be considered as a ferroelectric quadrupolar system.
$\overline{\overline{Q}}$ is then proportional to the orientational order parameter S of the
nematic, usually defined, for an uniaxial system, as: $S = \frac{3}{2}\ \langle\cos^2\theta - 1/3\rangle$,
with $\theta = (\vec{m},\ \vec{n}_o)$, $\vec{m}$ a unit vector (as $\vec{n}$) placed along the long axis of one
individual molecule, and the bracket means an averaging over molecules
around the point of interest. One can then write: $e = e_o S$ where $(-e_o)$ is
the quadrupolar density of a completely aligned nematic phase. It is now
obvious that the flexo polarization is just one component of a more
general vector quantity which should have been written as $\vec{P} = e_o\ \vec{\nabla}.[S(\vec{n}\vec{n}$
$-\overline{\overline{I}}/3)]$. P can be considered as the sum $\vec{P}_f + \vec{P}_o$ of the previously defined
flexo-polarization and of a new order-electricity contribution: $\vec{P}_o =$
$e_o\vec{\nabla}S\colon (\vec{n}\vec{n} = \overline{\overline{I}}/3)$. On the opposite of the flexo polarization, $\vec{P}_o$ exists
when n is uniform, but S varies in space (Fig. 1b). With this simple argu-
ment, one expects the order-electric coefficient to be the same as the
flexo coefficient e in the range of 10^{-3} cgs. In fact, there are two flexo
coefficients, e_1 and e_3, the sum of which is e. There are also two order
electricity coefficients. A detailed discussion of their correspondence
is given in Ref. [1].

2. THEORY

     To understand how order electricity could influence the behavior of a
liquid crystal, let us recall for instance the argument given in Ref. [1]
to explain the spontaneous molecular tilt close to a nematic-isotropic
interface (Fig. 2). A gradient of S, of order S/ξ, appears normally

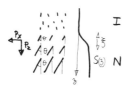

Fig. 2. Sketch of a nematic (N) to isotropic (I) interface. The
gradient of S builds the order polarization $\vec{P}(x, z)$. The
molecules tilt by an angle θ to decrease the dielectric energy
associated with P_z.

(along z) to such an interface. ξ is the coherence length at the
transition temperature. ξ is of the order of 500 A, and does not diverge
since the transition is (weakly) first order. For molecules tilted by a
constant angle θ at the interface, there must exist an order polarization
with P_x and P_z components. For a uniform interface, P_z only creates
charges, and generates a field $\vec{E}$ such that: $\varepsilon E_z + 4\pi P_z = 0$ (ε is the
assumed isotropic dielectric constant. This results in a dielectric self
energy of the order

$$\frac{1}{2} \left(\frac{4\pi}{\varepsilon}\right) P_z^2 = \frac{1}{2} \left(\frac{4\pi}{\varepsilon}\right) \left(\frac{S}{\xi}\right)^2 (\cos^2\theta - 1/3)^2 .$$

Note that the charges $\rho = -\vec{\nabla}.\vec{P}$ are not screened by ionic conductivity,
since the Debye screening length (d $\sim$ 1 μm) is usually much larger than ξ.
Assuming no coupling between the S(z) profile and θ, the dielectric energy
due to order electricity is minimum for the "magic" tilt angle $\theta \sim 53°$, such
that $\cos^2 \theta = 1/3$. In the absence of mechanical anisotropy, as in cyanobi-
phenyls compounds, for instance, there is indeed a director tilt at the
nematic-isotropic interface [6] of $\theta = 52°$. Note that a free energy term
in P^2 is a "non-analytic" term. It can be written as $\frac{(\vec{P}.\vec{\nabla}S)^2}{(\vec{\nabla}S)^2}$. As $\vec{P}$
already implies a gradient term, P^2 appears as a ratio $\sim (\nabla)^4/(\nabla)^2$ and
cannot be represented as the usual expansion in $(\nabla)^2$ and $(\nabla)^4$ terms currently used for the bulk free energy.

     For the case of nematic or cholesteric liquid crystals, order elec-
tricity should also be considered to describe the behavior of disinclina-
tion lines, in the core of which the order parameter also goes through
zero on the short scale ξ. This results in a charge per unit length of
the order of c, of same sign on each disclination core. These charges
should result in an additive repulsive force between disclination lines,
when their distance is smaller than d. In fact, this is not true, since
between the two lines, S goes through a maximum, creating charges -e, and
resulting in an attractive force. Such an electric attractive effect could
help to stabilize periodic arrays of disclinations as the ones occurring
in blue phases, where the period of dechiralization lines (the pitch) could

be shorter than d. It could also explain the attraction between a disclination and a wall, where the order parameter S could be larger than in the
bulk. The electric field effect on blue phases, taking into account order
electricity, has been sketched in Ref. [2].

To extend these ideas to smectic interfaces, we ignore any source of
electric polarization associated with the layer ordering, such as the
piezoelectric effect due to layer compression, assuming uniform undistorted
smectic layers. The first situation of interest in the behavior of thin
films of smectic A material (2 to 50 molecular layers, for example) which
one knows [7] to produce with two free smectic-air interfaces. It has been
observed that there is a strong tendency of the smectic A phase (molecules
normal to the layers) to become a smectic C phase (molecules tilted from
the normal layer). For thin layer films, for example, the $S_A \rightarrow S_C$ transition is observed at a temperature higher by twenty degrees C than in the
bulk. A model was made to describe this A $\rightarrow$ C transition [8], but without
a real physical description of what is going on. Order electricity, which
results in a spontaneous tilt on the nematic director from the normal to an
interface, can be invoked to explain this A $\rightarrow$ C surface-induced transition
since the surface layers themselves form the interface. For small θ, the
total free energy density in the surface layers must contain an order elec-
tric term like $e^2(\frac{S}{\xi})^2 (\cos^2 \theta - 1/3)^2 \sim - \frac{e^2S^2\theta^2}{\xi^2}$. This dielectric term is
destabilizing, i.e., it tends to induce a molecular tilt θ. It must be
compared to the elastic term $\frac{1}{2}B_\perp\theta^2$ which stabilizes the molecules along
the normal layer in the smectic A phase [9]. Close to a smectic A $\rightarrow$ C
phase transition, $B_\perp$ can be written in the Landau form as $B_\perp = \alpha(T - T_{CA})$,
where T_{CA} is the phase transition temperature for a second order A $\rightarrow$ C
transition. The order of magnitude for B is $\frac{kT_{CA}(T-T_{CA})}{m^3 \quad T_{CA}}$ where k is the
Boltzmann constant and m a molecular length. In most compounds, the quad-
rupole density eS is of the order of 10^{-3} cgs, i.e., e^2S^2 compares with
the curvature Frank elastic constant K $\sim 10^{-6}$ cgs. As K $\sim \frac{kT_{NI}S^2}{m}$, where
T_{NI} is the nematic isotropic transition temperature, and $\xi^{-2} \sim m^{-2} \frac{T_{NI}-T}{T_{NI}}$
from a crude mean field model of the nematic-isotropic transition, one
sees that the destabilizing order electric term $- \frac{e^2S^2}{\xi^2} \theta^2$ is equivalent
to an increase of T_{CA} comparable to $T_{NI} - T_{CA}$, very large indeed.

A direct observation of the order electricity in thin smectic layers
could be the following: if molecules tilt to suppress the normal P_z component of the order-polarization, close to the free boundary surfaces (Fig.
3), one is left with a P_x order electric component. This P_x component must
be distinguished from the ferroelectric polarization which occurs in chiral
smectic C^*; P_x is oriented along the molecular (director) projection C on
the smectic layers (and not perpendicular, as the ferroelectric SC^*

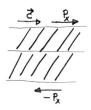

Fig. 3. A thin smectic bilayer, tilted at the magic angle, with a
resulting antiferro electric ordering $+p_x$ and $-P_x$.

polarization) and exists even with non-chiral molecules. Point defects of
the director $\vec{C}$ have been observed in thin tilted smectic layers [10].
Around these points, divergence of P_x builds a charge q (Fig. 4). Note
that, by symmetry, if the upper layers possess the polarization P_x, the
lower must show an opposite polarization $-P_x$, i.e., will build a -q charge
around the defect. A point defect must then possess an electric dipole
normal to the layers. It should be simple to observe such a dipole by
applying a nonuniform electric field on the smectic layers, with, for
example, two electrodes placed on the insulating holder. The experiment
is currently under way [11].

Finally, we can use order-electricity to explain the recently observed
[12] change of smectic ordering close to a smectic-air interface. A compound presenting an SA_1 ordering is oriented to present the smectic layers
parallel to the air-smectic interface. The A_1 phase contains only one
molecule per layer. The material is a cyano compound, with a strong end
dipole along the molecular axis. In the SA_1 phase, these dipoles are up
and down oriented at random, so that the A_1 phase is not ferroelectric.
An X-ray analysis shows that at the free surface, an additional SA_2 ordering appears, with a spatial period of two molecules. The SA_2 phase possesses an anti-ferroelectric ordering, with a regular alternation of aliphatic
tails and polar heads. The explanation for this local change in smectic
ordering could be the following: close to the interface there is a gradient of order parameter S, i.e., a surface layer of order polarization. As
the system is far from a possible $S_A \rightarrow S_C$ phase transition, it cannot
decrease the associated dielectric energy by a simple molecular tilt, as
previously suggested. It will change its ordering to build the observed

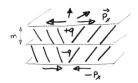

Fig. 4. A point defect in a smectic bilayer possesses an electric
dipole q m.

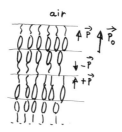

Fig. 5. The SA$_2$ antiferroelectric ordering in response to the order electric polarization $\vec{P}_0$ at the air-smectic interface.

antiferroelectric layers, i.e., it gives rise to the SA$_2$ phase at the air-smectic interface (Fig. 5.).

In conclusion, order-electricity, i.e., the electric polarization associated with the gradient of the orientational order parameters, exists in principle in all liquid crystals. Its influence is visible when the spatial variation of the order appears on a scale shorter than the Debye screening length in the practically used material. This is generally the case for all liquid crystal-air interfaces. In this paper we have discussed some observed, but not really understood, properties of free-standing smectic thin films. We have shown that the existence of an order electric polarization close to the smectic-air interfaces could explain the observed spontaneous tilt of the molecules inside the thin smectic films. Order electricity could also explain the antiferroelectric ordering observed at a free smectic-air interface.

We acknowledge interesting discussions with N. Clark, G. Barbero, Ph. Martinot-Lagarde and J. F. Palierne.

3. REFERENCES

1. G. Barbero, I. Dozov, J. Palierne, and G. Durand, Phys. Rev. Lett. 56, 2056 (1986).
2. H. R. Brandt, Mol. Cryst. Liq. Cryst. (Lett.) 3, 147 (1986).
3. M. Mondake, Ph. Martinot-Lagarde, and G. Durand, Europhysics Lett. 2, 299 (1986).
4. R. B. Meyer, Phys. Rev. Lett. 22, 319 (1969).
5. J. Prost and J. Marcerou, J. Physique (Paris) 38, 315 (1977).
6. S. Faetti and V. Palleschi, Phys. Rev. A30, 3241 (1984); J. Physique-Lettres (Paris) 45, 313 (1984).
7. C. Y. Young, R. Pindak, N. A. Clark, and R. B. Meyer, Phys. Rev. Lett. 40, 773 (1978).
8. S. Heinekamp, R. A. Pelcovits, E. Fontes, E. Yi Chen, R. Pindak, and R. B. Meyer, Phys. Rev. Lett. 52, 1017 (1984).
9. See for instance P. G. de Gennes, "The Physics of Liquid Crystals" (Oxford University Press, 1975).
10. R. Pindak, C. Y., Young, R. B. Meyer, and N. A. Clark, Phys. Rev. Lett. 45, 1193 (1980).
11. N. Clark, Private communication.
12. E. F. Gramsbergen, W. H. de Jeu, and J. Als Nielsen, J. Physique (Paris) 47, 711 (1986)

240

MACROSCOPIC VARIABLES IN COMMENSURATE AND INCOMMENSURATE CONDENSED PHASES,

QUASICRYSTALS AND PHASMIDS

Harald Pleiner

Department of Physics
University of Colorado
Boulder, Colorado

Abstract   Macroscopic variables, necessary for a useful dynamical
description of many condensed phases, are discussed. The nature of
these macroscopic variables, their origin and their implications for
the dynamics are considered. The somewhat special case of phasmids is
treated separately.

1. RIGOROUS HYDRODYNAMICS

Hydrodynamics has proven to be a powerful description of the dynamics
of various condensed phases. It is a rigorous theory making use only of
conservation laws, symmetries and thermodynamics. It is applicable if
local thermodynamic equilibrium holds, i.e., for long wavelength and low
frequency excitations. This approach was used to describe simple fluids
[1,2], superfluid ^{4}He [3,4], magnetic systems [5]. It was introduced to
treat the superfluid phases of ^{3}He by Ref. 6-9, crystals and liquid crystals by Ref. 10-12 (a generalization and application to more recent liquid
crystalline phases is given in Refs. 13-18).

The first step of this method consists of choosing the hydrodynamic
variables: i) all densities (e.g., mass density, etc. [19]) connected to
conservation laws (which are related to specific symmetries of the underlying Hamiltonian), are hydrodynamic. Being conserved quantities they
cannot be created or destroyed instantly (e.g., by molecular collisions),
they can only be redistributed spatially, which involves transport. This
is a slow process and makes the dynamics of these hydrodynamic variables
much slower as and quite distinct from the large number of microscopic
degrees of freedom. ii) All variables describing spontaneously broken
continuous symmetries [20] are hydrodynamic (Goldstone variables); the

phase connected with broken gauge symmetry in superfluid systems is an example. Describing symmetry operations, under which the Hamiltonian is invariant, homogeneous excitations of these Goldstone variables do not feel any restoring force. Only inhomogeneous excitations cost energy and provoke restoring forces, which vanish if the inhomogeneity vanishes. Thus the frequency ω of such excitations goes to zero, when their wave vector $\vec{k}$ goes to zero. This, however, is just the definition of a hydrodynamic excitation. It is obvious that the number of hydrodynamic variables equals the number of hydrodynamic excitations. For a more rigorous discussion of this connection between symmetry and hydrodynamics cf. Ref. 21.

The second step of the hydrodynamic method consists of series expansions in powers of the hydrodynamic variables and their thermodynamic conjugates (or the gradients thereof), both for the statics and dynamics. The coefficients of these expansions are phenomenological parameters, the generalized susceptibilities and (reversible or irreversible) transport parameters, respectively. The number of these parameters is strongly restricted by symmetry and thermodynamic requirements. Usually, nothing is known about the existence of these power series, nor about their limits of applicability. In this note, however, I will mainly be concerned with the first step of the method, the choice of variables.

2. MACROSCOPIC DYNAMICS

There are several cases where the rigorous hydrodynamic treatment does not lead to a complete macroscopic description of the dynamics of a system. There may be a few microscopic degrees of freedom, which become slow under certain circumstances, or there may be some almost hydrodynamic variables. Both types of non-hydrodynamic variables, which will be discussed in detail in the following, lead to non-hydrodynamic modes, whose frequencies show a gap in the homogeneous limit, i.e., $\omega(\vec{k}\to 0) \to \omega_0 \neq 0$. If ω_0^{-1} is much larger than the microscopic time scale τ (the collision time) relevant for the huge number of microscopic modes, then it is useful to add these few non-hydrodynamic (but slow) variables to the list of variables, by which the dynamics of a given phase is described. The resulting dynamic description will be called macroscopic dynamics. A rigorous hydrodynamic description would be restricted in this case to be valid only for frequencies ω smaller than ω_0, which could be a rather limited and uninteresting regime.

There is, however, no general procedure how to choose these macroscopic, but non-hydrodynamic variables. On the other hand, it is believed that the occurrence of those variables is connected with special properties

of a given phase or with special external circumstances the phase is subject to. Thus, it should be possible to find all macroscopic variables
relevant for the given phase, under the external constraints, and for the
frequency regime one is actually interested in. The non-universal character of these macroscopic variables deserves a special treatment for any
case.

Slow Microscopic Variables

There are two major situations where microscopic variables become
slow, phase transitions and defects. The complete order parameter of a
condensed system (simple fluids excluded) usually consists of two parts,
one which describes the strength of ordering (the "modulus"), and one which
contains the structure of ordering, i.e., describes the nature of the broken symmetry. The latter contains the Goldstone variables, while the former, the modulus, usually a scalar quantity, is a microscopic variable,
constant on a macroscopic time (and length) scale. Near a (second order
or nearly second order) phase transition, where the order vanishes, the
modules of the order parameter also vanishes and its dynamics become slow
("critical slowing down"). Then it has to be considered as a macroscopic
variable giving rise to a soft mode, whose gap vanishes only at the transition point, but may be small already in the vicinity of it.

A similar situation occurs when defects with a core are present. In
the core region of the defect the modulus of the order parameter tends to
zero, in order to avoid unphysical singularities in the ordered structure
(which would require a diverging energy). Again the modulus has to be
treated as a macroscopic variable, when the dynamics of condensed phases
with defects is considered. Early examples for the two cases having
included the modulus in the macroscopic dynamics are in Refs. 22-25 and
Refs. 26-29, respectively.

Almost Hydrodynamic Variables

The application of external fields (electric, magnetic, temperature
gradients, etc.) can destroy the hydrodynamic character of some Goldstone
variables. These fields break some symmetries [30] externally and the
appropriate Goldstone variables are no longer hydrodynamic. Even a homogeneous symmetry operation costs energy due to the presence of the external
field and the dispersion relation $\omega(k)$ shows a gap. If the interaction
energy between (former) Goldstone variable and external field is small,
the gap is small, too. Then it is sensible to keep the (former) Goldstone
variable as a macroscopic variable. An example is a nematic liquid crystal

in an external magnetic field $\vec{H}$, where the magnetic anisotropy interaction energy [31] between the director and $\vec{H}$ leads to gaps in the Goldstone modes (the director-shear modes). In this example the interaction is small and usually the director is kept as macroscopic variable (without further notice). If the interaction were stronger and the external field extremely strong, we would discard the director as macroscopic variable and would end up with a hydrodynamic description isomorphic to simple liquids.

The role of the external field played in the preceding example can also be played by a small internal interaction. E.g., in the superfluid phase ^{3}He-A, the very tiny nuclear magnetic dipole interaction couples the preferred direction $\hat{n}$ in spin space to that ($\hat{\ell}$) in orbit space. Thus, fluctuations of $\hat{n}$ (relative to $\hat{\ell}$) are not hydrodynamic, but nearly hydrodynamic because of the smallness of the interaction [7,32]. A rather similar situation occurs in the tilted smectic phases with bond orientational order [33] (smectics F, I [34,35] and C [36]), where rotational symmetry is spontaneously broken by both, the tilt direction and the bond ordering directions. However, the two appropriate Goldstone variables (rotations about a common axis, i.e., the layer normal) are locked together and a relative rotation (tilt direction versus bond ordering directions) is not hydrodynamic, but may still be looked at as a macroscopic variable [37,38].

A different kind of interaction occurs if long-range forces are involved. The homogeneous translation of a constant electric charge density (against a background of opposite charge density) leads to uncompensated charges at the surfaces. Due to its long-ranged nature, the resulting attracting (restoring) force cannot be neglected in the bulk and leads to a gap in the dispersion relation. The plasma frequency in superconductors is an example of this.

Interactions of Goldstone variables with walls can act in the same way as external fields discussed above. If the interaction is small, however, one keeps the hydrodynamic character of the Goldstone variable and meets the interactions with walls in the form of boundary conditions.

3. COMMENSURATE, INCOMMENSURATE SYSTEMS, QUASICRYSTALS

Commensurate phases are those where a symmetry is spontaneously broken more than once and where the appropriate Goldstone variables are coupled energetically [39]. The examples of the preceding subsection belong to this class. However, contrary to these cases where the interaction is small, very often the coupling energies are high (the gap large) and the appropriate variables are not kept as macroscopic variables. The smectic A phase is an example, where the layer normal and the director (the averaged direction of the long molecular axes) both break rotational symmetry.

Since they are strongly coupled together, relative fluctuations of one direction against the other are discarded in the macroscopic dynamics of smectics A [40]. The same holds with respect to the broken, translational symmetry in smectics A_1, A_2 and A_d, since the mass density modulation and the polarization density modulation [41] are strongly coupled in these systems.

I will call phases incommensurate, if the Goldstone variables connected with a more than once broken symmetry are unlocked and not coupled energetically [42]. In that case these variables keep their hydrodynamic character and there are no gaps in the excitation spectrum. Relative fluctuations of these variables are usually called phasons. One example for such an incommensurate phase is smectics C^*, where the (one-dimensional) translational symmetry is broken twice, by the layer structure and by the helix. Relative translations of the helix with respect to the layers (performed simply by rotating the helix about its axis) constitute therefore a gapless phason mode [17]. Further examples are the incommensurate smectic A_{ic} phase [43], where the mass density wave and the polarization density wave are unlocked, the--so far hypothetical--incommensurate discotic phases, where in one or two dimensions two different kinds of unlocked, spatial modulations may exist [44], and some crystalline structures which allow the (one-dimensional) transport of chains of one atomic species relative to the other atoms [45]. The latter systems, however, can acquire a gap in the phason spectrum, under certain conditions, due to pinning or friction [46].

True phason modes (without gap) can be either diffusive, propagative or a mixture of both [47] depending on how the appropriate linear combination of Goldstone variables couples to the other hydrodynamic variables. If the Goldstone mode involves effective mass transport, as in the case of the incommensurate crystalline structures mentioned above [48], one can expect a propagative behavior. In other cases, where only rotations (as in smectics C^*) or polarization fluctuations (as in smectics A_{ic}) are involved, the phason is diffusive. Diffusive or propagative behavior of a phason mode can also depend on the direction of its wave vector [49].

Quasicrystals [50] can be considered as having 6 (4) independent mass density waves in a 6 (4) dimensional hyperspace for the 3 dimensional, icosahedral (2 dimensional, pentagonal) case [51-54]. Thus, all translational symmetries are broken twice and 3 (2) additional macroscopic variables are to be included. In the real physical space, however, the Goldstone variables cannot be interpreted as relative displacements of different kinds of atoms. Thus, it is not obvious whether the appropriate modes are true phasons or whether they have a gap [54].

4. PHASMIDS

Phasmids [56] are basically discotic liquid crystals showing a two-dimensional regular lattice (in the x-z plane) of liquid-like columns (columnar axes parallel to z-axis). The columns are made by piling up molecules of rod-shaped cores (pointing in y-direction) with three flexible chains at each end. In the case of a rectangular lattice the cores and the end chains all lie in alternating planes parallel to the x-z plane. In this respect the structure resembles a smectic-like structure. Since the interaction between end chains of different molecules is small along these planes, it is easy to move the planes against each other. This is a new degree of freedom, not present in ordinary discotics, which may influence the macroscopic dynamics. Contrary to the case of smectics, these layers are coupled energetically, albeit probably very weakly. This reminds one of the case of commensurate phases. However, in phasmids the translational symmetry (in x-direction) is not broken twice. Thus the variable describing the new degree of freedom (relative layer displacement) is not independent from the Goldstone variable u_x, describing smooth lattice displacements around the equilibrium structure. To describe the displacement of one layer over several lattice constants one would have to go to infinite order in the hydrodynamic gradient exanpsion of u_x. Of course, this is not feasible and, instead, the introduction of a new macroscopic variable is proposed. In analogy to the treatment of single defects [29] a discrete version is chosen, i.e., $R_\alpha(z)$ may describe displacements of the α-th layer along the x-direction; it can still depend on z. In the absence of lattice vibrations an isolated layer will follow the dynamical law

$$m_\alpha \ddot{R}_\alpha = \frac{\delta \varepsilon_I}{\delta R_\alpha} - \eta \dot{R}_\alpha + \bar{\nu} \nabla_z^2 R_\alpha \tag{1}$$

with the potential energy ε_I provided by the neighboring layers

$$\varepsilon_I = D(\cos 2_\pi \frac{R_\alpha}{a} - 1) \tag{2}$$

which reflects the lattice periodicity, a, and the commensurability. The inertial force due to the layer mass m_α has been kept and a friction force has been added. If the displacement is inhomogeneous in the z-direction (the liquid-like direction), a diffusion takes place.

If lattice vibrations are taken into account, the lattice displacement $u_x(x, y, z)$ enters the dynamical equation for R_α

$$m\ddot{R}_\alpha = \frac{\delta\varepsilon_I}{\delta R_\alpha} - \eta(\dot{R}_\alpha - \langle u_x \rangle_{y=\alpha}) + \bar{\nu}\nabla_z^2 R_\alpha \tag{3}$$

with

$$\varepsilon_I = D < \cos\frac{2\pi}{a}(R_\alpha - u_x) - 1 >_{y=\alpha} \tag{4}$$

where $\langle\ldots\rangle_{y=\alpha}$ denotes the mean over the x-direction at $y=\alpha$. Generalizations to the case of more than one moving layer are straightforward; eventually one can then switch to a continuum description using $R(y,z)$. Details will be given elsewhere.

In conclusion, the macroscopic variable introduced in the description of phasmids turns out to be completely different in nature from those macroscopic variables discussed in the preceding chapters.

5. ACKNOWLEDGMENT

Support by the Deutsche Forschungsgemeinschaft is gratefully acknowledged.

6. REFERENCES

1. L. D. Landau and E. M. Lifshitz, "Fluid Mechanics" (Pergamon, London), 1959.
2. L. P. Kadanoff and P. C. Martin, Ann. Phys. 24, 419 (1963).
3. P. Hohenberg and P. C. Martin, Ann. Phys. 34, 291 (1965).
4. I. M. Khalatnikov, "Introduction to the Theory of Superfluidity" (Benjamin, New York, 1965).
5. B. I. Halperin and P. Hohenberg, Phys. Rev. 188, 898 (1968).
6. R. Graham, Phys. Rev. Lett. 23, 1431 (1974).
7. R. Graham and H. Pleiner, Phys. Rev. Lett. 24, 792 (1975).
8. R. Graham and H. Pleiner, J. Phys. C9, 279 (1976).
9. M. Liu, Phys. Rev. Lett. 43, 1740 (1979).
10. P. C. Martin, O. Parodi and P. S. Pershan, Phys. Rev. A6, 2401 (1972).
11. D. Forster, Ann. Phys. 84, 505 (1974).
12. T. C. Lubensky, Phys. Rev. A6, 452 (1972).
13. H. Brand and H. Pleiner, J. Physique 41, 553 (1980).
14. H. Pleiner and H. Brand, J. Physique Lett. 41, L491 (1980).
15. H. Brand and H. Pleiner, Phys. Rev. A24, 2777 (1981).
16. H. Pleiner and H. Brand, Phys. Rev. A25, 995 (1982).
17. H. Brand and H. Pleiner, J. Physique 45, 563 (1984).
18. H. Pleiner and H. R. Brand, Phys. Rev. A29, 911 (1984).
19. Except the angular momentum density, which is dynamically not an independent variable (cf. Ref. 10).
20. Except when long-ranged forces are present.
21. D. Forster, "Hydrodynamic Fluctuations, Broken Symmetry, and Correlation Functions" (Benjamin, Reading, 1975).
22. F. Jähnig and H. Schmidt, Ann. Phys. 71, 129 (1972).
23. W. L. McMillan, Phys. Rev. A9, 1720 (1974).
24. J. Swift, Phys. Rev. A13, 2274 (1976).
25. M. Liu, Phys. Rev. A19, 2090 (1979).
26. B. Julia, and G. Toulouse, J. Physique Lett. 40, L395 (1979).

27. I. E. Dzyaloshinskii and G. E. Volovik, Ann Phys. 125, 67 (1980).

28. K. Kawasaki, Ann. Phys. 154, 319 (1984).

29. H. R. Brand and K. Kawasaki, J. Phys. A17, L905 (1984).

30. Usually rotational symmetry; but if a field is modulated in space, it also breaks translational symmetry.

31. P. G. de Gennes, "The Physics of Liquid Crystals" (Clarendon, Oxford, 1974).

32. H. Pleiner, J. Phys. C10, 4241 (1977).

33. D. R. Nelson and B. I. Halperin, Phys. Rev. B21, 5312 (1980).

34. R. Pindak, D. E. Moncton, S. C. Davey and J. W. Goodby, Phys. Rev. Lett. 46, 1135 (1981).

35. J. J. Benattar, F. Moussa and M. Lambert, J. Chim. Phys. 80, 99 (1983).

36. J. D. Brock, A. Aharony, R. J. Birgeneau, K. W. Evans-Lutterodt, J. D. Litster, P. M. Horn, G. B. Stephenson and A. R. Tajbakhsh, Phys. Rev. Lett. 57, 98 (1986).

37. H. Pleiner, Mol. Cryst. Liq. Cryst. 114, 103 (1984).

38. H. Pleiner and H. R. Brand, Phys. Rev. A29, 911 (1984).

39. This definition is not restricted to translational broken symmetries.

40. Except near the smectic A-nematic phase transition (ref. 25).

41. J. Prost in "Liquid Crystals of One- and Two-Dimensional Order," W. Helfrich and G. Heppke, eds. (Springer, Berlin, 1980), p. 125.

42. This definition avoids the physically somewhat unsatisfactory distinction between rational and irrational numbers; on the other hand, it may be difficult to distinguish between no interaction and extremely small interaction.

43. J. Prost and P. Barois, J. Chim Phys. 80, 65 (1983).

44. H. R. Brand and H. Pleiner (to be published).

45. H. Brand and P. Bak, Phys. Rev. A27, 1062 (1983).

46. In that case, I would call these systems commensurate.

47. The real and imaginary parts of the dispersion relation vanish with equal power of the wave vector.

48. The masses of the different atomic species are different.

49. J. D. Axe and P. Bak, Phys. Rev. B26, 4963 (1982).

50. D. Schechtman, I. Blech, D. Gratias and J. W. Cahn, Phys. Rev. Lett. 53, 1951 (1984).

51. P. Bak, Phys. Rev. Lett. 54, 1517 (1985).

52. T. C. Lubensky, S. Ostlund, S. Ramaswamy, P. J. Steinhardt, and J. Toner, Phys. Rev. Lett. 54, 1520 (1985).

53. N. D. Mermin and S. M. Troian, Phys. Rev. Lett. 54, 1524 (1985).

54. M. Kleman, Y. Gefen and Y. Pavlovitch, Europhys. Lett. 1, 61 (1986).

55. The latter is proposed in T. Janssen, these proceedings.

56. J. Malthete, A. M. Levelut and Nguyen Hu Tinh, J. Physique Lett. 46, L876 (1985).

HOW THE SMECTIC A PHASE ADAPTS TO TWO INCOMMENSURATE PERIODS IN ASYMMETRIC

LIQUID CRYSTALLINE SYSTEMS

G. Sigaud, F. Hardouin, M. F. Achard, and H. T. Nguyen

Centre de Recherche Paul Pascal
Université de Bordeaux I
Domaine Universitaire
33405 Talence Cedex, France

Abstract   Asymmetric polar rod-like mesogens induce antiferroelectric
associations which play a special role in the structure of layered
liquid crystalline phases introducing a second period most often
incommensurate with the basic layering factor which remains connected
to the length of the molecules. The fluid smectics (mainly smectic A)
use different ways to overcome this incommensurability. Other kinds
of asymmetric liquid crystalline materials such as side-chain polymers
are now expected to present similar behaviors.

1.   INTRODUCTION

The idea of the role of incommensurability in liquid crystalline
phases generates from the special structural properties of layered smectic
phases of polar compounds. Among the smectic phases the smectic A one
(S_A) has been long considered as the simplest. A common picture is the
stacking of liquid layers along a given direction, the thickness of each
layer (d) being close to the length of the molecule (L_1) considered as a
rod [1]. However, d in the S_A phase of some polar compounds was clearly
observed to be larger than L_1 [2] and antiferroelectric associations were
proposed [3] to explain the value of the ratio $d/L_1 = 1.3$ usually measured
with compounds such as the alkoxycyanobiphenyls: RO—⟨O⟩—⟨O⟩—CN

Nevertheless, with these substances no difference was ever evidenced
between S_A with $d/L_1 = 1.3$ and S_A with $d/L_1 = 1$ since both are in all
cases totally miscible and thus are considered as perfectly isomorphous.

The real intervention of incommensurability arose from the discovery
of bilayered S_A[4] and of phase transitions connected to a need for the

system to change largely its layer thickness in the range T_1 to $2T_1$. We observed this special behavior with species which possess a typical chemical structure [5]:

$$\left.\begin{array}{c} RO \\ or \\ R \end{array}\right\} -\!\!\!\bigotimes\!\!\!- X -\!\!\!\bigotimes\!\!\!- Y -\!\!\!\bigotimes\!\!\!- \left\{\begin{array}{c} CN \\ or \\ NO_2 \end{array}\right.$$

and this paper is a summary of the experimental work performed to date showing the consequences of two competing lengths, the length of the molecule and the length of antiparallel pairs, to impose a smectic layer spacing.

2. THE DIFFERENT KINDS OF SMECTIC A

The first result of this competition is the occurrence of different S_A that we were able to classify according to three groups (Fig. 1):

S_{A1}: for monolayer S_A, i.e., the layer spacing represents the length of one molecule.

S_{Ad}: for partial bilayer S_A, i.e., the layer spacing is comprised between one and two molecular lengths.

S_{A2}: for bilayer S_A, i.e., the layer spacing is twice the molecular length. This S_{A2} phase is the final step for uniaxial fluid smectics toward low temperature which seems normal since a commensurate lock-in takes place between the two periods.

We know that the S_{A2} phase is likely to occur in compounds of the following series:

$$\left.\begin{array}{c} RO \\ or \\ R \end{array}\right\} -\!\!\!\bigotimes\!\!\!- O -\!\!\underset{\substack{|| \\ O}}{C}\!\!- \bigotimes\!\!\!- O -\!\!\underset{\substack{|| \\ O}}{C}\!\!- \bigotimes\!\!\!- CN$$

to which we will refer as DBnO or DBn in the figures (DB for DiBenzoate, n(O) being the number of carbon atoms in the alkoxy (RO) or alkyl (R) chain).

For short tails (n < 6, Fig. 2), these compounds undergo a direct nematic (N) to S_{A2} transition. The addition of a nonpolar compound of similar size modifies the system in such a way that a S_{A1} phase appears between N and S_{A2}. Thus, a S_{A1} - S_{A2} transition is induced, the observation of which led us to deny the uniqueness of the smectic A state in 1978 [4]. Since that time it has been widely studied both theoretically and experimentally [6,7]. High resolution techniques have shown that this transition becomes second order beyond a tricritical point [8,9].

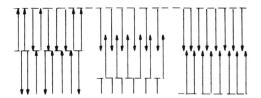

Fig. 1. Schematic representation of the three smectic A classes for asymmetric molecules. Left, monolayer SA (S_{A1}); middle, partial bilayer SA (S_{Ad}); right, bilayer SA (S_{A2}).

For larger n, the commensurate lock-in is not as easily established since a S_{Ad} separates now N and S_{A2} (Fig. 2). Of course, this means that a S_{Ad}-S_{A2} transition is observed [10]. Theoretical calculations expect a first order transition which by some analogy with the liquid-gas transition would present somewhere a critical end point (J. Prost, these proceedings). Two fundamental properties, the variation of the layer thickness and of the heat excess from DSC, as a function of the length of the aliphatic chain show indeed a strong evolution which could sustain this point of view [10].

Now the possible transitions among the three types of smectic A have not been all considered.

Only S_{A1}-S_{A2} and S_{Ad}-S_{A2} cases have been discussed. What about the S_{Ad}-S_{A1} change (Fig. 3)?

The situation is more complex since a direct transition is not the most usual. The most usual is surprisingly the passage from S_{Ad} to S_{A1} through a reentrant nematic (N_{re}), i.e., the reappearance above S_{A1} of the more symmetric N phase at lower temperature than a first N state followed by the S_{Ad}.

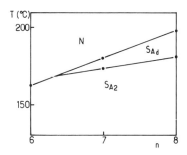

Fig. 2. Evolution of the polymorphism in the series of the alkoxy-dibenzoates (DBnO) as a function of the number of carbons in the aliphatic chain.

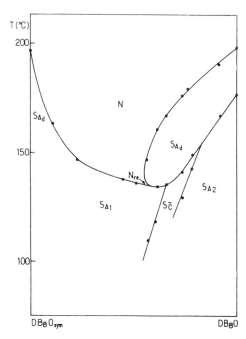

Fig. 3. This phase diagram exhibits the three different ways to change from S_{Ad} to S_{A1}. From the left to the right, continuous change, passage through a reentrant nematic, first order transition (note the vicinity of the N_{re}-S_{A1}-$S_{\tilde{C}}$ triple point).

This behavior has been widely reported in these systems since its first observation [11,12].

Only later has the direct transition from S_{Ad} to S_{A1} been definitely characterized [13] which was not obvious owing to the shortness of the line in every example known to date.

It is shrunk between a N_{re}-S_{Ad}-S_{A1} point predicted to be a peculiar bicritical point and a S_{Ad}-S_{A1}-$S^{\sim}$ triple point where arises one of the bidimensional states which will be discussed further (see section 3). This N_{re} seems to always accompany the S_{Ad}-S_{A1} transition in binary mixtures of polar compounds. One must add that no S_{Ad}-S_{A1} transition has been observed in a pure compound so far.

A third way is made possible from S_{Ad} to S_{A1}, a continuous evolution of the layer thickness without transition [14].

However, one might wonder whether the S_{Ad} phases are similar in all cases of S_{Ad}-S_{A1} transition. But Cladis and Brand recently observed a cholesteric "island" [15 and these proceedings] by mixing such a long polar rod with a nonpolar chiral mesogen. We have shown that this isolated part is not basically due to the cholesteric nature of the phase and, following the same idea of mixing a less polar compound with a polar one, we have described a nematic closed loop and shown that the topology of

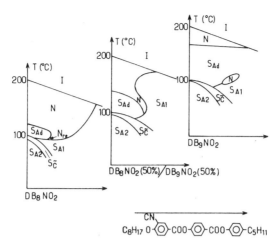

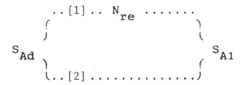

Fig. 4. Transformation of the bicritical N_{re}-S_{A1}-S_{Ad} point into a nematic closed loop. Both S_{Ad} on the left and on the right are isomorphous.

the bicritical N_{re}-S_{Ad}-S_{A1} point changes gradually into this closed loop (Fig. 4) [16]. As a consequence, we demonstrate that the S_{Ad} which led to S_{A1} through a phase transition is isomorphous to the S_{Ad} which changes to S_{A1} continuously since in this space it is possible to switch from one to the other without crossing any phase line.

To conclude this part, we can remark that the manner according to which one changes from a nematic to S_{A1} or S_{A2} through a S_{Ad} depends strongly upon the sense of variation of the layer thickness in the S_{Ad} phase with decreasing temperature.

Either d decreases smoothly, leading to S_{A1} through a N_{re} [1] or without phase change [2]:

$$S_{Ad} \quad \overset{\ldots [1] \ldots \; N_{re} \; \cdots}{\underset{\ldots [2] \ldots}{}} \quad S_{A1}$$

Or d increases, in this case the S_{Ad} transforms in S_{A1} through a N_{re} [4] or a phase line [3] or S_{Ad} transforms in S_{A2} with [6] or without [7] phase transition.

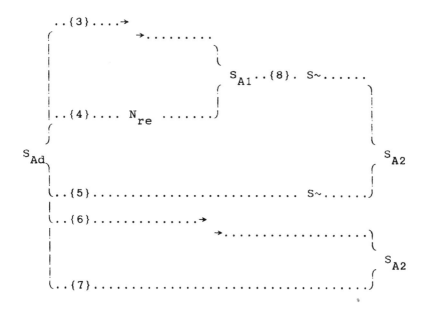

3. THE BIDIMENSIONAL ESCAPES TO INCOMMENSURABILITY (fig. 5)

On these ways to the low temperature S_{A2} state we already have mentioned the "tilde" phases which appear in the previous drawing [5], [8]. We propose now to explain their nature and polymorphism.

These mesophases possess a bidimensional structure due to a long range in plane modulation represented by the "tilde" sign ($\sim$). Nevertheless, the local order remains smectic with liquid-like disorder of the molecules inside the portions of layers. For this reason we keep the letter S to name this type of mesophase although their symmetry group is equivalent to the symmetry group of a columnar phase. Moreover, these mesophases separate in every case a fluid smectic from another fluid smectic.

The first example known of a phase of this type presents a centered rectangular lattice of "columns" with S_{A2} local order. By analogy with some structures of metallic alloys it has been named a fluid antiphase by A. M. Levelut [17]. The stability of such a phase has promoted extensive works, and the Prost's theory describes it as an escape to macroscopic incommensurability.

In this $S_{\tilde{A}}$ the "columns" whose shape is better then one of a board are of equivalent size. The domains of S_{A2} structure are separated by real walls as demonstrated by freeze fracture experiments [18]. This $S_{\tilde{A}}$ appears between S_{A1} and S_{A2}, but the sequence

$$S_{A1} \; -- \; S_{\tilde{A}} \; -- \; S_{A2}$$

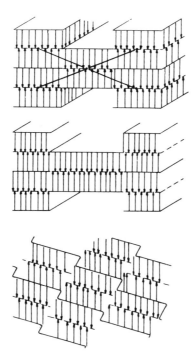

Fig. 5. Schematic representation of the three fluid bidimensional
smectics. Up, centered rectangular lattice for the $S_{\tilde{A}}$;
middle, simple rectangular lattice for the $S_{\tilde{A}cre}$; down,
oblique lattice for the $S_{\tilde{C}}$.

is in fact usually more complex; a second bidimensional phase takes place
between S_A and S_{A2} in a narrow range of temperature, as discovered by
A. M. Levelut [18]. The local order inside the domains still remains S_{A2}
but two adjacent domains are of unequal size. The centered rectangular
lattice is consequently transformed into a simple rectangular lattice.
We named it $S_{\tilde{A}cre}$ because of its "crenelated" profile.

The last kind of 2D phase with S_{A2} local order corresponds to an

We have performed microscopic observations in several (x, T) diagrams
in order to follow the evolution of the stability of this phase [19].
However, it has been experimentally impossible to determine where the $S_{\tilde{A}}$
- $S_{\tilde{A}cre}$ line ends and whether a first order $S_{\tilde{A}}$ - S_{A2} line actually exists.

The last kind of 2D phase with S_{A2} local order corresponds to an
oblique lattice which we have labeled $S_{\tilde{C}}$ although we do not know to what
extent the molecules could be tilted [20]. At short range the $S_{\tilde{C}}$ is certainly close to a S_{A2} structure but the stacking is periodically interrupted by oblique defects. How does this $S_{\tilde{C}}$ locate with regard to the other
2D phases already described? From observations at the polarizing microscope, the sequence of phases with decreasing temperature would be the
following [21]:

$$S_{A1} \text{ -- } S_{\tilde{C}} \text{ -- } S_{\tilde{A}} \text{ -- } S_{\tilde{A}cre} \text{ -- } S_{A2}$$

At last it seems that all systems could not find an escape to incommensurability since the theory describes a possible S_A state in which the two longitudinal wave vectors are simultaneously condensed [22]; such a phase is incommensurate with a double stacking corresponding to two independent incommensurate periods. An experimental study of a mixture of two polar compounds of very different lengths supports this idea [23 and these proceedings]. The confirmation of the stability of this incommensurate S_A should now be found in a pure compound.

4. CONCLUSION AND PERSPECTIVES

One may wonder whether this polymorphism is restricted to the special class of polar mesogens described previously. Regarding the length, a recent experimental work has reported a bilayer S_{A2} according to our classification with shorter compounds of this type [24]:

$$RO \longrightarrow \hspace{-1em}\bigcirc\hspace{-1em}\longrightarrow N = CH \longrightarrow \hspace{-1em}\bigcirc\hspace{-1em}\longrightarrow (CH2)_n \longrightarrow CN$$

We showed previously that very different systems induce surprisingly S_{A1}, $S_{\tilde{A}}$ and S_{A2} phases. In this case some association between very different molecules, the cyanobicyclohexyls and a certain class of aromatic amines should occur [25]. However, the experimental investigation seems to indicate that the dipolar interactions are of chief importance for the appearance of a smectic A polymorphism; all these compounds entail a strong polar function.

Are these interactions absolutely necessary? According to the theory, no [26,27]. Experimentally the S_A phases of some side-chain polymers have revealed partial bilayer structures although the mesogenic side groups do not possess a strongly polar part [28,29]. The asymmetry inherent to the linkage on the backbone on one side and to the freedom on the other end seems now sufficient to obtain antiparallel arrangements and variations of the layer thickness which remind what occurs in the low molecular weight compounds.

$$\left[O \underset{\underset{(CH_2)_n}{|}}{\overset{\overset{CH_3}{|}}{Si}} \right]_{35}$$ main chain

$(CH_2)_n \longrightarrow O \longrightarrow \hspace{-1em}\bigcirc\hspace{-1em}\longrightarrow O \longrightarrow \overset{O}{\overset{||}{C}} \longrightarrow \hspace{-1em}\bigcirc\hspace{-1em}\longrightarrow O \longrightarrow C_m H_{2m+1}$

spacer   side-group   tail

One can finally add that antiphase fluctuations have been yet detected by several authors in these polymeric systems [30,31] and that we have observed a reentrant nematic phase by mixing a nonpolar side-chain polymer with a nonpolar low molecular weight compound [32].

5. REFERENCES

1. S. Diele, P. Brand, and H. Scakmann, Mol. Cryst. Liq. Cryst. 16, 105 (1972).
2. J. E. Lydon and C. J. Coakley, J. Phys. 36, C1-45 (1975).
3. A. J. Leadbetter, J. C. Frost, J. P. Gaughan, G. W. Gray, and A. Mosley, J. Phys. 40, 375 (1979).
4. G. Sigaud, F. Hardouin, M. F. Achard, and H. Gasparoux, Int. Liq. Cryst. Conf. Bordeaux (1978); J. Phys. 40, C3-356 (1979).
5. Nguyen Huu Tinh, J. de Chimie Phys., 80, 83 (1983), and the references therein.
6. J. Prost, J. Phys. 40, 581 (1979); J. Wang, T. C. Lubensky, J. Phys. 45, 1653 (1984).
7. F. Hardouin, A. M. Levelut, J. J. Renattar, and G. Sigaud, Solid State Commun. 33, 337 (1980).
8. K. K. Chan, P. S. Pershan, T. R. Sorensen, and F. Hardouin, Phys. Rev. Lett. 54, 1694 (1985).
9. C. Chiang, C. W. Garland, Mol. Cryst. Liq. Cryst. 122, 25 (1985).
10. F. Hardouin, M. F. Achard, Nguyen Huu Tinh, and G. Sigaud, J. Phys. Lett. 46, L1-123 (1985).
11. F. Hardouin, G. Sigaud, M. F. Achard, and H. Gasparoux, Solid State Commun. 30, 265 (1979).
12. F. Hardouin, A. M. Levelut, M. F. Achard, and G. Sigaud, J. de Chimie Phys. 80, 53 (1983).
13. H. T. Nguyen, G. Sigaud, M. F. Achard, H. Gasparoux, and F. Hardouin, "Advances in Liquid Crystal Research"(Pergamon Press, Oxford, 1980), p. 147.
14. P. E. Cladis and H. R. Brand, Phys. Rev. Lett. 52, 2261 (1984).
15. F. Hardouin, M. F. Achard, H. T. Nguyen, and G. Sigaud, Mol. Cryst. Cryst. Lett. 3, 7 (1986).
16. G. Sigaud, F. Hardouin, M. F. Achard, A. M. Levelut, J. Phys. 42, 107 (1981).
17. G. Sigaud, M. Mercier, H. Gasparoux, Phys. Rev. A32, 1282 (1985).
18. A. M. Levelut, J. Phys. Lett. 45, L1-603 (1984).
19. G. Sigaud, F. Hardouin, M. F. Achard, Phys. Rev. A31, 547 (1985).
20. F. Hardouin, H. T. Nguyen, M. F. Achard, and A. M. Levelut, J. Phys. Lett. 43, L1-327 (1982).
21. G. Sigaud, M. F. Achard, F. Hardouin, J. Phys.Lett. 46, L1-825 (1985).
22. P. Barois, Phys. Rev. A33, 3632 (1986).
23. B. R. Ratna, R. Shashidar, V. N. Raja, Phys. Rev. Lett. 55, 1476 (1985).
24. F. Barbarin, M. Dugay, D. Guillon, A. Skoulios, J. Phys. 47, 931 (1986).
25. G. Sigaud, M. F. Achard, F. Hardouin, H. Gasparoux, J. Phys. Lett. 46, L1-321 (1985).
26. J. Prost and P. Barois, J. de Chimie Phys. 80, 65 (1983).
27. F. Dowell, Phys. Rev. A31, 3214 (1985).
28. R. M. Richardson and N. J. Herring, Mol. Cryst. Liq. Cryst. 123, 143 (1985).
29. M. Mauzac, F. Hardouin, H. Richard, M. F. Achard, G. Sigaud, and H. Gasparoux, Eur. Polym. J. 22, 137 (1986).
30. P. Davidson, P. Keller and A. m. Levelut, J. Phys. 46, 939 (1985).

31. P. Keller, B. Carvalho, J. P. Cotton, M. Lambert, F. Moussa, G. Pepy, J. Phys. Lett. $\underline{46}$, L1-1065 (1985).

32. G. Sigaud, F. Hardouin, M. Mauzac and H. T. Nguyen, Phys. Rev. $\underline{A33}$, 789 (1986).

AN INCOMMENSURATE SMECTIC A PHASE

B. R. Ratna, R. Shashidhar and V. N. Raja

Raman Research Institute
Bangalore 560080, India

Abstract   An incommensurate smectic A phase (A_{ic}) existing between
partially bilayer A_d and bilayer A_2 phases was discovered recently by
us [1]. Here we report detailed X-ray and DSC studies as well as
dielectric constant measurements on this new phase.

1.   INTRODUCTION

Several types of smectic A phases are known [2] to be exhibited by
compounds whose constituent molecules possess a strongly polar end group.
These phases, which have been characterized by X-ray diffraction, are (i)
the monolayer A_1 phase which gives a reflection corresponding to a wavevector $2q_0 = 2\pi/\ell$, ℓ being the length of the molecule; (ii) the partially
bilayer A_d phase giving a reflection at $q_0' = 2\pi/\ell'$ where $\ell < \ell' < 2\ell$; and
(iii) the bilayer A_2 phase characterized by two reflections, the fundamental at q_0 corresponding to twice the molecular length and its harmonic at
$2q_0$. In addition, such compounds show "antiphases" ($\tilde{A}$) with X-ray diffraction peaks off the Z-axis [3] (the layer normal). It is now generally
accepted that the $\tilde{A}$ phases are not strictly smectic A phases at all [4].

Recently we have observed yet another kind of smectic A with two collinear incommensurate density modulations. In this paper we present
detailed X-ray, DSC and dielectric measurements on this new phase.

2.   EXPERIMENTAL

X-ray studies were performed on magnetically oriented samples contained in 0.5 mm diameter Lindemann glass capillaries using monochromatic copper K_α radiation and a flat photographic film. The sample was cooled
extremely slowly (less than 1°C per hour) in order to obtain a mono-domain

of the incommensurate phase. The constancy of temperature during any expo-
sure was ±0.1°C. The relative accuracy in the determination of the temperature variation of the layer spacing is ±0.1 Å or better. Differential
scanning calorimetry runs were taken with a Perkin-Elmer DSC-4 calorimeter
in conjunction with the Thermal Analysis Data Station.

The static dielectric constants ($\varepsilon_\parallel$ and $\varepsilon_\perp$) were measured using a
Hewlett-Packard Impedance Analyzer (4192A). The sample, typically 50-100
μm thick, was aligned in the nematic phase by a 1.5 Tesla magnetic field
and cooled to the smectic phase in the presence of this field.

3. RESULTS AND DISCUSSION

The incommensurate phase (A_{ic}) was observed in a binary mixture of
4-n-octyloxy-4'-cyanobiphenyl (8OCB) and 4-n-heptyloxyphenyl-4'-cyanobenzy-
loxy benzoate (DB7OCN). The molecular structures of the two pure compounds
are given in Fig. 1. In Fig. 2 we present the relevant part of the temperature concentration (T-X) diagram. For X > 24%, the A_{ic} phase intervenes
between the A_d and A_2 phases. The A_d-A_{ic} and A_{ic}-A_2 transitions were
clearly seen by optical microscopy. In Figs. 3a-d we show the textures of
the A_d, A_{ic} and A_2 phases obtained on cooling the focal conic texture of
the A_d phase. Zig-zag folds appear on the back of the focal conic fans at
the A_d-A_{ic} transition (Fig. 3b), become very prominent well in the A_{ic}
phase (Fig. 3c) and disappear at the A_{ic}-A_2 transition (Fig. 3d). On cooling the homeotropically aligned A_d phase, the A_{ic} phase makes a dramatic
appearance in the form of loops as shown in Fig. 4a. These defects do not
completely clear off in the A_2 phase. But on heating the sample back into
the A_{ic} phase a burst of parabolic focal conic-like defects [5] (see Fig.
4b) appears at the A_2-A_{ic} transition.

The thermal evolution of the layer spacing corresponding to the A_d,
A_{ic} and A_2 phases for 34.8% mixture is shown in Fig. 5. In the A_{ic} phase,
three reflections are observed corresponding to wavevectors q_0, q_0' and
$2q_0$, with $2\pi/q_0$ and $2\pi/q_0'$ decreasing with decrease of temperature. It
should be mentioned that we did not see any reflections corresponding to
combinations of q_0 and q_0' even with long exposures. (The setup did not

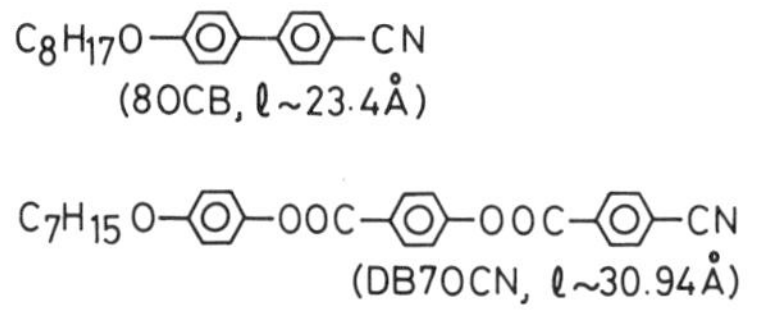

$$C_8H_{17}O-\bigcirc-\bigcirc-CN$$
$$(8OCB, \ell \sim 23.4 Å)$$

$$C_7H_{15}O-\bigcirc-OOC-\bigcirc-OOC-\bigcirc-CN$$
$$(DB7OCN, \ell \sim 30.94 Å)$$

Fig. 1. Molecular structures of 8OCB and DB7OCN. Their molecular lengths
measured using Dreiding model are also given in brackets.

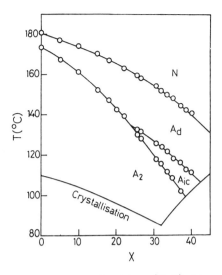

Fig. 2. Partial temperature-concentration (T-X) diagram for varying mole percent (X) of 80CB in DB70CN.

allow very low angle reflections ($\theta < 0.5°$) to be recorded.) We have also verified from the high angle diffraction maximum that the in-plane order is liquid like in all the three phases.

Figure 6 gives the intensity contour diagram (obtained with an X-Y microdensitometer – Joyce-Loebl Scandig-3, in conjunction with an on-line computer) of the photograph taken for the 34.8% mixture at 115.5°C. Typical widths of the spots are 0.8×10^{-2} Å^{-1} in the Z-direction and 1.7×10^{-2} Å^{-1} in the X-direction. The larger widths in the X-direction arises from the geometry of the X-ray monochromator setup. However, it is evident that any displacement of the reflections along the X-axis arising from a lateral periodicity of several hundred angstroms would at once be revealed in the contours. We therefore conclude that the three wave vectors are <u>collinear</u> along the Z-axis.

Microdensitomer scans along the Z-axis of a series of representative photographs for the above mixture are given in Figs. 7a-7f. Starting from the A_d phase at 110°C (Fig. 7a), we see a sharp peak at q_0'. On cooling, a second sharp peak is seen at q_0 close to q_0' (Fig. 7b). This signifies the onset of the A_{ic} phase. On further decrease of temperature the intensity of the reflection at q_0' decreases while that at q_0 increases with an accompanying increase in the intensity of the second harmonic at $2q_0$. The switchover of the relative strengths of the q_0' and q_0 reflections is clearly seen in Figs. 7c and 7d. Finally at 108°C the peak at q_0' has disappeared altogether (Fig. 7(f)) leaving a clear signature of the A_2 phase - strong reflections at q_0 and $2q_0$. It must be emphasized that regardless of their amplitudes, the sharpness of these reflections remains the same throughout at all temperatures.

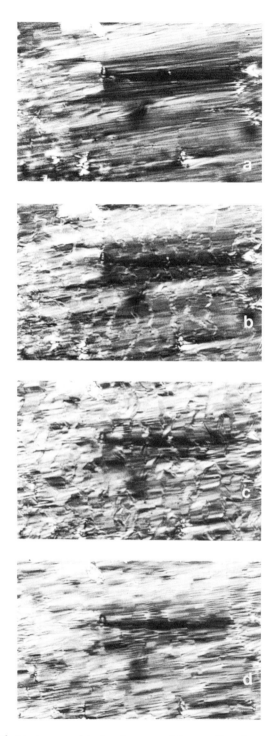

Fig. 3. Optical textures obtained on cooling a focal conic region of the A_d phase: (a) A_d phase, (b) at the A_d-A_{ic} transition, (c) A_{ic} phase and (d) A_2 phase.

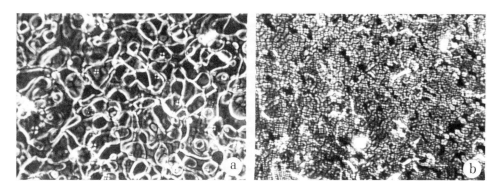

Fig. 4. Optical textures in the A_{ic} phase obtained (a) on cooling from a homeotropic A_d phase and (b) on heating from the A_2 phase.

It can be argued that the A_{ic} phase is a two-phase regime consisting of A_d and A_2 regions, each giving rise to its own X-ray diffraction pattern. In such a case let us consider a tie line MN (as shown in Fig. 8a) – a constant temperature line parallel to the concentration axis and connecting the A_d-A_{ic} and A_{ic}-A_2 boundaries (marked 1 and 2 in the figure). It is well known [6] that for all concentrations along such a

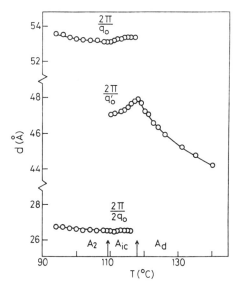

Fig. 5. Thermal variation of the layer spacing (d) in the A_d, A_{ic} and A_2 phases of the $X = 34.8\%$ mixture.

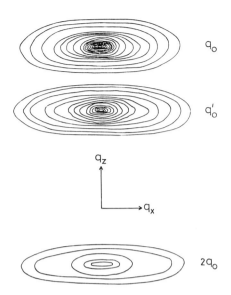

q_z

q_x

Fig. 6.   Intensity contour map of an X-ray diffraction photograph taken
          for the X = 34.8 mole % mixture at 115.5°C. Widths of the
          spots are discussed in the text. The spot at $2q_0$ has been
          displaced closer to the other two spots for convenience.

tie line the layer spacing corresponding to the A_d and A_2 modulations is
given by the values at M and N respectively. In other words, for any concentration the temperature variation of the layer spacing in the two-phase
region is uniquely determined by the variations along the boundaries 1 and
2. Using the experimental data the expected layer spacing variations, if
the A_{ic} phase were to be a two-phase region can be evaluated and these are
shown in Fig. 8(b). It is clear that for every concentration, the data
for $2\pi/q_0'$ in the two-phase region should fall on the same curve. This
argument holds for $2\pi/q_0$ also. On the contrary, the experimental data
shown in Fig. 9 for five concentrations in the A_d and A_{ic} phases, show
that at any temperature in the A_{ic} phase there is clear change in the layer
spacing with concentration. In fact, we find that the layer spacing corresponding to the q_0 modulation at any temperature in the A_{ic} phase is
higher for a higher concentration of the shorter molecular species, viz. -
80CB. Thus the possibility of the A_{ic} phase being a two-phase region is
ruled out.

        We present the differential scanning calorimetry runs for pure DB70CN
in Fig. 10 and for mixtures of it with 80CB in Fig. 11 after normalizing to
unit weight of the sample so that they can all be directly compared. All
the exotherms were recorded at 0.5°C/min cooling rate and 0.5 mcal/sec
sensitivity. The A_d-A_2 transition shows a rapid decrease in the strength
of the signal with increasing 80CB concentration (Figs. 11a-11c). In fact,

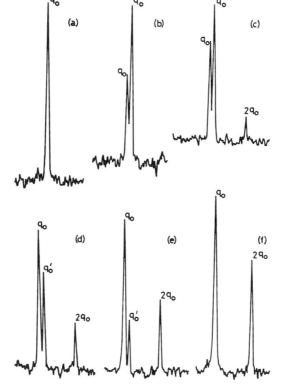

Fig. 7. Microdensitometer scans of the X-ray diffraction photographs
taken along the Z-axis for the X - 34.8 mole % mixture at
(a) 119°C in the A_d phase, (b)-(e) at 117, 116, 114.5 and
112°C in the A_{ic} phase and (f) at 108°C in the A_2 phase.

at 20% which is only a few percent away from the concentration at which
A_{ic} makes its appearance, the transition is seen as a barely perceptible
baseline change. Even in the region where A_{ic} phase exists (see Figs. 11d-
11e), no signals were observed corresponding to the A_d-A_{ic} and A_{ic}-A_2 transitions.

The static dielectric constants, $\varepsilon_\parallel$ and $\varepsilon_\perp$, as well as the dielectric
anisotropy $\Delta\varepsilon = \varepsilon_\parallel - \varepsilon_\perp$ are plotted in Fig. 12 for the 30 mole % of 80CB mixture. At the A_d-A_{ic} transition, $\varepsilon_\parallel$ starts decreasing at a faster rate
while $\varepsilon_\perp$ shows a still stronger increase. This can be correlated to the
appearance of the bilayer modulation at this transition. In the A_2 phase
both $\varepsilon_\parallel$ and $\varepsilon_\perp$ tend to stabilize since in this phase the lock-in of the dipoles is almost complete.

Prost [7,4] developed a phenomenological model to understand the different smectic phases that arise as a result of a competition between two
incommensurate lengths, víz. the length of the molecule (ℓ) and the length
of the antiparallel pair (ℓ'). The model predicts two incommensurate
phases depending on the ratio ℓ'/ℓ; A_{i1} when ℓ'/ℓ is close to 1 and A_{i2}

265

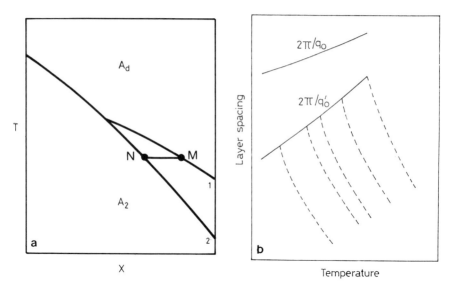

Fig. 8.   a. Schematic T-X diagram of the A_d-A_{ic} and A_{ic}-A_2 phase boundaries
(marked 1 and 2 respectively) showing a 'tie-line' which
intersects these boundaries at M and N.

   b. Schematic representation of the expected temperature variation of the layer spacings (solid curves) for different
concentrations if the A_{ic} phase were a two-phase region
consisting of A_2 and A_d. The dashed curves show the experimental variation in the A_d phase.

when ℓ'/ℓ is close to 2. The incommensurate phase observed by us is of
the A_{i2} kind. The only other observation of two coexistent incommensurate
density modulations has been in 4-octyl-4'-cyanoterphenyl[8], in a three-
dimensionally ordered smectic E phase. Ours is the first observation of
an <u>incommensurate fluid phase</u>.

According to Prost's theory, the competition between the elastic and
lock-in terms in the free energy can lead to an incommensurate phase. When
the elastic term barely wins over the lock-in term, the A_{ic} phase consists
of regions of A_2 modulation periodically separated by phase solitons. When
the lock-in term decreases further, i.e., in the weak coupling limit, the
two modulations coexist percolating through each other. The lack of any
satellite reflections in the A_{ic} phase is probably indicative of the weak
coupling between the two modulations. Recently Barois [9] has computed a
theoretical phase diagram by including fourth order terms in the free
energy expression of Prost. The stability of the incommensurate phase
depends on the coefficients of the fourth order terms. He showed that the
phase diagram (Fig. 13) when represented in the temperature-incommensur-
ability (t-z) plane (the incommensurability parameter z is a measure of

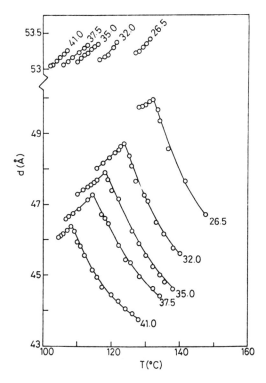

Fig. 9. Experimental variation of the layer spacing with temperature for different concentrations of 80CB in the A_d and A_{ic} phases.

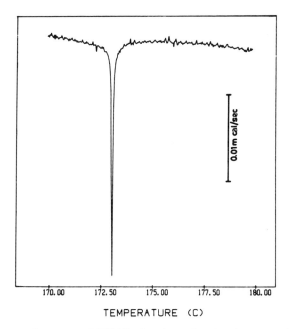

Fig. 10. DSC scan for pure DB70CN showing the A_d-A_2 transition, taken in the cooling mode at 0.5°C/min rate.

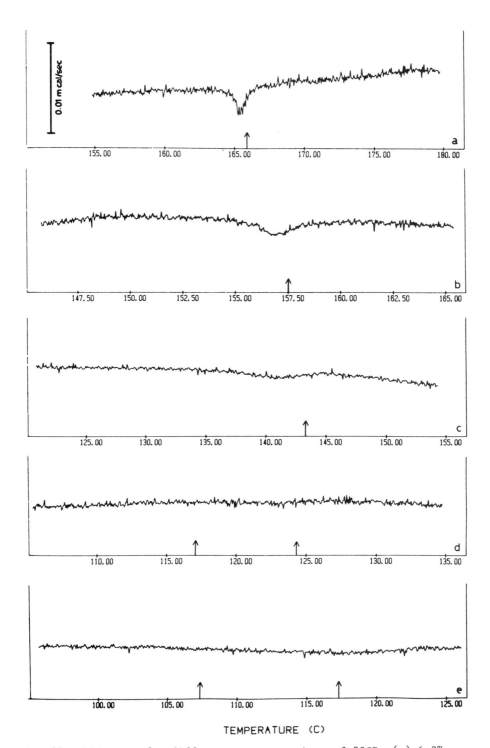

Fig. 11. DSC scans for different concentrations of 80CB: (a) 6.3%,
(b) 12% and (c) 20% exhibit only the A_d-A_2 transition; (d) 31%
and (e) 35.2% have A_d,A_{ic} and A_2 phases. The transition
temperatures are shown by the arrows. All the runs were
taken at 0.5°C/min cooling rate.

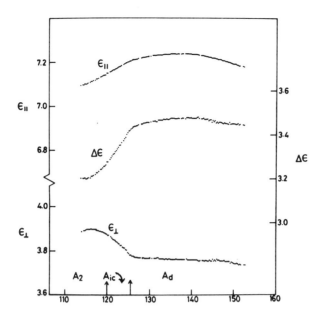

Fig. 12. Temperature variation of the static dielectric constants and and the dielectric anisotropy $(\Delta\varepsilon = \varepsilon_{\parallel} - \varepsilon_{\perp})$ in the A_d, A_{ic} and A_2 phases for 80CB concentration of 30 mole%.

the mismatch between q_o and q_o') is fully compatible with our experimental topology in the T-X plane (see Fig. 1). Finally, we would like to mention that we have observed the A_{ic} phase in a few other binary systems also, which we will be reporting in a separate publication.

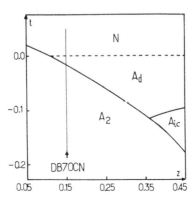

Fig. 13. Theoretical plot of the temperature (5) versus incommensurability (z) taken from Ref. 9.

269

4. ACKNOWLEDGMENT

We are highly indebted to Professor S. Chandrasekhar for many important suggestions and valuable discussions.

5. REFERENCES

1. B. R. Ratna, R. Shashidhar and V. N. Raja, Phys. Rev. Lett. $\underline{55}$, 1476 (1985).
2. See e.g., F. Hardouin, A. M. Levelut, M. F. Achard and G. Sigaud, J. Chim. Phys. $\underline{80}$, 53 (1983).
3. G. Sigaud, F. Hardouin, M. F. Achard and A. M. Levelut, J. Physique $\underline{42}$, 107 (1981).
4. J. Prost and P. Barois, J. Chim. Phys. $\underline{80}$, 65 (1983).
5. C. S. Rosenblatt, R. Pindak, N. A. Clark and R. B. Mayer, J. Physique $\underline{38}$, 1105 (1977).
6. See e.g., C. S. Barrett and T. B. Massalski, "Structure of Metals," McGraw-Hill Series Material Science and Engineering, 1966.
7. J. Prost, "One and Two-dimensional Order," W. Helfrich and G. Heppke, eds. (Springer Verlag, Berlin, New York, 1980), p. 125.
8. G. J. Brownsey and A. J. Leadbetter, Phys. Rev. Lett., $\underline{44}$, 1608 (1980).
9. P. Barois, Phys. Rev. A. $\underline{33}$, 3632 (1986).

THE NEMATIC AND SMECTIC-A$_1$ PHASES IN DB7NO$_2$: HIGH RESOLUTION X-RAY STUDY AND SYNTHESIS

C. R. Safinya and L. Y. Chiang

Corporate Research Science Laboratories
Exxon Research and Engineering Company
Annandale, NJ 08801

Abstract   We report a high resolution x-ray study of the fluctuations in the nematic and smectic-A$_1$ phases in DB7NO$_2$. As a function of decreasing temperature, the fluctuations evolve from an incommensurate region, through an intermediate region of coexisting (competing) incommensurate and smectic-$\tilde{C}$ fluctuations, to a region dominated by pretransitional smectic-$\tilde{A}$ fluctuations which exhibit a phase lock-in smectic $\tilde{A}_1$-smectic-$\tilde{A}$ transition. This behavior directly confirms the current theoretical description of the smectic-$\tilde{A}$ and smectic-$\tilde{C}$ phases as alternatives to the incommensurate phase. In DB7NO$_2$, although the average structure of the smectic-$\tilde{A}$ phase appears to be consistent with earlier findings, high resolution x-ray work reveals additional weaker (possibly satellite) peaks indicating a more complicated under-lying structure. Finally, we outline a novel new synthetic route for preparation of high purity samples of DB7NO$_2$ in large quantities.

1.   INTRODUCTION

The smectic-A(SmA) phase of liquid crystals consists of oriented molecules segregated into stacks of two-dimensional liquid layers with spacing d. Recently, considerable experimental [1-3] and theoretical [4-5] effort has been concentrated on understanding the nature of the phases and transitions of an entirely new class of polar liquid crystals with compe-tition between two different layer spacings in SmA phases. In the SmA$_1$ (monolayer) phase, the polar molecules are randomly up and down in each layer and the one-dimensional density wave has periodicity $d_2 = \ell$ = molecu-lar length. The SmA$_2$ bilayer (antiferroelectric) phase consists of layers in which the molecules are preferentially up or down and the density wave has period $d_1=2\ell$ commensurate with the molecular length. Previous [1] low resolution x-ray work has also revealed novel $Sm\tilde{A}$ and $Sm\tilde{C}$ antiphase structures in which the direction of the dipolar molecule is modulated in the plane of the layers (Top; Figs. 2 and 4).

2. THEORY

To describe these phases within a unified framework, Prost and Barois [4] (P-B) proposed a phenomenological model with Ψ_2 and Ψ_1, characterizing the density and the antiferroelectric order parameters. As essential ingredients, the model incorporates two types of important terms. First, the elastic terms, $|(\nabla^2 + k_1^2)\ \Psi_1|^2$ and $|(\nabla^2 + k_2^2)\ \Psi_2|^2$, which describes spatial modulations with preference to order at incommensurate wave vector moduli $|\vec{k}_1| = 2\pi/\ell'$ ($\ell' < 2\ell$) and $|\vec{k}_2| = 2\pi/\ell$ and second, the coupling term $\Psi_1^2\Psi_2^*$ which favors the lock-in (commensurability) of $\vec{k}_1$ and $\vec{k}_2$. These opposing tendencies lead to frustration and the system responds with the formation of various SmA phases. In the model, for one-dimensional modulations, either a commensurate phase is stabilized when the lock-in term dominates, or an incommensurate (I) phase when the elastic terms dominate. To date, however, while the SmA_2 commensurate phase has been observed in numerous systems [1], there is only one very recent report [3] of the possible existence of an ordered (I) phase in a binary mixture in which two density waves with incommensurate period coexist. To explain this scarcity, Prost and Barois hypothesize that in most systems, the smectic antiphases ($\tilde{A}$ and $\tilde{C}$) will be preferred to the incommensurate phase because the phases compromise between the elastic and commensurability energies. For example, in the $Sm\tilde{A}$ phase, the antiferroelectric incommensurate modulation at $\vec{k}_1$ tilts away from the z-axis (developing an in-the-layer component $\vec{k}_1$ = $k_{1z} \hat{z} + k_{1\perp} \hat{\imath}$; (Fig. 2 (top)) with k_{1z} locked ($2k_{1z} = 2q_0$) onto the monolayer modulation at $\vec{k}_2 = 2q_0\hat{z}$. This configuration simultaneously minimizes the commensurability energy and preserves modulus $|\vec{k}_1|$ favored by the elastic terms. Wang and Lubensky [5] (W-L) studied the density fluctuations in the monolayer SmA_1 phase using the (P-B) model as a starting point. Significantly, they predict that the competition between the incommensurate (I) phase and the $Sm\tilde{A}$ antiphase may be directly observable as a region of coexisting (I) and $Sm\tilde{A}$ fluctuations.

To elucidate the problem and understand the nature of the onset of antiphase ordering, we carried out a high resolution x-ray study of the density fluctuations in the nematic (N) and SmA1 phases (T_{NA_1} = 96.90°C) associated with the SmA_1-$Sm\tilde{A}$ transition ($T_{A_1\tilde{A}}$ = 92.21°C) in the compound 4-heptylphenyl 4-(4-nitrobenzoyloxy) benzoate (1) (DB7NO$_2$). Our most

272

important result is that the crossover from incommensurate (I) to SmÃ
antiphase fluctuations in DB7NO$_2$, does in fact occur through a region of
coexisting incommensurate and antiphase fluctuations directly confirming
the competition between these fluctuations.   This then strongly supports
the Prost-Barois hypothesis that the smectic antiphases are alternatives to
the (I) phase.

3. EXPERIMENT

The compound of 4-heptylphenyl 4-(4-nitrobenzoyloxy) benzoate ($\underset{\sim}{1}$)
used in this study was synthesized and purified by a new procedure as
illustrated in the scheme in Table 1.   Upon treatment of p-hydroxybenzoic
acid with 2.3 equivalents of sodium hydride in DMF, the initially formed
disodium salt was allowed to react with benzyl chloride to afford benzyl
4-benzyloxybenzoate as an intermediate.   Hydrolysis of this benzoate in
alkaline solution gave 4-benzyloxybenzoic acid ($\underset{\sim}{2}$) and p-heptyl with
N-(N,N'-dimethylaminopropyl)-N'-ethyl-carbodiimide hydrochloride in the
presence of 1-hydroxybenzotriazole and triethylamine in DMF solution.   The
coupled product of ester $\underset{\sim}{3}$ was isolated in 69% yield.   Catalytic hydroge-
nation of 3 in the presence of 5% Pd/C in acetic acid gave a 90% yield of
heptylphenyl 4-hydroxybenzoate ($\underset{\sim}{4}$).   Treatment of compound 4 with p-nitro-
benzoyl chloride in the presence of triethylamine afforded DB7NO$_2$ ($\underset{\sim}{1}$) in
88% yield.

Table 1

Scheme

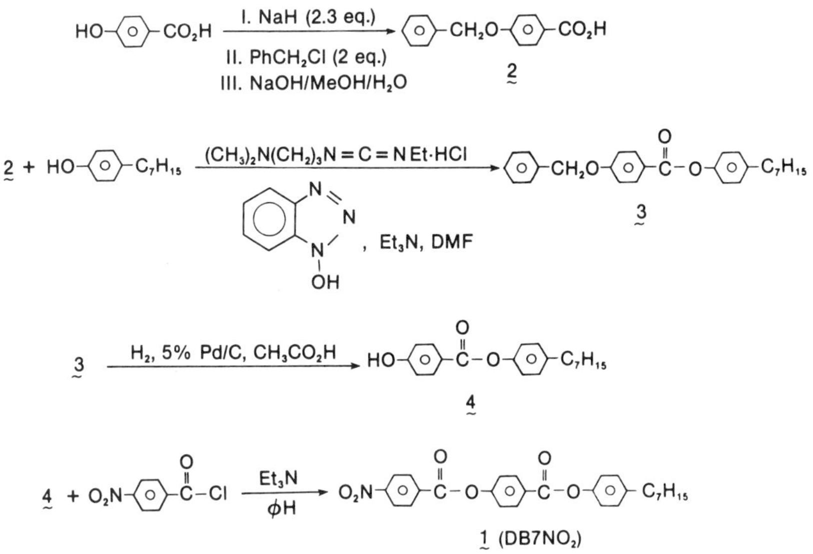

$\underset{\sim}{1}$ (DB7NO$_2$)

The experiments were carried out on a rotating anode x-ray source using Ge(111) crystals as monochromator and analyzer elements. The resolution expressed in terms of the half width at half maximum (HWHM) was 4.2×10^{-4} Å^{-1} in the longitudinal direction $< 2 \times 10^{-5}$ Å^{-1} in the transverse in-plane and 5×10^{-3} Å^{-1} in the transverse out-of-plane direction. An applied magnetic field of 3.0 kG provided alignment of the nematic director in the scattering plane. A two-stage oven provided temperature control of ± 2.5 mK.

We begin with the behavior in the nematic phase. The critical scattering associated with the onset of a one-dimensional density wave with monolayer periodicity at the nematic-SmA$_1$ transition is centered about $\vec{k}_2$ $= (0, 0, 2q_0 = 2\pi/\ell = 0.229\text{Å}^{-1})$ in reciprocal space. Simultaneously, a second incommensurate density fluctuation with period less than the anti-parallel pair length is centered about $\vec{k}_1 = (0, 0, k_1 = (1 + \varepsilon_1)q_0)$ with the incommensurability parameter ε_1 decreasing from 0.23 to 0.16 in the N phase. The scattering is shown schematically at the top of Fig. 1 with the nematic director along $\hat{z}$. For each temperature, we carried out a complete series of longitudinal ($q_\parallel$ varied, $\parallel = \hat{z}$) and transverse ($q_\perp$ varied) scans mapping out the entire scattering plane in the vicinity of $\vec{k}_1$ and $\vec{k}_2$. We show in Fig. 1A longitudinal scans in the nematic phase through

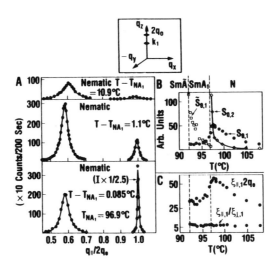

Fig. 1. Top Panel: Nematic phase scattering spectrum in reciprocal space with diffuse spots at incommensurate $\vec{k}$ and $\vec{k}_2 = 2q_0\hat{z}$. (A) longitudinal scans through $\vec{k}_1$ and $\vec{k}_2$. The solid lines are the results of fits of Eq. (1) to the data. (B) Susceptibilities for the monolayer $S_{0,2}$, incommensurate $S_{0,1}$ and antiphase $\tilde{S}_{0,1}$ fluctuations. (C) Correlation length $\varepsilon_{\parallel,1}$ and the ratio $\varepsilon_{\parallel,1}/\varepsilon_{\perp,1}$ for the incommensurate fluctuation.

$(0, 0, k_1)$ and $(0, 0, k_2)$ at three temperatures $T - T_{NA_1} = 10.90°C$, $1.10°C$, and $0.085°C$ ($T_{NA_1} = 96.90°C$) as the SmA_1 phase is approached. Quantitatively the scattering may be described by an x-ray structure factor [6] centered at $(0, 0, k_1)$ and $(0, 0, k_2)$:

$$S(\vec{q}) = \sum_{i=1}^{2} \frac{S_{o,i}}{1 + \varepsilon_{\parallel,i}^2 (q_\parallel - k_i)^2 + \varepsilon_{\perp,i}^2 q_\perp^2 + d\varepsilon_\perp^4 q_\perp^4} \tag{1}$$

The solid lines through the data of Fig 1A are the results of a fit of Eq. (1) convoluted with the instrumental resolution, to longitudinal and transverse scans through $\vec{k}_1$ and $\vec{k}_2$. This yields susceptibilities $S_{o,i}$ and correlation lengths $\xi_{\parallel,i}$, $\xi_{\perp,i}$ for the incommensurate (i=1) and monolayer (i=2) density fluctuations. At the $N=SmA_1$ transition, the monolayer density wave at $\vec{k}_2$ condenses with diverging susceptibility $S_{o,2}$ (Fig. 1B) and correlation lengths similar to previous N-SmA [6] transitions studies. Very importantly, the incommensurate density wave at $\vec{k}_1$ exhibits significant temperature dependence in the nematic phase (Fig. 1A) with the susceptibility $S_{o,1}$ and correlation lengths $\xi_{\parallel,1}$, $\xi_{\perp,1}$ exhibiting a broad maximum about one degree above the NA_1 transition as plotted in Fig. 1B and 1C. Thus, as the temperature is decreased in the N phase, this behavior at $\vec{k}_1$ which indicates the approach towards an ordered incommensurate phase, is preempted (presumably because of the cost in lock-in energy), just above the SmA_1 phase by a crossover to a region of coexisting incommensurate and anti-phase fluctuations. Before describing this region, we first briefly discuss the limiting $Sm\tilde{A}$ fluctuation behavior near the SmA_1-$Sm\tilde{A}$ transition.

In the SmA_1 phase, the scattering exhibits a Bragg spot at $\vec{k}_2 = (0, 0, 2q_0)$ and diffuse scattering near q_0. In the vicinity of the SmA_1-$Sm\tilde{A}$ transition, the diffuse scattering is dominated by $Sm\tilde{A}$ fluctuations. We show schematically at the top of Fig. 2 the real space centered rectangular $Sm\tilde{A}$ unit cell [1]. The $Sm\tilde{A}$ fluctuation in the SmA_1 phase manifests itself as a ring of scattering at $\vec{k}_{\tilde{A}} = (k_\perp \cos\psi, k_1 \sin\psi, k_\parallel)$, $0<\psi<2\pi$ in reciprocal space. Shown in Fig. 2 are the results of four (series of $q_\parallel$ and $q_\perp$) mesh scans near the bilayer wave vector q_0 plotted as equal intensity contours in the nematic phase (A) and in the SmA_1 phase (B, C, D). For symmetry reasons we show only $q_\parallel > 0$ and $q_\perp > 0$. We see that the single spot pattern characteristic of incommensurate fluctuations at $\vec{k}_1 = (0,0, k_1 = (1 + \varepsilon_1)q_0; \varepsilon_1 \sim 0.2)$ in the nematic phase (A) crosses over to a single ring pattern indicative of pretransitional $Sm\tilde{A}$ fluctuations near the SmA_1-$Sm\tilde{A}$ transition [2D] at $k_{\tilde{A}} = (k_\perp, 0, k_\parallel = q_0 (1 + \varepsilon))$ with very small incommensurability $\varepsilon \sim 0.01$.

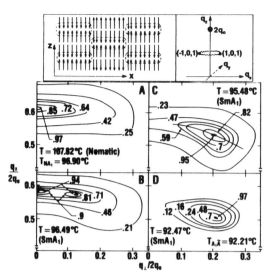

Fig. 2. Top Panel (left) Centered rectangular unit cell associated with
short range antiferroelectric Sm-Ã order (arrows indicate polar
end) and (right) resulting diffuse ring in reciprocal space in
the SmA₁ phase. Lower panel (A, B, C, D): Contour plots showing
the evolution of scattering near the bilayer q_o position in the
nematic and smectic-A₁ phases as discussed in the text.

The evolution of the fluctuations from incommensurate to SmA behavior
occurs through an intermediate region as we now discuss. At the incommensurate susceptibility $S_{o,1}$ decreases (Fig. 1B), antiphase fluctuations set
in near the N-SmA₁ transition. We show in Fig. 2B the contour plot of
<u>coexisting incommensurate and antiphase fluctuations</u> at T = 96.49°C = T_{NA}
- 0.41°C, where the scattering into the spot and antiphase ring are comparable. At this temperature, the precise nature of the antiphase fluctuation is masked by the presence of strong incommensurate scattering. As the
temperature is further decreased, we see that at T = 95.48°C = T_{NA_1} -
1.42°C, the contour plot of Fig. 2C shows the ring pattern clearly. Qualitatively, two features become apparent. First, the fluctuations in the
SmA₁ phase appear to occur on a constant $|\vec{q}|$ radius in reciprocal space
(solid·radial line, Fig. 2C). Second the contour lines appear asymmetri-
cally about this radial line; that is, the distance from the peak position
to equal intensity contours above and below this line along $q_∥$ is not
equal. A single ring pattern which is the behavior near the SmA₁-SmÃ transition shown in Fig. 2D, does not exhibit this asymmetry. This second
feature is elucidated in Fig. 3 where we show longitudinal ($q_∥$) scans
through the peak $q_⊥/2q_o$ = 0.2 and center $q_⊥$ = 0 of the ring pattern of
Fig. 2C. The scan through the ring maximum (Fig. 3B) reveals a double
peak structure (arrows mark the peak positions). In this temperature range
the antiphase fluctuation is manifest as two closely spaced rings which is
characteristics of SmC [4,5] fluctuations. We show schematically at the

276

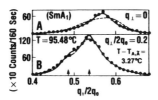

Fig. 3. Longitudal scans through the center ($q_\perp = 0$) and peak ($q_\perp/2q_0 = 0.2$) of the ring pattern of Fig. 2C. The dashed lines are fits to Eq. (2) and the solid lines are fits to a sum of Eq. (2) and Eq. (1) (for i=1) (see text).

top of Fig. 4 the real space unit cell of the oblique centered $Sm\tilde{C}$ lattice which can be thought of simply as a "sheared $Sm\tilde{A}$." This structure gives rise to $(1,0,1)$ and $(-1,0,1)$ reflections with unequal wave vector moduli, which when azimuthally averaged gives rise to two rings in reciprocal space (Top, Fig. 4). The double peaked $Sm\tilde{C}$ scattering may be quantitatively described by the sum of two rings

$$S(\vec{q}) = \sum_{i=1}^{2} \frac{\tilde{S}_{0,i}}{1 + \xi_\parallel^2 q_{\parallel,i}'^2 + \xi_\perp^2 q_\perp'^2 + d_\parallel q_{\parallel,i}'^4 + d_\perp q_{\perp,i}'^4 + c q_{\parallel,i}' \dot{q}_\perp'}$$

(2)

with $q_{\parallel,i}' = (q_\parallel - k_{\parallel,i})$, $q_\perp' = (q_\perp - k_\perp)$ and where $k_{\parallel,1}$ and $k_{\parallel,2} \equiv k_{\parallel,1} - \Delta$ and $\tilde{S}_{0,1}$, $\tilde{S}_{0,2}$ are the longitudinal peak positions, and the amplitudes respectively, of the upper and lower ring ($\vec{k}(101)$ and $\vec{k}(-101)$ reflections shown schematically at top of Fig. 4) separated by Δ. The data require nonnegligible quartic terms. The lowest order cross term is needed to characterize the approximately radial nature of the fluctuations (Fig. 2C, 2D). This finding supports one of the important assumptions of the (P-B) and (W-L) models which introduce elastic terms of the type $|(\nabla^2 + k_1^2)\psi_1|^2$ to indicate the preference of ψ_1 to order at $|\vec{k}_1|$; these terms, in turn, lead to the prediction [5] of the radial nature of $S(\vec{q}) \sim \langle \psi_1(\vec{q}) \psi_1(-\vec{q}) \rangle$. Equation (2) which characterizes the double ring scattering cannot simultaneously describe the scattering away from the ring maxima near the incommensurate peak at $\vec{k}_1 = (0,0,k_1)$. This is quantitatively evident from Fig. 3 where the dashed line is a fit of Eq. (2) to simultaneous longitudinal scans around the ring maximum at $q_\perp/2q_0 = 0.2$ which describes the double ring accurately (3B) but not the scattering through $q_\perp = 0$ (3A) which is significantly narrower. The solid line is a best fit ($\chi^2 \sim 2$) of $q_\parallel$ and $q_\perp$ scans spanning the entire fluctuation range to a sum of Eq. (2) and Eq. (1) (for i = 1) characterizing an intermediate region of coexisting SmC and incommensurate (I) fluctuations. (I) fluctuations appear to persist even near the SmA_1-$Sm\tilde{A}$ transition.

277

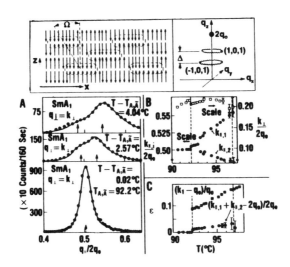

Fig. 4. Top panel: (left) centered oblique unit cell (arrows indicate polr
end) associated with short range Sm-C̃ order and (right) resulting
diffuse double ring in reciprocal space in the SmA₁ phase.
(A) Longitudinal scans through the peak of the double ring Sm-C̃
(top and center) and single ring Sm-Ã (bottom) diffuse scattering
profile. Solid lines are result of fits (see text). (B) Longi-
tudinal ($k_{\parallel,1}$ and $k_{\parallel,2}$) and transverse ($k_\perp$) peak positions of
the double ring (Sm-C̃) followed by single ring (SmÃ) scattering.
(C) Incommensurability parameter for the incommensurate (solid
square) and antiphase SmC̃ and SmÃ (solid circle fluctuations).
The vertical dashed lines give the N-SmA₁ and SmA₁-SmÃ phase
boundaries.

As the temperature is decreased in the SmA₁ phase, we find that the
SmC̃ fluctuations evolve continuously into SmÃ fluctuations. We show in
Fig. 4A longitudinal scans through the double ring maximum at $q_\perp = k_\perp$ which
indicate that the separation $\Delta = k_{\parallel,1} - k_{\parallel,2}$ between the longitudinal peak
positions (marked by arrows) decreases with an eventual collapse ($\Delta = 0$)
of the rings above the weakly first order SmA₁-SmÃ transition. We plot
the peak positions $k_\perp, k_{\parallel,1}$ and $k_{\parallel,2}$ of the antiphase scattering as a function of temperature in Fig. 4B which corresponds to SmC̃ fluctuations with
an oblique angle $\Omega = \tan^{-1}\Delta/2k_\perp$ (top, Fig. 4) that continuously shear to
the SmÃ ($\Delta,\Omega = 0$) configuration (top, Fig. 2) about 0.75°C above the SmA
phase. In the (P-B) model the coupling term $\sim\psi_1^2\psi_2^*$ leads to a commensurability energy which is minimum for $\vec{k}_2(= 2q_0\hat{z}) = \vec{k}(101) + \vec{k}(-101)$. The SmÃ
represents the symmetric case when $|\vec{k}(101)| = |\vec{k}(-101)|$ (top, fig. 2).
For the SmC̃, $|\vec{k}(101)| \neq |\vec{k}(-101)|$ (top Fig. 4). <u>Significantly we find that the
incommensurability parameter</u> $\varepsilon = (k_{\parallel,1} + k_{\parallel,2}-2q_0)/2q_0$ <u>plotted in Fig. 4C
associated with antiphase SmC̃ and SmÃ fluctuations is small (<0.02) over
the entire SmA₁ phase and locks-in ($\varepsilon = 0$) across the SmA₁-SmÃ transition.</u>
We also show $\varepsilon_1(I) = (2k_1 - 2q_0)/2q_0 \gtrsim 0.10$ associated with incommensurate
fluctuations at $\vec{k}_1$ which are large only in the nematic phase. Thus,

278

antiphase fluctuations, which set in this intermediate region, occur with $\varepsilon \ll \varepsilon_1$ reflecting the importance of the coupling term in the (P-B) and (W-L) models.

To summarize the behavior in the nematic and smectic-A_1 phase we find that the approach of incommensurate (I) fluctuations with large incommensurability $\varepsilon_1 > 0.16$ towards ordering in the nematic phase, is preempted by a crossover, through an intermediate region of coexisting (I) and $\widetilde{SmC}$ anti-phase fluctuations, to a region dominated by $\widetilde{SmA}$ antiphase fluctuations with very small incommensurability $\varepsilon \sim 0.01$ and which exhibits a lock-in SmA_1-$\widetilde{SmA}$ transition. This intermediate region directly confirms the competition between (I) and $\widetilde{SmC}$ antiphase fluctuations and strongly supports the (P-B) hypothesis that the Smectic Antiphases are alternative to the (I) phase. Although the present (W-L) model predicts a coexisting region of (I) and $\widetilde{SmA}$ fluctuation rather than (I) and $\widetilde{SmC}$ which is what is observed in DB7NO$_2$, a more complete analysis which goes beyond the harmonic approximation should yield either types of antiphase fluctuations in the coexistence region.

Before concluding, we discuss the scattering in the lower temperature smectic($\widetilde{A}$) phase in DB7NO$_2$. We show in Fig. 5 the results of mesh scans in the $(q_{\parallel},0,q_{\perp})$ plane plotted as equal intensity contours at 91.65°C in the smectic($\widetilde{A}$) phase. This contour plot should be compared to that shown in Fig. 2D which was data taken just above the smectic-A_1smectic-$\widetilde{A}$ phase transition. The first immediate difference is that the primary peak at $(q_{\parallel} = q_o, 0, q_{\perp} = 0.41q_o)$ is extremely sharp in the smectic($\widetilde{A}$) phase as compared to the antiphase fluctuations in the smectic(A_1) phase. This indicates that the antiphase modulation shown schematically in the top of Fig. 2 is quite long range at least on the order of several microns. However, in addition to the primary peak, our high resolution capability has revealed weaker satellite peaks with intensities about 1/30 of that of the primary peak. As is clear from Fig. 5, the extra peaks have no obvious symmetry associated with them. We point out that the Bragg peak at $q_{\parallel} = 2q_o$ evident in the smectic-A_1 phase (shown schematically in the top of Fig. 2) remains unchanged in the smectic($\widetilde{A}$) phase with no evidence of any additional peaks. However, we find significant structureless diffuse scattering near the Bragg peak at $q_{\parallel} = 2q_o$. One possible explanation for the extra weaker peaks in the lower temperature phase is that the sample may consist of coexisting smectic($\widetilde{A}$) and smectic($\widetilde{C}$) phase regions. The additional weaker peaks can be indexed on an oblique smectic(C) lattice if one assumes that the oblique angle is not the same through the smectic($\widetilde{C}$) regions but may actually have certain preferred discrete values. For example, a pair of weaker peaks (see Fig. 5) may be indexed as the primary

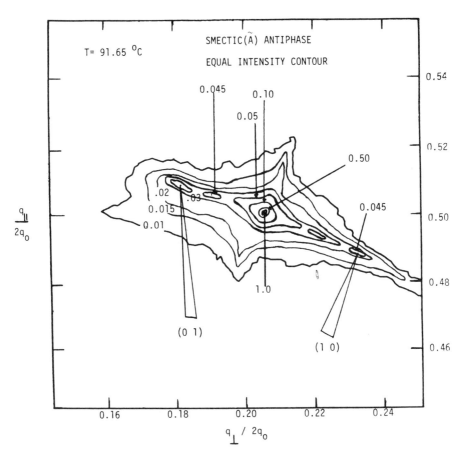

Fig. 5. Equal intensity contour plot of the scattering near the off-axis peak position ($q_\parallel = q_0$, $0, q_\perp = 0.4q_0$) (that is the (1,0,1) peak shown schematically in the top of Fig. 2) in the smectic (A) antiphase. In addition to the primary intense antiphase peak, weaker satellite peaks are also evident.

(1,0) and (0,1) peaks of a two-dimensional smectic($\tilde{C}$) structure (following the identical notation used by A. M. Levelut in reference [1]). More work is clearly required to unambiguously determine the precise structure in this phase.

We wish to thank S. Alexander, R. J. Birgeneau, N. A. Clark, C. C. Huang, and R. Pindak for stimulating discussions. One of us (C.R.S.) is particularly grateful to T. C. Lubensky and J. Prost for extensive conversations.

4. REFERENCES

1. A. M. Hardouin, M. F. Levelut, G. Archard, and G. Sigaud, J. Chim. Phys. 80, 53 (1983); C. C. Huang, S. C. Lien, S. Dumrograttana, and L. Y. Chiang, Phys. Rev. A30, 965 (1984).
2. K. K. Chan, P. S. Pershan, L. B. Sorenson, and F. Hardouin, Phys. Rev. Lett. 54, 1694 (1985).

3. B. R. Ratna, R. Shashidhar, V. N. Raja, Phys. Rev. Lett. $\underline{55}$, 1476 (1985).
4. J. Prost and P. Barois, J. Chim. Phys. $\underline{80}$, 65 (1983); J. Prost, Advances in Physics, $\underline{33}$, 1-46 (1984).
5. J. Wang and T. C. Lubensky, J. Phys. $\underline{45}$, 1653 (1984).
6. R. J. Birgeneau, C. W. Garland, G. B. Kasting, and B. M. Oco, Phys. Rev. $\underline{A24}$, 2624 (1981).

X-RAY DIFFRACTION BY INCOMMENSURATE LIQUID CRYSTALS

A. M. Levelut

Laboratoire de Physique des Solides, Université Paris-Sud
91405 Orsay (France)

Abstract. The comparison between X-ray patterns of various liquid
crystalline phases can solve the problem of the existence of incommensurability in liquid crystals similar to the incommensurability in
crystals. Modulation occurs more easily over the liquid order than
over the remaining periodic molecular ordering which can be found in
smectic and columnar phases. Differences and similarities between the
behavior of modulated liquids and crystals are here reviewed.

1. INTRODUCTION

Among the liquid crystalline phases, the smectic A appears as the
simplest example of a 1D periodic fluid phase since this uniaxial liquid
undergoes a periodic density modulation parallel to its unique axis. This
modulation implies that the bulk phase has a layered structure. Generally, the periodicity of the modulation is of few nm and comparable to the
largest dimension of an extended molecule.

The diffraction patterns of smectics have been extensively studied
some years ago. Nevertheless, it appears that the simple representation
of the smectic A phase failed in some cases and especially when the
molecule has an intrinsic longitudinal dipole. This class of molecules
can show different values of the layer periodicity for the same composition, and therefore phase transitions involving a jump in the layer
thickness can occur in such systems. Moreover, it appears from the
analysis of the X-ray diffraction patterns that two periods can coexist in
a direction parallel to the director, at least with short coherence
length. The ratio between these two periods is not at all an integer or
a simple fractional number. Therefore, the term "discommensuration in

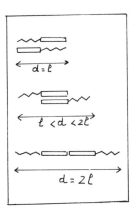

Fig. 1.   Three molecular associations of polar mesogens.   The layer
          periodicity d of the corresponding smectic A phase is compared
          to the molecular length ℓ.

smectic A" is used.   J. Prost [1] has developed a theoretical model of

the free energy of these systems and this model can describe the rich

polymorphism of smectic A involved by the existence of the discommensura-

tion.   The true 1D modulated smectic A phase predicted by the model has

not been yet put in evidence.   Nevertheless, the study of X-ray diffrac-

tion patterns gives evidence of short range modulation or of long range

modulation non-parallel to the director.   In this paper, we will discuss

the implication of the presence of such a modulation upon the diffraction

patterns of the liquid crystalline phases.

In the first part, we will discuss the X-ray pattern of a 1D modula-

ted layered liquid crystal.   The comparison between the experimental

results and the predicted pattern is developed in part II.   In the third

part, we will discuss the behavior of some modulated fluids.   Finally, in

the last part we will present modulated liquid crystals out from the frame

of the smectic structures.

2.   X-RAY DIFFRACTION BY A MODULATED SMECTIC A PHASE

Let us consider a lamellar system with two natural period L_1 and L_2

parallel to the normal Oz to the layer planes.   Since we speak of smectic

fluid phases, no translational periodic order takes place inside each

layer and we consider that modulations occur only along Oz.

If the two lamellar structures coexist in the same sample without

any coupling, the diffraction pattern will be the superposition of the two

sets of Bragg spots given by each periodicity.   In case of segregation of

the two structures, or by one set of Bragg spots corresponding to the mean

period in case of a complete ideal mixing of the two periods.

284

A coupling term between the two natural layer thicknesses will induce a modulated structure in which both the thickness and the nature of the layer can vary periodically. Let us look for the simplest form of modulation [2]. For the nth layer, the thickness is $L_n = c(1 + \varepsilon \sin \frac{2\pi nc}{M}$

Its coordinate

$$z_n = nc - \frac{M\varepsilon}{2\pi} \cos 2\pi nc \qquad (1)$$

and its form factor

$$F_n = F(1 + \eta \sin \frac{2\pi nc}{M}) \qquad (2)$$

c is the mean layer thickness and M the period of modulation; ε and η characterize respectively the amplitude of the displacement modulation and of the chemical modulation.

The diffracted amplitude is the Fourier transform F of the function $\sum\limits_{n} F_z \delta(z - z_n)$, where z_n is given by Eq. 1 and $F_z = F(1 + \eta \sin \frac{2\pi z}{M})$. The Fourier transform vanishes unless the wave vector q is parallel to the z axis.

$$\mathsf{F}(1 + \eta \sin \frac{2\pi z}{M}) = \delta q_z - i\eta/2 \, \delta(q_z - \frac{2\pi}{M}) + i\eta/2 \, \delta(q_z + \frac{2\pi}{M}) \qquad (3)$$

$$\mathsf{F}[\sum \delta(z - z_n)] = \sum_{-\infty}^{+\infty} 2\pi i^n \, \delta(q_z - \frac{2\pi \ell}{c} - \frac{2\pi m}{M}) \quad J_m(\frac{M\varepsilon q_z}{2\pi}) \qquad (4)$$

where J_m is the mth order Bessel function, m and ℓ are integers. Since ε and η are small, we keep only first order terms in ε and η, and therefore neglect high order Bessel functions $(m \geq 2)$ and cross term in $\varepsilon \eta$. The diffracted intensity is then measurable for $\ell \neq 0$, m = 0 and m = ± 1.

In the first case (fundamental Bragg reflexion),

$$I(\ell) = 4\pi^2 \, F^2 J_0^2 \, (\frac{\ell M\varepsilon}{c}) \qquad (5)$$

In the second case we have the first satellites ℓ, m = ± 1

$$I(\ell \mp 1) = 4\pi^2 F^2 \left\{ J_1 \left[M\varepsilon(\frac{\ell}{c} \mp \frac{1}{M}) \right] \pm \frac{1}{2} \eta \, J_0 \left[M\varepsilon(\frac{\ell}{c} \mp \frac{1}{M}) \right] \right\}^2 \qquad (6)$$

For m = 0, the intensity of the fundamental Bragg reflexions corresponding to the mean periodicity decrease with q_z. For m = ± 1, the first term is driven by the displacement modulation and its amplitude increases with q_z. The second one comes from the chemical modulation decrease with q_z. Moreover, the two satellites lying on each side of one

fundamental spot are of inequal intensity: the inner satellite is more
important if $\varepsilon/\eta > 0$. In our model, the high order satellites $|m| > 1$
have a very weak intensity. However, a correct description of the
diffraction pattern must take into account the higher harmonic components
of the modulation. These components can be derived from the term which
couples the two natural wavelengths of the system. Nevertheless, this
very crude model allows a qualitative description of the reciprocal
space: each fundamental Bragg spot is surrounded by satellite spots lying
on the same q_z axis. The distance between the first satellite and the
Bragg spot gives the period of the modulation. The ratio between the
intensity of two given spots of the same Brillouin zone varies smoothly
when going to the next zone.

In crystals if we except the case of alloys and non-stoichiometric
compounds, the modulation is purely displacive. In liquid crystals, the
existence of two periods in polar compounds corresponds generally to a
great change in the chemical content of the layer as it is shown in
Fig. 1, which represent schematically the three possible layer arrays
encountered in polar systems. Therefore, a large dissymmetry between the
two sides of a given fundamental Bragg spot is expected in LC system while
a 3D incommensurate crystal shows a complete symmetry of the pairs of
equivalent satellites. In Fig. 2, we give a schematic view of the
reciprocal space of a modulated smectic since the modulation is induced by
the presence of two natural lengths L_1 and L_2 of the system. The location
of the first fundamental Bragg peak $\ell = \pm 1$ and of the corresponding
satellite are close to the wave vector $q_z = \dfrac{2\pi}{L_1}$ and $\dfrac{2\pi}{L_2}$.

3. COMPARISON BETWEEN THE EXPERIMENTAL DIFFRACTION PATTERN AND THE
    PREDICTION

The main features of the diffraction pattern of a simple smectic A
phase are a set of one or two pairs of Bragg spots lying on the axis
parallel to the director and two broad crescents at $q \simeq 2\pi/4.5$ Å, the
maximum of which is in the plane perpendicular to the director. This

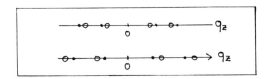

Fig. 2. X-ray diffraction pattern of a modulated smectic A phase. The
         fundamental Bragg peaks (open circles) are surrounded by
         asymmetric satellites (full circles). The visible satellites
         can be located either on the outer side or on the inner side
         of the main Bragg peaks.

286

outer diffraction ring is rather insensitive to the specificity of a given smectic A phase. Besides these main features one sees also maxima at large value of q_z (1.2Å^{-1}) originated from intramolecular interferences. In the following section, we will restrict our discussion to the small angle area surrounding the Bragg spots which characterize the layer structure. The Bragg peaks of SmA phase are surrounded by a weak diffuse scattering which originates from undulation modes of the layers. We admit that the width of the main peak is given by the resolution function of a conventional device. Near a nematic to smectic A transition in the nematic phase, the Bragg spot is replaced by a diffuse spot whose extension is related to the correlation length of smectic fluctuations parallel and perpendicular to the director.

With polar molecules the coexistence of the two layering periodicities will induce scattering intensity for q_z values distinct from those corresponding to the Bragg peaks. This intensity is not necessarily localized on the q_z axis but is maximum for small values of $q_\perp$ ($< 10^{-1}$Å^{-1}).

We can give a classification of the patterns from an analysis of the sequence of q_{nz} where q_{nz} is the z component of the scattering vector at the maximum of each peak (either δ function diffusive). A simple periodic system will be described by a sequence $q_{\ell z} = \dfrac{2\pi\ell}{c}$. Beside the simplest case, we have two kinds of complex sequences (Fig. 3).

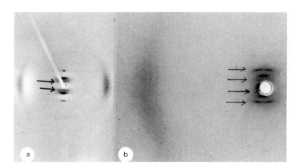

Fig. 3. Example of X-ray patterns of SmA phases in which diffuse spots coexist with the Bragg peaks:
    a) one sees a unique diffuse spot lying at $\left| q_z \right| = \dfrac{2\pi}{L_2}$ and two Bragg peaks at $q_{oz} = \dfrac{2\pi}{L_1}$ and $\dfrac{q_o}{2}$

    ($C_5H_{11}-\phi-CH=CH-\phi-O_2C-\phi-CN$ at 152°C)

    b) two diffuse spots lying at $q_z = \dfrac{2\pi}{L_2}$ and $q'_z = \dfrac{2\pi}{L_1} + \dfrac{2\pi}{L_2}$ beside a unique peak at $q_z = \dfrac{2\pi}{L_1}$

    ($C_{10}H_{21}-O-\phi-N=CH-\phi-O-CH_2-\phi-CN$ at 128.5°C)

    Note: the diffuse spots are pointed out by arrows.

- Beside a first set of q_ℓ, $q_\ell = \dfrac{2\pi\ell}{c}$ with $|\ell| < 2$, we observe a unique q'_z value $q'_z = \dfrac{2\pi}{c'}$. In such a case, it appears that the two periods are uncoupled ($\dfrac{c'}{c}$ is not a simple rational fraction).

- All the values of q_z can be described in a unique sequence

$$q_{nz} = \frac{2\pi\ell}{c} + \frac{2\pi m}{M} \; ; \; M/c \text{ is not an integer.} \quad |\ell| < 2 \quad |m| < 1$$

This second pattern can be related to the existence of modulation of a mean period c; the modulation period is M.

Let us give a detailed description of these two cases:

Uncoupled Periods

The coexistence of two sets of Bragg spots lying at $q_z = \dfrac{2\pi\ell}{c}$, $q_\perp = 0$ and $q'_z = \dfrac{2\pi m}{c'}$, $q_\perp = 0$ (where ℓ and m are integers, c and c' the two periods) have been seen in one case by Ratna et al. [3]. If no diffuse scattering surrounds these peaks, a segregation between two phases of different periods takes place at a scale larger than the inverse of the resolution width in the experience. Coexistence of one set of Bragg peaks at $q_z = \dfrac{2\pi\ell}{c}$ with a unique diffuse peak (or ring) at $q_z = \dfrac{2\pi}{c'}$ is seen in various smectic A phases of polar compounds mainly when the smectic A phase is unique [4] or when the two wavelengths of the two smectic phases become commensurate, i.e., transition between a monolayer paralectric smectics and a bilayer antiferroelectric one. Very often, the intensity-scattered in the plane $q_z = 2\pi/c'$ is not maximum for $q_\perp = 0$ but out of the q_z axis, in such a way that it is localized on a ring. This ring shape is indicative of a m plane ordering due to repulsive forces between small ordered zones of period c'.

Existence of Modulation Fluctuations

The transition between two smectic A phases of incommensurate periods is usually announced on the diffraction pattern by precursors. These precursors occur in case of direct transition and in case of transition occurring through intermediate phases such as the reentrant nematic. The precursor diffuse features lie at value of q_z specific of modulation fluctuations: if the Bragg spots are seen for $q_z = \dfrac{2\pi\ell}{c}$, $\ell = \pm 1$ ($\ell = \pm 2$), diffuse spots are seen for $|q'_z| = \dfrac{2\pi\ell}{c'}$ and $q_{z'} + q_z$ where c' is close to the second natural period of the system.

Therefore, the position of the diffuse scattering spots corresponds to the description of part 1 with $\dfrac{1}{M} = \left| \dfrac{1}{c} - \dfrac{1}{c'} \right|$. A lot of systems present such modulations, but in all cases, the maximum of intensity in the satellite reciprocal plane is out of the q_z axis. This fact implies an n-plane ordering of the modulated domains. Table 1 gives some characteristics of the diffraction pattern for two pure compounds. The first one has a reentrant sequence with two nematic and smectic A phases (4) and the modulated domains are present even in the reentrant nematic phase. In the second case, which corresponds to Fig. 3b, the SmA phase has a layer thickness equal to 1.7 times the molecular length. In general, when one of the two natural periods of the system is larger than 1.5 times the molecular length, the reentrant nematic phase is not seen and the coexistence of the two periods induces intra-layer modulations: a two-dimensional periodic structure in which double layers of antiferroelectric ordering are cut by parallel equidistant walls. Crossing the wall, the polarization of a single layer takes the reverse orientation (in other words, we have a 2D network of antiphase ribbons). The walls could be tilted with respect to the normal to the layer ($S_{m\tilde{C}}$) or normal to the layer ($S_{\tilde{A}}$). This last phase which has a centered rectangular lattice is obtained when the precursors lie near $q_z = \frac{1}{2} q_{oz}$, q_{oz} corresponding to the Sm A_1 monolayer ordering (Fig. 4). The occurrence of such phases with 2D network has been explained by J. Prost et al. [5] as an escape to the incommensurability. The

Table 1.   Evidence of Modulated Fluctuation in Pure Compounds

Sample Characteristic Features	$\frac{2\pi}{q_z}$ (Å)	ℓ	m	Intensity
$C_8H_{17}-CO_2-\phi-CO_2-\phi-C=C-\phi-CN$	41.3	1	0	Very strong
Nematic T - 150°C	31.8	1	1	Weak
c = 41.3Å	18.0	2	1	Very weak
M = 139.0Å	-	2	0	-
$C_8H_{17}-CO_2-\phi-CO_2-\phi-C=C-\phi-CN$	45.35	1	1	Weak
Smectic A T = 101°C	31.30	1	0	Very strong
c = 31.30Å	18.50	2	1	Very weak
M = 101.3Å	15.65	2	0	Very weak
$C_{10}H_{21}-\phi-N=CH-\phi-OCH_2-\phi-CN$	88.75	1	1	Very weak
Smectic A T = 128.5°C	50.90	1	0	Very strong
c = 50.9Å	32.40	2	1	Weak
M = 119.3Å	-	2	0	-

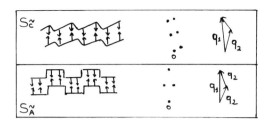

Fig. 4.   The structure (on left side) and the reciprocal space (on the
          middle) of the two $S_A^{\sim}$ and $S_C^{\sim}$ ribbon phases. On the right side,
          it is shown how a sum of wave vectors of length q_1 and q_2
          induces the $S_A^{\sim}$ and $S_C^{\sim}$ phases.

modulations of the liquid order of each layer are energetically favored by
comparison of a layer thickness modulation. The rule of sum over the wave
vector (Fig. 4) fits with the experimental diffraction patterns.

Since these modulated fluids occur in the polymorphism of smectics in
place of the incommensurate phases in the polymorphism of molecular
crystals, let us see if some similarities between these modulated phases
can be found.

3.   STUDY OF THE MODULATED FLUID PHASE

We focus our attention on two points:

- the transition from the paralectric phase to the modulated phase:
the evolution is similar for the S_A $S_A^{\sim}$ transition and the S_A $S_C^{\sim}$ one,

- the evolution of the modulated liquid towards an antiferroelectric
structure.

a.   Pretransitional Features of the Paraelectric S_A Phase

We have seen above that precursor diffuse spots (in fact, rings)
exist above the transition towards a 2D $S_A^{\sim}$ or $S_C^{\sim}$ phase. As we approach the
transition, the longitudinal component of the reciprocal vector which
characterizes the maximum of the diffuse spot goes smoothly to its equilibrium value for the low temperature phase. At the same time, the width of
the spot decreases. Figure 5 gives an example of contour lines of the
diffuse spots in the Sm_A half a degree above the transition towards the S_A
phase. The intensity profile along a line perpendicular to the director
is asymmetric and extended mainly on the inner side of the diffuse ring.
This effect can be understood as a geometrical one, due to the uniaxial
symmetry of the X-ray pattern: we have to compare energies diffracted in
a constant azimuthal range as $q_\perp$ varies. Therefore, the raw data have to
be multipled by $q_\perp$ (in other words we apply the suitable Lorentz correction
to our raw data). After this correction, the diffuse spot has a quasi-

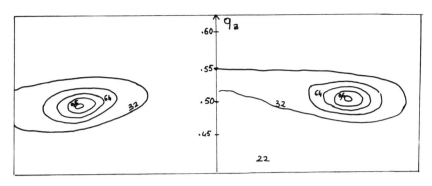

Fig. 5. Contour lines of the diffuse spots near a S_A $S_{\tilde{A}}$ transition. The raw data are shown on the right side and a Lorentz correction is applied on the left side. The unit in q_z corresponds to the Bragg peaks of the S_{A1} phase (mixture of $C_5H_{11}-\phi-CH=CH-\phi-O_2C-\phi-CN$ (55%) and $C_6H_{13}-\phi-O_2C-\phi-CN$ 45% at $133^\circ C$).

Lorentzian shape. Anyway, the symmetry axes of the spot are tilted versus the director. We can explain this unexpectable orientation and the slightly ovoid shape of the spot by adding orientational fluctuations of the mean wave vector to the exponential decay of the positional correlations parallel and perpendicular to the director. The disorientations of the layers (measured on the Bragg peaks) are less than $\pm 1^\circ$ while the wave vector for the local $S_{\tilde{A}}$ ordering undergoes $\pm 3.5^\circ$ fluctuations. At the same temperature (corresponding to Fig. 5), the transverse correlation length is 120 Å and the longitudinal one is 500 Å.

Similar observations can be done near the transition S_A $S_{\tilde{C}}$. In the example of Fig. 3b, half a degree above the $S_{\tilde{C}}$ transition the anisotropy between $\xi_{\parallel}$ and $\xi_{\perp}$ is similar to that near a $S_{\tilde{A}}$ phase. Orientational fluctuations of the wave vector are the same for the Bragg spot and for the diffuse ring. These orientational fluctuations near the transition towards a ribbon phase are consistent with the theory (5) since rotation of the wave vectors induces the transition. Nevertheless, microscopical observations [7] and X-ray data [8] seem to be in favor of the existence of a transient Sc phase in the case of some S_A $S_{\tilde{C}}$ sequences. In the vicinity of a S_A $S_{\tilde{A}}$ $S_{\tilde{C}}$ triple point fluctuations of both $S_{\tilde{A}}$ and $S_{\tilde{C}}$ phases can coexist [8].

Evolution of the Modulated Structure

The $S_{\tilde{C}}$ phase can transform itself either in an antiferroelectric double layer phase (S_{A2} or S_{C2}) or in a $S_{\tilde{A}}$ phase. These transitions have not been extensively studied by X-ray diffraction and moreover the transition $S_{\tilde{A}}$ $S_{\tilde{C}}$ is not at all evidenced by X-ray data. For the compound presented here in Fig. 3b and Table 1, a transition involving a molecular

291

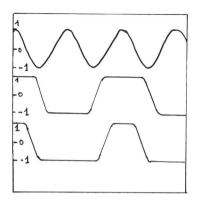

Fig. 6. The in-layer modulation of the electric polarization. For the centered rectangular phase $S_A^\sim$ close to the paralectric phase (upper curve) close to the transition towards the rectangular lattice (in the middle) and for the rectangular lattice itself (in the bottom).

intralayer ordering takes place before the other possible transformations [7]. In each layer, the molecules form a 2D hexatic array [10] characterized by a long-range bond orientational order and short-range hexagonal positional order. Nevertheless, walls of inversion of polarization built a second 2D network of infinite length ribbons. The two networks lie in perpendicular planes. At the intersection of the two planes the two periods are incommensurate, but very weakly coupled. Therefore, we have an incommensurate "twofold 2D crystalline structure." We have to notice that, starting at high temperature with a paralectric liquid, the modulation appears at higher temperature than the molecular ordering.

The $S_A^\sim$ $S_{A2}^\sim$ transition is more simple and therefore it has been studied in some detail [6]. This transition shows some similarities with the incommensurate-ferroelectric transition of some molecular compound such as thiourea [11].

A good insight of the evolution of the system can be obtained by following the evolution of the electric polarization inside one specific layer, keeping in mind the fact that for the same in-plane coordinates, the adjacent layers have the reverse polarization (Fig. 6). At high temperature, the modulation is sinusoidal with a wave vector of $\sim$ 100 Å. As the temperature decreases, the wavelength increases and harmonics of the in-plane modulation appear simultaneously. In other words, the walls corresponding to the inversion of the dipoles become narrower and the antiferroelectric ordered zones increase in size as the temperature decreases. Close to the S_{A2} phase transition, a symmetry change occurs: one kind of antiferroelectric domain increases in width while the other kind progressively disappears. In this second modulated phase, the wavelength is

292

constant (equal to 330 Å in our experiment); a breakdown of the symmetry is induced by a pairing of the walls of polarization inversion. Nevertheless, this symmetry breakdown does not induce immediately the S_{A2} phase. The symmetry change from simple rectangular network to a centered rectangular one occurs with some hysteresis on heating; at the same time, we have found an hysteresis for the in-plane period. We also notice that a model of this transition, which appears to be specific of the modulated liquid state, has been proposed by L. Benguigi [12]; he assumes that the simple rectangular phase is a modulated phase in which a constant term is superimposed to a sinusoidal one. This model does not take into account the higher harmonics which are clearly visible on the diffraction patterns (201) in the centered rectangular lattice, (201) and (301) in the simple rectangular one). We think that the presence of these harmonics is one of the main factors that must be taken into account in any model of this transition. Going down in temperature, the walls become narrower and more rigid and this may induce a change of their interaction forces. Electron microscopical observations by means of the freeze etching technique show also narrow and rigid walls [13].

4. OTHER EXAMPLES OF INCOMMENSURATE LIQUID CRYSTALS

The polymorphism of the polar thermotropic compounds shows some similarities with that of the lyotropic liquid crystals [14], and especially if we think to the 2D ribbon phases. Moreover, it has been shown recently that side chain liquid crystalline polymers [15] and phasmidic [16] compounds have the same trend to form 2D liquid. The formation of these phases can enter in the frame of the theory of J. Prost. Since the main feature that it is taken into account is the existence of a competition between two equilibrium periodicities, the physical reasons for such a competition can be found in electric interaction but also in some elastic constraints. Microscopical models have been developed on this basis (301) by J. O. Indekeu and A. N. Berker [17] and F. Dowell [18].

Elastic constraint can also induce modulation in non-smectic liquid crystalline phases and a theoretical model has been proposed by P. G. De Gennes [19]. This model applies to columnar phases and discusses the impact over the stacking inside a column of the mismatching which could occur between the inter core and inter chains spacing. This mismatch induces a modulation of the linear periodicity along the column axis (Fig. 7a). This model is based upon X-ray data of D. Guillon et al. [19] obtained on powder samples and in fact, information about the direction and the polarization of the distortion cannot be obtained unambiguously.

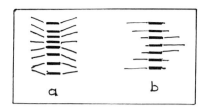

Fig. 7. Intracolumn modulations in columnar phases of disc-like molecules.
   a) a longitudinal modulation after De Gennes' model
   b) a helicoidal modulation found in some triphenylene derivatives

In most of the columnar phases of the disc-like compounds the core to core distance is of 3.6 Å, while a typical chain radius reaches 4.5 Å. A good linear packing is obtained by the rotation of two adjacent molecules which avoid the superposition of paraffinic chains. This rotation can induce a helicoidal modulation [20] (Fig. 7b). The more straightful evidence of a helicoidal array is shown by the columnar phase of a conical molecule [21] (Fig. 8). In this phase, the paraffinic chains which are in a melted state form a 3D crystal similar to a regular array of parallel imbricated screws. The cores are regularly stacked in each column with a period of 4.82 Å. A longitudinal distortion of wave length equal to the helix pitch (20.96 Å) is superimposed over the mean period, but no intercolumn correlation exists between the cores. This absence of correlation could be induced by an up-down orientational disorder of the cores. The paradoxal feature of this phase is that the 3D imperfect networks come from the flexible part of the molecule which is in a melted state.

   In fact, liquids can undergo periodic modulation in three dimensions. The blue phase is one of the most famous examples [23]. Cubic and tetragonal phases of smaller unit cell size (100 to 200 Å) have been found in lyotropic systems [24] but also in thermotropic liquid crystals [25,26, 16].

5. CONCLUSION

   X-ray diffraction experiments on incommensurate liquid crystals are usually in good agreement with a simple phenomenological theory.

   First put in evidence more than twenty years ago in lyotropic liquid crystals, periodic modulation superimposed over a liquid ordering appears to cover an important field of the thermotropic liquid crystalline polymorphism.

   On the one hand, the microscopic investigation of these systems cannot be pursued easily with some accuracy since investigations of the correlations in a liquid are not easy in general and a fortiori for liquids

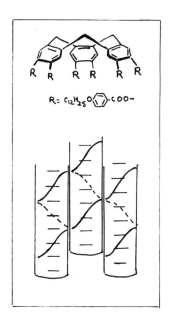

Fig. 8. The structure of the low temperature mesophase of a conical molecule shown on the upper part of the figure ($R=C_{12}H_{25}-O-\phi-CO_2-$). The paraffinic chains form a 3D imperfect network of helices. The symmetry is $P_{2_1}22_1$. The cores are regularly stacked in each column but there is no correlation of their position along the column axis and of their orientation (up or down) between neighboring columns.

made of high molecular weight and flexible molecules. But, on the other hand, the fluid character erases the specific differences and perhaps this is at the origin of the success of the models.

6. ACKNOWLEDGMENTS

The X-ray pattern has been obtained partly at the synchrotron facility LURE (Orsay) with the support of S. Megtert.

The contour lines of Fig. 5 have been drawn by the "Service de Microdensitometrie du CNRS" (Orsay).

7. REFERENCES

1. J. Prost, "Proceeding of the Conf. on Liq. Cryst. of One and Two Dimensional Order at Garmisch Partenkirschen" (Springer Verlag, Berlin, Heidelberg, New York, 1980).
2. A. Guinier, "X-ray Diffraction" (Freeman, San Francisco, 1963).
3. R. Ratna, S. Shashidhar and V. N. Raja, Phys. Rev. Lett. 55, 1476 (1985).
4. F. Hardouin, A. M. Levelut, G. Sigaud, J. de Phys. 42, 71 (1981).
5. J. Prost and P. Barois, Journal de Chimie Physique 80, 65 (1983).
6. A. M. Levelut, J. Phys. (Paris) Lett. 45, L603 (1984).
7. A. M. Levelut and N'Guyen Huu Tinh, 11th international liquid crystal conference, Berkeley (1986).

8. E. J. Fontes, P. A. Heiney, J. N. Haseltine, A. B. Smith, 11th international liquid crystal conference, Berkeley (1986).

9. C. R. Safinya, L. Y. Chiang, W. A. Varady, P. Dimon, 11th international liquid crystal conference and this book.

10. B. I. Halperin and D. R. Nelson, Phys. Rev. Lett $\underline{41}$, 121 (1978); R. Pindak, D. E. Moncton, S. C. Davey and J. W. Goodby, Phys. Rev. Lett. $\underline{46}$, 1135 (1981).

11. F. Denoyer and R. Currat, "Incommensurate Phases in Dielectrics 2," R. Blinc and A. P. Levanyuk (eds.) (Elsevier Science Publishers New York, 1986).

12. L. Benguigui, Phys. Rev. $\underline{A33}$, 1429 (1986).

13. G. Sigaud, F. Hardouin and M. F. Achard, Phys. Rev. $\underline{A31}$, 547 (1985).

14. J. Charvolin, Journal de Chimie Physique $\underline{80}$, 15 (1983).

15. P. Davidson, P. Keller, A. M. Levelut, Journal de Physique $\underline{46}$, 936 (1985).

16. N'Guyen Huu Tinh, C. Destrade, A. M. Levelut, J. Malthete, Journal de Physique $\underline{47}$, 553 (1986).

17. J. O. Indekeu and A. N. Berker, Phys. Rev. $\underline{A33}$, 1158 (1986).

18. F. Dowell, Phys. Rev. $\underline{A31}$, 2464 (1985).

19. P. G. De Gennes, J. Physique Lettres $\underline{44}$, L657 (1983).

20. D. Guillon, A. Skoulios, C. Piechocki, J. Simon, P. Weber, Mol. Cryst. Liq. Cryst. $\underline{100}$, 275 (1983).

21. A. M. Levelut, Journal de Physique Lettres $\underline{40}$, L81 (1979).

22. A. M. Levelut, J. Malthete, A. Collet, Journal de Physique, $\underline{47}$, 351 (1986).

23. S. Meiboom, M. Sammon, Phys. Rev. Lett. $\underline{44}$, 882 (1980); D. L. Johnson, J. H. Flack, P. P. Crooker, Phys. Rev. Lett. $\underline{45}$, 641 (1980).

24. V. Luzzati, in "Biological Membranes," D. Chapman (ed.) (Academic Press, London, 1968).

25. S. Diele, P. Brand and H. Sackmann, Mol. Cryst. Liq. Cryst $\underline{17}$, 163 (1972).

26. A. M. Levelut, C. Germain, P. Keller, L. Liebert, J. Billard, J. Phys. $\underline{44}$, 623 (1983).

ANOMALOUS HEAT CAPACITY ASSOCIATED WITH THE INCOMMENSURATE SmA PHASE

IN DB_7OCN + 80CB

C. W. Garland and P. Das

Department of Chemistry and Center for Materials Science
and Engineering
Massachusetts Institute of Technology
Cambridge, Massachusetts 02139

Abstract   A high-sensitivity ac calorimeter has been used to determine
the anomalous heat capacity associated with the SmA_d-SmA_{ic}-SmA_2 phase
transitions in a smectic liquid crystal mixture of DB_7OCN and 80CB.
SmA_d is a partial bilayer smectic A phase (L < d < 2L) and SmA_2 is a
bilayer smectic (d = 2L), where L is the molecular length. SmA_{ic} is an
incommensurate smectic phase exhibiting the simultaneous presence of
two collinear mass density modulations with periods whose ratio is not
a rational fraction. In addition, heat capacity and x-ray data on pure
DB_7OCN indicate that the direct SmA_d-SmA_2 transformation in that
material occurs continuously without undergoing a thermodynamic
transition.

1. INTRODUCTION

An incommensurate smectic-A liquid crystal phase should exhibit two
collinear mass density modulations for which the ratio of wavevectors is
not an integer or rational fraction. Such a phase, denoted as SmA_{ic}, has
been reported [1] in binary mixtures of heptyloxyphenyl-cyanobenzoyloxyben-
zoate (DB_7OCN) and octyloxycyanobiphenyl (80CB). The significant portion
of the phase diagram is shown in Fig. 1.

X-ray data [1] show the simultaneous presence in the SmA_{ic} phase of
quasi-Bragg peaks at q_0' and at q_0. Thus there exists a SmA_d-like (partial
bilayer smectic) modulation with wavelength $2\pi/q_0' \cong 1.75L$ and a SmA_2-like
(bilayer smectic) modulation with wavelength $2\pi/q_0 = 2L$ (and also the SmA_2
second harmonic $2\pi/2q_0 = L$), where L is the effective molecular length.
Detailed analysis of the x-ray patterns and the temperature dependences
of q_0' and q_0 shows this is an incommensurate phase and not merely a coexis-

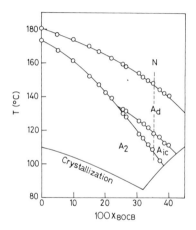

Fig. 1.   Part of the T-X phase diagram for mixtures of DB₇OCN and 80CB
          [taken from Ref. 1]. X_{80CB} denotes the mole fraction of 80CB.
          The vertical dashed line shows the temperature range studied
          for the 35 mole percent mixture.

tence of two separate phases [2]. The amplitude of the q_0' modulation
decreases on cooling below the SmA_d-SmA_{ic} transition, finally reaching
zero at $T_{A_{ic}A_2}$.

DSC measurements have been made on pure DB₇OCN and three mixtures that
transform directly from SmA_d to SmA_2 as well as on two mixtures exhibiting
the SmA_{ic} phase [2]. DB₇OCN exhibits a sharp, well-resolved SmA_d-SmA_2
peak, and mixtures containing 6 to 20 mole percent 80CB show small
SmA_d-SmA_2 peaks also. However, there are absolutely no indications of DSC
thermal anomalies at the SmA_d-SmA_{ic} or SmA_{ic}-SmA_2 transitions in mixtures
with 31 or 35 mole percent 80CB. These measurements were carried out on
cooling at 0.5K/min, which is quite a slow scan rate for the SC technique.

2. EXPERIMENT

High-sensitivity heat capacity measurements on pure DB₇OCN and a mixture containing 35 mole percent 80CB have now been carried out with an ac
calorimeter using scan rates in the range 0.5 to 1.5 K/h [3]. Figure 2
shows an overview of the C_p variation for DB₇OCN. Note the small N-SmA_d
peak at 452.6K as well as the large peak associated with the SmA_d-SmA_2
transformation (at 445.7K on the first run). The integrated enthalpy
$\delta H = \int \Delta C_p\ dT$ for the latter is ~5.4 J/g. A discussion of the thermal
behavior of pure DB₇OCN will be given later in this paper. Figure 3 presents the C_p variation observed in a mixture with $X_{80CB} = 0.35$ on cooling
and warming through the SmA_{ic} region. The differences shown are not due
to ordinary hysteresis effects. The cooling run yields the true heat
capacity, whereas the "apparent" C_p values obtained on warming are

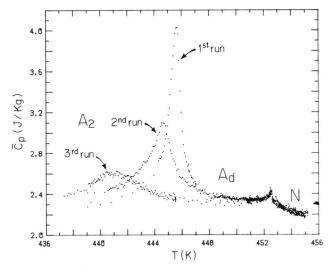

Fig. 2. Specific heat $\bar{C}_p$ for pure DB7OCN. All three runs were taken on cooling from ~ 456K.

artificial in the SmA$_{ic}$ range 383.5K-394.0K. This is indicated by the phase shifts ϕ in the observed $T_{ac}(\omega)$ signal shown in Fig. 4. The phase of $T_{ac}(\omega)$ is independent of temperature on cooling, in agreement with many observations for diverse systems near second-order transitions with small ΔC_p peaks. During warming runs, there is a systematic distortion in the sine-wave temperature oscillation, leading to an appreciable phase shift and an apparent dip in C_p that is an artifact of the analysis procedure.

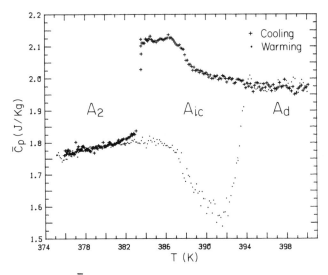

Fig. 3. Specific heat $\bar{C}_p$ in the SmA$_{ic}$ region for a 35 mole percent mixture of 8OCB + DB7OCN. See text.

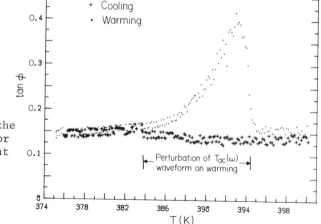

Fig. 4. Phase shift of the $T_{ac}(\omega)$ signal for a 35 mole percent mixture of $CB_7OCN + 80CB$.

The reason for this wave-form distortion on warming is not yet clear. It has subsequently been discovered that the transition at 383.5K on cooling is due to freezing of the SmAic phase into a metastable plastic crystal phase that is not the same as the SmA_2 phase obtained on rapid cooling. Details will be given in Ref. 9.

The behavior shown in Figs. 3 and 4 is reproducible. We have seen behavior for two cycles of warming and cooling on each of two samples. The absence of any appreciable thermal decompositon during these long runs (66 h for two cycles) is confirmed by the reproducibility of the crystal-SmA_d melting peak. Furthermore, the observed SmA_{ic} range agrees well with that reported in Ref. 1 (382.7K–391.5K). The excess or transitional enthalpy δH associated with the SmA_2-SmA_{ic}-SmA_d region is ~2.3 J/g, which is smaller than, but comparable to, the direct SmA_2-SmA_d enthalpy in pure DB_7OCN.

3. ANALYSIS

Recent theoretical work [4] on the Landau theory of frustrated smectics has shown that such a SmA_{ic} phase can be stable for a certain range of model parameters. A theoretical phase diagram equivalent to Fig. 1 has been given in Ref. 4. The exact nature (first-order or second-order) of the SmA_2-AmA_{ic} and SmA_{ic}-SmA_d transitions is uncertain theoretically since this feature is sensitive to the magnitude of a third-order coupling term. However, the direct SmA_d-SmA_2 transition at small values of X_{80CB} is predicted to be first order, which seems to be in conflict with the experimental data on pure DB_7OCN.

Both alkylphenyl-cyanobenzoyloxybenzoates (e.g., DB_6CN) and alkoxy-pehnylcyanobenzoyloxybenzoates (e.g., DB_7OCN) exhibit the phase sequence N →SmA_d →SmA_2 on cooling. The SmA_d range is quite narrow (T_{NA_d}-$T_{A_dA_2}$ ≅ 1.0K for DB_6CN and 6.9K for DB_7OCN), but the nematic range is very broad (T_{NI}-T_{NA_d} ≅ 96K). According to the theoretical phase diagram [4] for

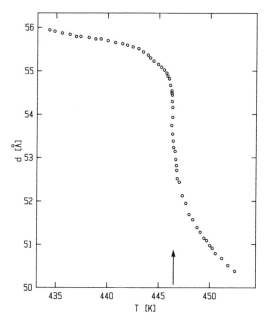

Fig. 5. Temperature dependence of the smectic layer spacing d̲ in pure
DB₇OCN (results of R. Shashidhar [5]). The arrow indicates
the apparent temperature of the SmA$_d$-SmA$_2$ transformation.

these families of liquid crystals, the N-SmA$_d$ transition is second-order
but the SmA$_d$-SmA$_2$ transition is described by a first-order line ending at a
critical point C (much like the familiar liquid-gas coexistence line in
a pure fluid). As shown in Fig. 2, heat capacity measurements have been
made near both the N-SmA$_d$ and the SmA$_d$-SmA$_2$ transitions in pure DB₇OCN.
Due to the high temperatures, some thermal decomposition occurred and T_{NA}
decreased slowly with time. The $T_{NA\,d}$ drift rate averaged approximately
-75 mK/h over the first 30 h (runs 1 and 2) and was zero subsequently.
The temperature scale on Fig. 2 corresponds to the first run; data for
other runs have been shifted slightly to compensate for shifts in $T_{NA\,d}$.
Note the good agreement in the size and shape of the N-SmA$_d$ peaks for three
different runs.

The most important feature to note in Fig. 2 is the broad rounded C_p
peaks associated with the SmA$_d$-SmA$_2$ transformation. Such peaks do not
correspond to those seen for either second-order or first-order phase transitions. The data strongly suggest a continuous evolution from SmA$_d$ to
SmA$_2$ without going through a thermodynamic phase transition. This certainly is allowed since SmA$_d$ and SmA$_2$ have the same symmetry. (Note that the
transition must be first-order if any transition actually does occur.)

Recent x-ray data [5] on DB₇OCN support this conclusion in that the
thermal expansion of the layer thickness below $T_{NA\,d}$ is qualitatively very

301

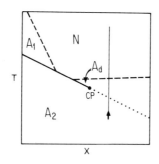

Fig 6. Phase diagram for the DB$_n$OCN system, slightly modified from that given in Ref. 4. Measurements on DB$_7$OCN correspond to a scan along the light vertical line marked with an arrow. No thermodynamic transition occurs: SmA$_2$ on heating evolves continuously and nonsingularly into SmA$_d$.

similar to our observed C_p behavior. As shown in Fig. 5, no discontinuity is observed in the thickness of the smectic layers. Instead, the layer thickness evolves continuously from $T_{NA_d} \cong$ "$T_{A_dA_2}$" + 7K to "$T_{A_dA_2}$" - 7K, where "$T_{A_dA_2}$" denotes the point of maximum thermal expansion.

The currently available results lead to the revised phase diagram shown in Fig. 6. The dotted line is not a transition line but merely the locus of maxima in the thermodynamic response functions in the one-phase region beyond the SmA$_d$-SmA$_2$ critical point C. This is a direct analogy of the situation in a pure fluid beyond its liquid-gas critical point. We propose that impurities are formed by slow thermal decomposition during the long heat capacity runs. They have no significant effect on the N-SmA$_d$ transition except for a small shift in T_c. However, they have a large effect on the "SmA$_d$-SmA$_2$" anomaly. As the system is shifted away from the critical point C, the anomalous peaks associated with the dotted line decrease substantially (again in direct analogy with pure fluids). This conclusion about the "SmA$_d$-SmA$_2$" transformation in DB$_7$OCN is supported by analogous results from heat capacity [6] and x-ray [7] investigations of DB$_6$N and DB$_6$OCN.

The above conclusion that the SmA$_d$-SmA$_1$ transformation in DB$_7$OCN occurs continuously without a thermodynamic transition raises some interesting questions with regard to the Landau model for such frustrated smectic liquid crystals [4,8]. The model parameters that gave rise to a phase diagram like Fig. 1 with a SmA$_{ic}$ phase led to the prediction of a first-order SmA$_d$-SmA$_2$ transition. It seems that experimentally one has no formal transition between SmA$_d$ and SmA$_2$ (merely a locus of maxima in C_p and thermal expansion) for DB$_7$OCN and dilute mixtures with 80CB up to a concentration $X_{8OCB} \simeq 0.25$, above which a SmA_{ic} phase is observed with unusual properties. Further work on this system is in progress as part of a collaboration with R. Shashidhar, and the detailed results will be published later [9].

4. ACKNOWLEDGMENT

This work was supported by Grant DMR 83-15637 of the National Science Foundation.

5. REFERENCES

1. B. R. Ratna, R. Shashidhar and V. N. Raja, Phys. Ref. Let. <u>55</u>, 1476 (1985).
2. B. R. Ratna, R. Shashidhar and V. N. Raja, J. Phys. (Paris), in press.
3. The ac calorimeter experimental technique is described in C. W. Garland, Thermochim. Acta <u>88</u>, 127 (1985).
4. P. Barois, Phys. Rev. <u>A33</u>, 3632 (1986).
5. R. Shashidhar, private communication.
6. C. W. Garland and P. Das (to be published).
7. K. Chan, P. S. Pershan, L. B. Sorensen and F. Hardouin, Phys. Rev. (in press).
8. P. Barois, J. Prost and T. C. Lubensky, J. Phys. (Paris) <u>46</u>, 391 (1985).
9. P. Das, C. W. Garland, R. Shashidhar (to be published).

THERMAL CONDUCTIVITY STUDIES AND FREE-STANDING LIQUID-CRYSTAL FILM

CALORIMETRY AS TWO APPLICATIONS OF THE AC CALORIMETRIC TECHNIQUE

C. C. Huang, G. Nounesis, and T. Pitchford

School of Physics and Astronomy
University of Minnesota
Minneapolis, MN 55455

Abstract   As extensions of the original ac calorimetric technique by
Sullivan and Seidel, we have developed suitable experimental techniques and have successfully carried out thermal conductivity studies of
bulk samples and heat-capacity measurements of free-standing liquid-
crystal films.

1. INTRODUCTION

In 1968, Sullivan and Siedel [1] invented the quasiadiabatic ac calorimetric technique to continuously measure heat-capacity with high resolution as a function of temperature. Since then, many types of critical
phenomena have been revealed and investigated by employing this experimental technique to obtain heat-capacity anomalies near the given phase transitions. During the past ten years, our understanding of various phase
transitions between different liquid-crystal mesophases has been greatly
enhanced through the heat-capacity anomalies obtained by the ac calorimetric technique. Here we will present two extensions of this experimental
technique, namely, heat-capacity studies on free-standing liquid-crystal
films and thermal conductivity studies on bulk samples which have been
successfully developed in our laboratories.

Now let us review briefly the basic idea behind the quasiadiabatical ac
calorimetric technique. Figure 1(a) shows schematically a liquid-crystal
sample of thickness h as used in the calorimeter. Of the heat energy ΔQ
applied to region X, a part ΔQ_1 is dissipated with a time constant τ_1 primarily through an exchange gas to the heat reservoir and another part ΔQ_2
goes to region Y with a time constant τ_2. In the limit of a two-dimension-

al system with primarily one-dimensional heat-flow through the sample, Sullivan and Seidel have shown that [1]

$$\Delta T = \frac{\Delta P_A}{\omega C_A}[1 + \frac{1}{(\omega \tau_1)^2} + (\omega \tau_2)^2 + 2(10)^{1/2}\frac{\tau_2}{\tau_1}]^{-1/2} \qquad (1)$$

Here ΔT is the temperature-oscillation amplitude in region Y, ΔP_A the power-oscillation amplitude per unit area applied to region X, ω the angular frequency of oscillation, C_A the heat capacity per unit area, $\tau_2 = h^2/[3(10)^{\frac{1}{2}}D_T]$, D_T is the thermal diffusivity of the sample, and $\tau_1 = C/g$ with C the heat capacity of the sample and g the sample-to-reservoir thermal conductance. For operation independent of the time constants τ_1 and τ_2, such as when the heat capacity is measured, then one chooses ω such that

$$\omega \tau_1 \gg 1 \text{ and } \omega \tau_2 \ll 1.$$

Under these conditions, Eq. (1) becomes

$$\Delta T = \Delta P_A/(\omega C_A) \qquad (2)$$

Physically, $\omega \tau_1 \gg 1$ ensures negligible ac heating loss to the heat reservoir over one oscillation cycle (quasiadiabatic condition) and $\omega \tau_2 \ll 1$ ensures sample thermal equilibrium.

Our ac calorimeter[2] [2] for both heat-capacity and thermal conductivity studies consists of a sample cell containing the liquid crystal, an oven to control the average sample temperature, and associated electronics. Figure 1(b) shows the sample cell with liquid crystal L, glass slides G, ac heater H, ac thermocouple A to measure ΔT, and dc thermocouple D to monitor average sample temperature. Now let us discuss thermal conductivity studies first and then the free-standing film calorimeter next.

2. THERMAL CONDUCTIVITY STUDIES

From Eq. (1), instead of measuring ΔT with a fixed ω such that $\omega \tau_1 \gg 1$ and $\omega \tau_2 \ll 1$, at a given temperature, one can measure $(\omega \Delta T)/(\Delta P_A)$ as a function of ω (= $2\pi f$), namely, the frequency response curve, in the region $\omega \tau_1 \gg 1$ and including the region $\omega \tau_2 \cong 1$. Then the high cut-off frequency, $f_H = \omega_H/2\pi$, ($\omega_H \tau_2 = 1$) obtained from this frequency response curve bears the information directly related to the sample thermal conductivity (K). Here $\omega_H = 1/\tau_2 = 3(10)^{\frac{1}{2}}D_T/h^2$. Figure 2 shows one typical frequency

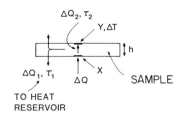

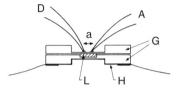

Fig. 1. (a) Schematic heat flow diagram of a thin sample in an ac
calorimeter. (b) Schematic liquid-crystal sample-cell cross
section for our ac calorimeter.

response curve [3] for one microscope cover glass slide with a thickness of
170 μm. The electronic apparatus used to measure ΔT have a low cut-off frequency around 0.2 Hz. This accounts for the slight decrease in $(\omega \Delta T)/(\Delta P_A)$
in the region $f \lesssim 1$ Hz. Actually the value of τ_1 has been measured for
our calorimeter to be 5 sec. Then two important features in the region
$f \gtrsim 1$ Hz provide two important properties of the sample. First, the flat
portion of the frequency response curve gives the sample heat capacity
(C). Secondly, the high cut-off frequency is directly related to the thermal diffusivity (D_T) of the sample. Then the thermal conductivity K =
$D_T C$. Careful measurements on three microscope cover glass slides with
different thicknesses have been reported by us [3]. This investigation gives
us results on the heat capacity and thermal conductivity of glass slides
which are in fairly good agreement with previously reported values. Also
we conclude that the contribution from thermocouple junction, GE varnish
and exchange gas can be neglected in our data reduction.

In the case of a liquid crystal, the sample is placed inside a sample
cell which is confined by one pair of chemically etched glass slides with
equal thickness in the sample cell area (see Fig. 1(b)). Then, the expression for ω_H becomes [3]

$$\frac{1}{\omega_H} = \frac{1}{3(10)^{1/2}} C_A (\frac{h_L}{K_L} + \frac{2h_G}{K_G}). \tag{3}$$

Here, $C_A = 2h_G C_G + h_L C_L$ is the total heat-capacity per unit area of the
sample and the sample cell and h, C, and K are thickness, heat-capacity per
unit volume, and thermal conductivity, respectively. The subscripts G and

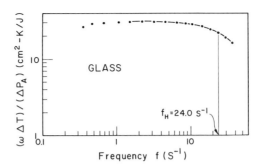

Fig. 2. Frequency response curve [i.e., $(\omega \Delta T)/(\Delta P_A)$ vs. f] for one microscope cover glass slide with thickness of 170 μm.

L denote the glass and liquid crystal, respectively. Again our measured results of heat-capacity and thermal conductivity of MBBA (p-methoxy benzylidene p-n-butylaniline) [3] at room temperature are in good agreement with previously reported values.

Because of the high resolution that can be achieved in both thermal conductivity (about 1%) and temperature about 3 mK), this new experimental technique enables us to investigate the temperature variation of thermal conductivity in the vicinity of various phase transitions. This has been

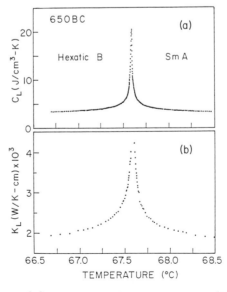

Fig. 3. Heat capacity (a) and thermal conductivity (b) vs. temperature near the SmA-hexatic-B transition of 650BC.

done near the smectic-A (SmA)-hexatic-B transition of 650BC (n-hexyl-4'-n-pentyloxybiphenyl-4-carboxylate) [4]. Figure 3(a) shows our temperature dependence of measured heat-capacity near the SmA-hexatic-B transition of 650BC. Details of obtaining frequency response curves have been published [4]. The frequency response curve at $T-T_c$ = -0.120 K is delayed in Fig. 4. The calculated thermal conductivity through Eq. (3) from this kind of frequency response curves is shown in Fig. 3(b) as a function of temperature. Comparing the anomalies in Fig. 3(a) and 3(b), the sharper anomaly in heat-capacity than that in thermal conductivity suggests that thermal diffusivity D_T(=K/C) should have a dip at the transition temperature. The results are illustrated in Fig. 5. Our thermal diffusivity data clearly demonstrate the critical slow down in thermal fluctuations near the continuous SmA-hexatic-B transition. The existing hydrodynamic theory [5] in the hexatic-B phase does not address the nature of the SmA-hexatic-B transition and fails to offer detailed comparison with our results.

Judging from the magnitude of thermal conductivity for typical liquid-crystal compound (about 2×10^{-3} W/K-cm) which is about five times smaller than that of glass, we still can measure the thermal conductivity anomaly near the phase transition and investigate the dynamic critical phenomena. This demonstrates that this experimental technique is a viable one for many applications.

3. HEAT-CAPACITY STUDIES ON FREE-STANDING LIQUID-CRYSTAL FILMS

We have employed ac calorimetric technique to investigate the evolution of heat-capacity anomaly as the thickness of the free-standing liquid-crystal film changes [6]. Liquid-crystal free-standing films have the following unique properties: They are substrate free. Films with wide range of thickness (two layers up to a few thousand layers) can be prepared easily and x-ray experiments have been carried out on two-layer films without rupture [7]. Thus far x-ray diffractions [7], mechanical studies [8], and

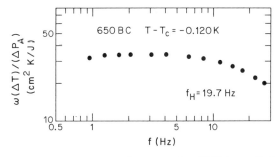

Fig. 4. Frequency response curve for sample 650BC at $T-T_C$ = -0.120 K.

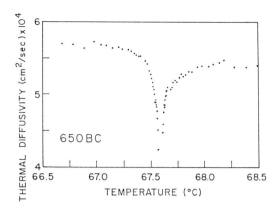

Fig. 5. Thermal diffusivity vs. temperature in the vicinity of the SmA-
hexataic-B transition of 650BC.

optical measurements [9] have been performed on many different free-standing liquid-crystal films. Here we have carried out the first heat-capacity
studies on the free-standing films near the SmA-hexatic-B transition of
460BC (n-butyl-4'-n-hexyloxybiphenyl-4-carboxylate). The large heat-capacity anomaly associated with the SmA-hexatic-B transition which has intrinsic
relationship with two-dimentional melting theories [10] makes this transition
the best candidate for heat-capacity studies on free-standing liquid-crystal
films. The reasons we choose 460BC are that it has large temperature ranges
for SmA and hexatic-B phase and no enhanced surface crystallization association with the lower temperature crystalline phase occurs.

       A schematic diagram of our free-standing film calorimeter is shown in
Fig. 6. To minimize the addendum heat-capacity contribution, we employ
chopped laser radiation (at 3.4 μm), which the liquid-crystal molecules
strongly absorb, as the heating source and a very small thermocouple junction made of 12.5 μm thermocouple wires as the temperature sensor for the
ac temperature oscillation of the sample. The separation between this
thermocouple and the films was 0.1 mm which is much smaller than one thermal diffusion length of argon exchange gas (about 1.2 mm) at the chopping
frequency employed (5.6 Hz). The average sample temperature was measured
with a thermistor which was located beneath and near the edge of the film
support to eliminate its contribution to the measured heat-capacity. From
Eq. (2) any laser intensity change will affect the measured C_A through the
variation in ΔP_A. Consequently, inside our temperature regulated oven,
another thermocouple junction was mounted well above the film to monitor
the laser intensity variation. After making the corresponding correction,
C_A (in relative units) can be measured with deviation less than 0.5%.
Although in the case of thick films (thickness about 5 μm) the transmission

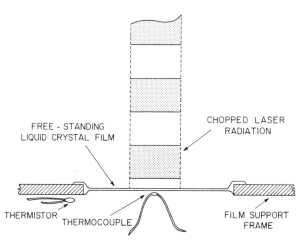

Fig. 6. Schematic diagram of our quasi-adiabatic free-standing film
calorimeter. Relevant dimensions are the following (approximate):
film thickness 0.03 - 7.0 μm, film diameter 5 mm, laser $1/e^2$
diameter 6.0 mm, thermocouple film separation 100 μm.

through the film is undetectable, in the case of thinner films (say 0.5
μm) the transmission becomes significant. Thus, a third thermocouple
junction was placed well below the film to detect the transmitted laser
intensity. The major reason that this calorimeter is an extension of the
original one by Sullivan and Seidel [1] is the following. Even though all
effort has been made to minimize the addendum heat-capacity contribution,
the effective heat-capacity per unit area from 0.5 atm. pressure Ar
exchange gas and the small thermocouple junction becomes larger than the
heat-capacity per unit area of the liquid-crystal film provided that the
film thickness is less than 0.1 μm [11]. Similar to the calculation by
Baloga and Garland for ac calorimetry at high pressure [12], the one-dimensional heat-flow equation has to be solved with proper boundary conditions.
Details of this calculation will be published in the future [11]. Here
let us show our primary findings. Figures 7(a), (b) and (c) show the measured heat-capacity anomalies with film thickness approximately 7 μm, 0.3
μm and 0.06 μm, respectively. In reducing the film thickness, two prominent features emerge. The heat-capacity anomaly becomes more asymmetric
with a stronger variation with temperature of the measured heat-capacity
developing on the low temperature side of the anomaly. Meanwhile, the difference between the background heat capacity on the high temperature side
and on the low temperature side of the heat-capacity anomaly grows larger
as the film thickness decreases. Qualitative explanations of our results
are given in Ref. 6. The solid line in Fig. 7(a) is the convoluting fitting to the power-law heat-capacity expression [6]. Because of the
finite laser beam size and without sufficient laser power output to expand

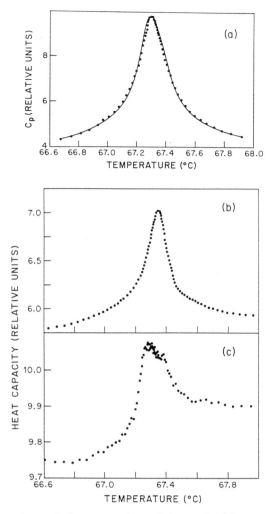

Fig. 7. Heat capacity of free-standing films of 460BC near the SmA-
hexatic-B transition. The results shown are for films
approximately 2,000 (a), 100 (b), and 20 (c) layers thick.

the beam size, the existence of Gaussian shape temperature gradient on the
sample requires a convoluting fitting in our heat-capacity data analyses.

Further decrease in the film thickness results in the heat-capacity
anomaly being buried in noise. This is due to the fact that the absorption of the film decreases dramatically and the heat-capacity contribution
from the liquid-crystal film towards the total measured heat-capacity
reduces. Thus more clever ways to provide ac heating or detect the ac
temperature oscillation, e.g., remote optical detectors, are necessary.
Finally, a much more powerful laser source is definitely desirable.

The work reported here was supported in part by the National Science
Foundation - Solid State Chemistry - Grant DMR85-03419.

4. REFERENCES

1. P. F. Sullivan and G. Seidel, Phys. Rev. $\underline{173}$, 679 (1968).
2. J. M. Viner, D. Lamey, C. C. Huang, R. Pindak, and J. W. Goodby, Phys. Rev. $\underline{28}$, 2433 (1983).
3. C. C. Huang, J. M. Viner, and J. C. Novack, Rev. Sci. Instrum. $\underline{56}$, 1390 (1985).
4. G. Nounesis, C. C. Huang, and J. W. Goodby, Phys. Rev. Lett. $\underline{56}$, 1712 (1986).
5. H. Pleiner and H. R. Brand, Phys. Rev. $\underline{A29}$, 911 (1984).
6. T. Pitchford, C. C. Huang, R. Pindak, and J. W. Goodby, Phys. Rev. Lett., $\underline{57}$, 1039 (1986).
7. S. C. Davey, J. Budai, J. W. Goodby, R. Pindak, and D. E. Moncton, Phys. Rev. Lett. $\underline{53}$, 2129 (1984) and references found therein.
8. D. J. Bishop, W. O. Sprenger, R. Pindak, and M. E. Neubert, Phys. Rev. Lett. $\underline{49}$, 1861 (1982) and references found therein.
9. D. H. Van Winkle and N. A. Clark, Phys. Rev. Lett. $\underline{53}$, 1157 (1984).
10. R. J. Birgeneau and J. D. Litster, J. Phys. (Paris) Lett. $\underline{39}$ L399 (1978).
11. T. Pitchford and C. C. Huang (unpublished).
12. J. D. Baloga and C. W. Garland, Rev. Sci. Instrum. $\underline{48}$, 105 (1977).

QUASIPERIODIC PATTERNS WITH ICOSAHEDRAL SYMMETRY

A. Katz and M. Duneau

Centre de Physique Théorique de l'Ecole Polytechnique
91128-Palaiseau, France

1. INTRODUCTION

This lecture deals with the theory of quasiperiodic tilings, and more generally with quasiperiodic patterns. We present the ideas introduced in [9] and developed in [10].

The interest for non-periodic tilings came first from problems of mathematical logic. However, since the invention by R. Penrose of his well known aperiodic tilings of the plane [2,3], the motivations have changed to the study of the geometrical properties of these patterns. After the first theoretical works by R. Penrose, J. Conway [2,3] and N. G. de Bruijn [6], A. L. Mackay [4] and Mosseri and Sadoc [5] have drawn attention to the possible implications of these structures for solid state physics. A 3-dimensional generalization of the Penrose tilings (using two rhombohedra instead of two rhombs) has been devised by R. Ammann and theoretically described by Kramer and Neri [7] in the context of the de Bruijn approach.

The special properties of the Fourier transform of the Penrose patterns, which are now identified as their quasiperiodicity, have been empirically observed by A. L. Mackay, using optical methods. D. Levine and P. Steinhardt have suggested in ref. [11] to take quasiperiodicity as the basis to build icosahedral quasicrystals.

The experimental discovery of icosahedral symmetry together with some kind of long-range spatial order in solids by D. Schechtman, I. Blech, D. Gratias, and J. W. Cahn [1] has aroused an intense activity among theorists, as well on the structural problem as on stability questions which will not be discussed here (see [13] for further references).

The theory described here, which takes into account both the icosahedral symmetry and the quasiperiodicity of the tilings, is based on an idea independently introduced in this context by V. Elser [12], P. A. Kalugin et al. [8] and us [9,10] (following N. G. de Bruijn).

2. A ONE-DIMENSIONAL MODEL

Let us begin by a description of this method in the simplest one-dimensional case (see Fig. 1): to tile of a straight line with two types of intervals, we start with the square lattice Z^2 in the plane R^2, in which the line E to be tiled is drawn. Consider now the strip obtained by shifting the unit square K of the lattice along the line. The claim is that, for almost all positions of the line, this strip contains an unique broken line (made of edges of the lattice), which joins exactly all the vertices falling inside the strip. It just remains to project this broken line orthogonally on the given direction to get the announced tiling, the two tiles a and b being clearly the projections of the vertical and horizontal edges of the unit square K.

One can understand the essential features of our tilings with this simple example. Consider first the choice of the unit square to build the strip. It is clear that one has to take a unit cell of the lattice in order to obtain a tiling (by the projections of the edges of the cell): with a narrower strip, one would get "holes" in the broken line, and with a larger strip, one would get extra points. However, any unit cell of Z^2 is possible, and not only the canonical square. This is closely related to the self-similarity properties of these tilings.

Secondly, consider the slope of the line E with respect to the canonical basis of R^2. It is clear that the tiling is periodic if and only if this slope is a rational number: in fact, the slope is just the ratio of the relative abundances of each type of tiles, and this ratio is certainly rational for a periodic tiling. Now, given an irrational slope, consider what happens when the strip is translated in R^2: each time a point leaves the strip, another point enters into the strip. These two points are the vertices of the diagonal of some square in the lattice, in such a way that the broken line "jumps" from one side of the square to the other. This is the only elementary change of the tiling, but for any finite translation of the strip (transverse to its direction) such jumps occur infinitely many times, in such a way that for each irrational slope, this construction generates a non-countable set of different tilings (of course, each time the translation belongs to Z^2, one gets merely a translation of the initial tiling).

Notice that we can slightly generalize the construction and distinguish between the line along which is built the strip and the line on which is projected the broken line to obtain the tiling. It is clear that the orientation of this last line (the tiled one) merely defines the relative sizes of the tiles, and that the orientation of the strip defines all the important properties, for instance the relative abundance of each type of tile. If we vary the orientation of the strip while maintaining unchanged the tiled line, we get different tilings with the same pair of tiles. Each time the slope is rational, the tilings are periodic, in such a way that the non-periodic tilings obtained for irrational slopes can be thought of as "interpolations" between periodic lattices.

Let us now explain why any finite patch of tiles that belongs to a tiling appears infinitely many times in any tiling defined through a strip with the same given irrational slope (Penrose's local isomorphism property, see [3]). Consider such a tiling, pick any finite patch of tiles in it, and consider the finite broken line that projects on this patch. Since the slope is irrational, it is clear that the projection of this broken line on E' (see Fig. 1) is strictly smaller than the projection of the

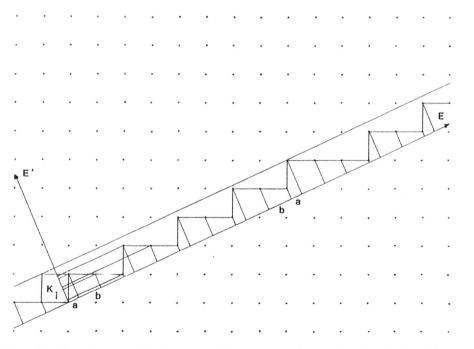

Fig. 1. The strip method in one dimension: the strip is obtained by translation of the unit square K along the line E. The unique broken line contained in the strip projects on E on a tiling by means of the two tiles a and b which are the projections of the two edges of K.

whole strip in such a way that there exists an open set (non-empty !) of
translations in E' that keep the projection of the finite broken line
inside the strip. Now consider the auxiliary strip that projects on E'
onto this set of translations; there are infinitely many vertices of Z^2
inside it, and each of them defines a translation that maps the finite
broken line inside our main strip onto a copy of it, which is by unicity a
part of the infinite broken line that projects on the tiling. This shows
that any finite patch which appears in a tiling, appears infinitely many
times. One can show with the same type of argument that, given any two
parallel strips and a finite broken line in one of them, there always
exists a translation in Z^2 that maps the finite broken line inside the
other strip, which means that any finite patch that appears in one tiling
appears in all the tilings associated with the same slope. There are
interesting questions concerning the distance between two copies of a given
patch. It can be shown that the rate of "rarefaction" of the copies of a
patch when its size increases depends on arithmetical properties of the
slope: for algebraic numbers, the mean distance between two copies is
proportional to the size of the patch, but for Liouville numbers, the mean
distance can grow arbitrary quickly.

Finally, let us describe the calculus of the Fourier transform of the
measure defined by putting one Dirac delta at each vertex of the tiling
(Fig. 2). Following our method, this measure is obtained in three steps:

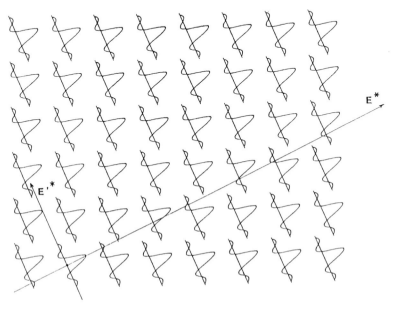

Fig. 2. Construction of the Fourier transform of the 1-dimensional tiling.

318

take the square lattice of Dirac deltas in the plane, multiply by the characteristic function of the strip, and then project on the line. To compute the Fourier transform, we have just to operate the "Fourier images" of these three steps in the reciprocal space. The first one is to take the Fourier transform of the square lattice, which is of course another square lattice of Dirac deltas. The second step is the convolution by the Fourier transform of the characteristic function of the strip, which is easy to compute since it is a product; in the direction E* dual to the direction of the strip, this Fourier transform is a Dirac delta and in the orthogonal direction E'*, it is the smooth function (sin k)/k, up to normalizations and phase factor. To complete the convolution, we have to translate a copy of this Fourier transform at each point of the reciprocal lattice. The third step (the Fourier image of the projection) is merely the restriction to the line E'*, as shown on Fig. 2.

Let us make a few comments about the result. The most important point is that the Fourier transform thus obtained is a sum of weighted Dirac deltas, carried by the projection of the whole lattice Z^2, i.e., by a two-generator Z-module, and this property of the support of its Fourier transform is the very definition of the quasiperiodicity of a measure. Observe that the quasiperiodicity follows directly from the strip being built on a straight line: one could produce more general tilings with a strip obtained by shifting the unit cell along a curved or undulating line (while maintaining the projection on a straight line) and they would not be quasiperiodic. The second point is that a translation of the strip changes the Fourier transform only by a phase factor (a different one for each Dirac delta) in such a way that all the tilings associated with the same irrational slope give the same diffraction patterns, as far as only intensities are measured.

Finally, observe that, although the support of the Fourier transform is a dense set, the intense peaks correspond to vertices which are near the line E* (see Fig. 2), and that the set of peaks whose intensity is greater than any strictly positive threshold is always discrete. In fact, the quantitative analysis of the intensities in the diffraction pattern may be used to reconstruct the geometry in the direct space (up to a translation of the strip) in the same sense and with the same limitations as in ordinary crystallography.

2. THE GENERAL CASE

We give now the generalization of this method to arbitrary dimensions. We generate quasiperiodic tilings of a p-dimensional space E as the projection, from a higher dimensional space R^n, of a p-dimensional surface made up of a suitable union of p-facets of a regular n-dimensional lattice L in R^n.

Actually, let $L = Z^n$ be the n-dim lattice in R^n generated by the natural basis $(e_1,...,e_n)$, and let γ_n be the unit cube. Let $E \subset R^n$ be a p-dimensional subspace of R^n, and assume that E does not contain any point of the lattice, except the origin.

There are $\binom{n}{p}$ different p-facets (the p-dimensional analog of an edge) of γ_n containing the origin. These facets project on E on a priori $\binom{n}{p}$ different p-volumes. A tiling of E by mean of these volumes is obtained in the following way: Let $S = E + \gamma_n$ be the open strip generated by shifting γ_n along E. Then the claim is that the union of all p-facets entirely contained in S is exactly a p-dimensional surface of R^n, the projection of which on E gives the announced tiling.

If E' is the orthogonal complement of E in R^n, the projection K of the strip on E' is just the projection of the unit cube γ_n. Moreover, if $E \cap L = \{0\}$, the projection of the lattice L in E' is one to one. Thus the set of vertices of the quasiperiodic tiling corresponds to the set of points of L that project in E' inside K.

Now, if E is invariant under the action of a subgroup G of the point group of the lattice L, the a priori $\binom{n}{p}$ different tiles fall into classes, in such a way that the tiles of each class have the same shape, and are permuted by G.

The set of p-volumes around a given vertex x of the tiling is completely specified by the set of corresponding p-facets falling in K around the corresponding point x' in E'. Note that no (p+1), (p+2),...,n-facet of L can project in E' strictly inside K.

3. THE ICOSAHEDRAL CASE

Let us now specialize to the case of the icosahedral symmetry. The simplest construction involves R^6 endowed with an orthogonal representation of the icosahedral group G, which permutes Z^6 and such that R^6 falls in two G-invariant 3-spaces, E and E' equipped with non-equivalent irreducible representations of G.

The 20 different 3-facets of $L = Z^6$ fall in E (and E') on two different rhombohedra, with the same facets (with angles Atn(2) and π-Atn(2)), each of which being repeated 10 times. These are the rhombohedra considered by A. L. Mackay in [4], P. Kramer and R. Neri in [7], and us in [9,10] (Fig. 3b and c).

The open strip is defined by $S = E + \gamma_6$ where γ_6 is the open unit cube of R^6. It can be seen that the projections (1) on E and $\pi'(L)$ on E' are Z-

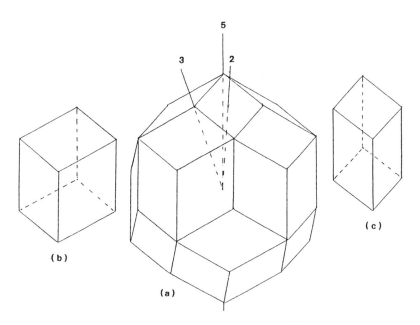

Fig. 3.   The projection (a) of the unit cube of R^6 in E and E' is a triacontahedron. A 5-fold axis, a 3-fold axis and a 2-fold axis are
represented. In (b) a flat rhombohedron and in (c) a thick
rhombohedron, which are the 2 tiles of the icosahedral tilings.

modules which are dense in E and E'. If $\varepsilon_1,\ldots,\varepsilon_6$, is the natural basis of
R^6, $\pi(\pm\varepsilon_1),\ldots,\pi(\pm\varepsilon_6)$ point to the 12 vertices of a regular icosahedron
centered at the origin of E and so do $\pi'(\pm\varepsilon_1),\ldots,\pi'(\pm\varepsilon_6)$ in E'.

The projection $\pi(S \cap L)$ in E is in one to one correspondence with the
other projection $\pi'(S \cap L) = K\cap\pi'(L)$ where $K = \pi'(\gamma_6)$ is a rhombic triacontahedron (Fig. 3a). the fact that no (closed) 4-facet of Z^6 can project
inside K insures that a non-periodic tiling of E by mean of the two types
of rhombohedra is obtained.

The vertex neighborhood around a given point x in E is completely
specified by the set of 3-facets around the corresponding point x' in E'.
Now observe that for a given rhombohedron (spanned by (e_i,e_j,e_k)) to be
entirely contained in the triacontahedron, it is necessary and sufficient
that each of its vertices fall inside another rhombohedron, spanned by
e_1,e_m,e_n, in such a way that (i,j,k,l,m,n) is a permutation of (1,2,3,4,
5,6), and located in the triacontahedron in one of eight canonical positions according to the vertex considered. Looking for the intersections
of these 160 rhombohedra yields a cell decomposition of the triacontahedron
in which each cell corresponds to a vertex neighborhood. One thus can see
that, up to the symmetry operations, there are only in these tilings 24
possible patterns of rhombohedra around a point. In particular, the
central cell of the triacontahedron corresponds to a 20-pronged star with
the icosahedral symmetry, made of 20 thick rhombohedra.

321

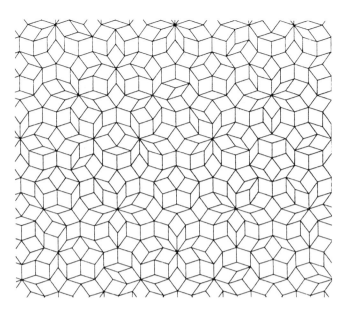

Fig. 4. A section of an icosahedral tiling, orthogonally to a 5-fold
         axis. A generalized Penrose tiling is obtained, with for
         instance a 10-pronged star.

    We present in Figs. 4, 5 and 6 three sections of this tiling, which
are associated to axes of order 5, 3 and 2 respectively. The cuts, which
are made along 2-dimensional surfaces taken from 2-facets of the tiling,
yield quasiperiodic tilings of the plane. Actually the first section,
carried out orthogonally to a 5-fold axis, projects on a generalized Penrose tiling. The two other sections project on quasiperiodic tilings of
the plane associated with 3-fold and 2-fold symmetries.

    As in the one-dimensional case, the strip can be translated, which
yields an uncountable set of non-isomorphic tilings.

    One of the most striking features of all these tilings is their quasi-
periodicity, in the sense that for any tiling, the measure defined by a
Dirac delta at each vertex, is quasiperiodic: its Fourier transform is a
sum of weighted Dirac measures with support in the $\mathbb{Z}$-module generated by
the projections of the basis vectors of the lattice. Moreover, all tilings
have the same Fourier transform, up to a phase at each point, in such a
way that they are identical from a "diffractional" point of view.

    In fact, the calculus of the Fourier transform runs exactly as in the
one-dimensional case. Here is a brief account of the proof: let $\nu = \chi_S \cdot$
$\sum_{\xi \in \mathbb{Z}^6} \delta_\xi$ be the measure associated to $S \cap \mathbb{Z}^6$, where χ_S is the characteristic
function of the strip in $\mathbb{R}^6$. Then $n = \sum_{\xi \in \mathbb{Z}^6} \chi_S(\xi) \cdot \delta_{\pi(\xi)}$ is the
measure associated to the tiling in E. If $\xi = (x, x')$ is the orthogonal
decomposition of $\xi \in \mathbb{R}^6$ in E and E', and if $\kappa = (k, k')$ is the corresponding
decomposition in the dual space, the Fourier transform of n is given by

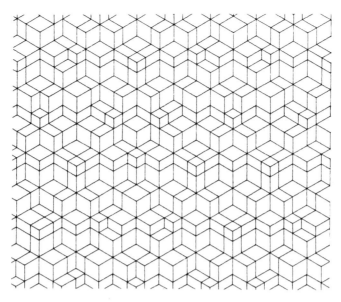

Fig. 5. A section of an icosahedral tiling, orthogonally to a 3-fold axis.

n*(k)=ν*(k,0).   Now $\nu^* = \kappa_S^* * \sum_{\lambda \in Z^6} \delta_\lambda$,   and since $\kappa_S(x,x') = \kappa_{TR}(x')$ where κ_{TR} is the characteristic function of the triacontahedron in E',

$\kappa_S*(k,k') = \delta(k).\kappa_{TR}*(k')$. Finally, $n*(k) = \sum_{\lambda \in Z^6} \delta(k-1).\kappa_{TR}*(-1')$ where $\lambda = (1,1')$ is the decomposition of λ in the dual space.

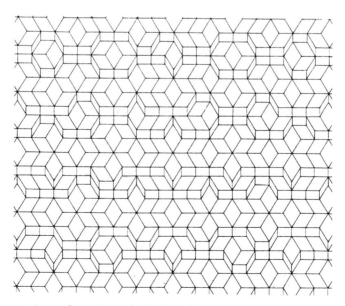

Fig. 6. A section of an icosahedral tiling, orthogonally to a 2-fold axis.

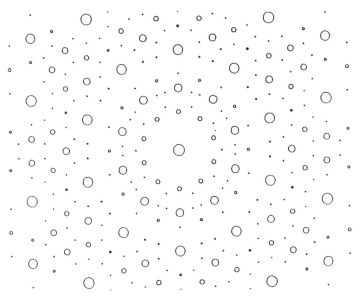

Fig. 7. The Fourier transform orthogonally to a 5-fold axis. The peaks above a given threshold are represented by circles with radii proportional to their amplitudes.

We give in Figs. 7, 8 and 9, the computed Fourier transforms of a tiling, in the 3 planes respectively orthogonal to symmetry axes of order 5, 3 and 2. These patterns are astonishingly similar to the electron

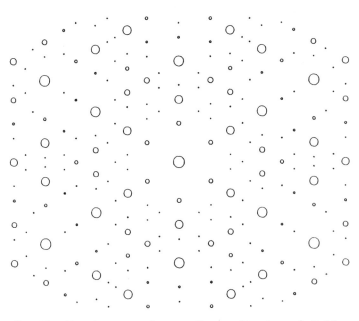

Fig. 8. The Fourier transform orthogonally to a 3-fold axis.

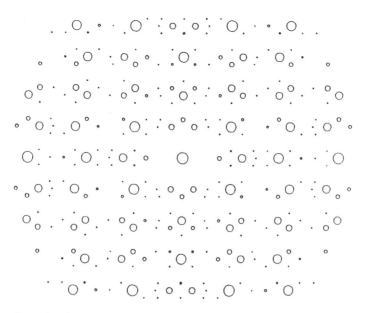

Fig. 9. The Fourier transform orthogonally to a 2-fold axis.

diffraction images of the quasicrystal Al-Mn. However, the exact significance of this fact is not known.

The general framework presented here can be generalized in many ways.
For instance, if the condition which ensures an exact tiling of the space
is removed, more general quasiperiodic patterns are obtained; the above
are special icosahedral patterns which are associated to special strips,
namely those generated by the translation of a unit cube of the hyperlattice. More general patterns can be obtained by replacing the triacontahedron by any other bounded subset K of E'. The corresponding pattern is
given by projecting on E the lattice points in the new strip $S = \pi'^{-1}(K)$.
However, as in the previous situation, such patterns exhibit a rigorous
long-range bond orientational order and are quasi-periodic.

The density of vertices in a pattern only depends on the lattice con-
stant $\alpha = |\varepsilon_i|$ of the hyperlattice in R^6 and on the volume $|K|$ in E': if V
is a bounded subset of E, the number N_V of points which fall in V is equal
to the number of lattice points in V×K, the part of the strip above V.
Then $N_V = |V| \cdot |K|/\alpha^6$ and the density is $|K|/\alpha^6$. One of the main geometri-
cal problems about these patterns is to understand which shape of K will
give rise to the densest icosahedral pattern with a given minimal distance
between any pair of points.

On the other hand, the projection space and the strip may have different orientations; the first one specifies the tiles while the second one
dictates their relative abundance. As in the one-dimensional model, the

quasiperiodic tilings thus obtained can be seen to interpolate between periodic ones, which correspond to rational orientations of the strip.

Notice that the quasiperiodicity of the tilings is independent of the fact that neither the pentagonal nor the icosahedral symmetries are compatible with periodic ordering. In fact, the sections corresponding to the 2-fold and 3-fold axes (given in Figs. 5 and 6), show quasiperiodic tilings of the plane while these symmetries are of crystal type.

As a final remark, let us stress on the idea that the construction presented herein works in any dimension and with any lattice. For instance, it is easy to see that another type of quasiperiodic tiling of the space with icosahedral symmetry can be obtained from R^{10}, the icosahedron being replaced by a dodecahedron.

5. REFERENCES

1. D. Schechtman, I. Blech, D. Gratias and J. W. Cahn, Phys. Rev. Let. - 53, 195 (1984).
2. R. Penrose, Math Intelligencer 2, 32 (1979).
3. M. Gardner, Sci. Am. 236, 110 (1977).
4. A. L. Mackay, Physica 114A, 609 (1982); Kristallografiya 26, 909 (1981), 909 (Sov. Phys. Crystallogr. 26, 517 (1981).
5. R. Mosseri and J. F. Sadoc, "Structure of Non-crystalline Materials 1982," Gaskell et al., eds. (London, Taylor and Francis, 1983).
6. N. G. de Bruijn, Nederl. Akad. Wetensch. Proc. A43, 39-66 (1981).
7. P. Kramer and R. Neri, Acta Crystallogr. A40, 580-587 (1984).
8. P. A. Kalugin, A. Yu. Kitayev and L. S. Levitov, J. Physique Lett. 46, 601 (1985); JETP Lett. 41, 145 (1985).
9. M. Duneau and A. Katz, Phys. Rev. Let. 54, 2688 (1985).
10. A. Katz and M. Duneau, J. de Physique 47, 181 (1986).
11. D. Levine and P. Steinhardt, Phys. Rev. Let. 53, 2477 (1984).
12. V. Elser, Acta Cryst. A42, 36 (1986).
13. Les Houces Workshop on Aperiodic Crystals, 11-22 March 1986, L. Michel and D. Gratias, eds., to be published in Journal de Physique (Colloque).

FRUSTRATION AND ORDER IN RAPIDLY COOLED METALS

Subir Sachdev [*]

AT&T Bell Laboratories
Murray Hill, New Jersey 07974

1. STATICS OF METALLIC GLASSES

The first metallic glass was produced in 1960 by Duwez and co-work-
ers [1] by rapidly cooling a molten alloy of gold and silicon. Metallic
glasses have since then been created in large numbers of simple metal,
transition metal and metalloid systems by a variety of ingenious methods
which achieve cooling rates of over a million degrees per second. The new
metallic materials so produced have proved to be of considerable importance
for their unique magnetic, mechanical and corrosion-resistance properties.

In spite of their importance, a complete microscopic understanding
of the static and dynamic properties of these materials is lacking. An
important reason for this shortcoming is that the properties of these
materials are complicated and very sensitive to the microscopic chemical
and electronic properties of their constituents. Unlike the case of ordinary metals, there is no simplifying feature like crystalline periodicity
which clears the way for further progress. It is the purpose of this pre-
sentation to convince you that it is possible to identify structural and
dynamic features of these materials which will be a simplified abstraction
of reality. However, because of its simplicity this model will allow us
to make definite predictions of the properties of metallic glasses.

The possibility that metals may condense into a glassy state arose
out of some experiments of Turnbull [2] on the nucleation of solid crystalline mercury from supercooled liquid mercury. By carefully excluding
sources of heterogeneous nucleation, Turnbull found a rather low rate of
homogeneous nucleation, implying that the short range order in the supercooled liquid and the crystal was rather different. This was followed
immediately by a suggestion by Frank [3] that the short range order in the

[*]Current Address: Sloane Physics Laboratory, Yale University, CT 06511

327

supercooled liquid was icosahedral and thus incompatible with crystalline order. A common feature of a large fraction of the metastable states of less than 200 particles, interacting via a simple Lennard-Jones like pair potential, is that the particles group rather well in sets of four, each set forming a tetrahedron [4]. The topology of the clusters can therefore be described as the packing of tetrahedra. Five tetrahedra packed face to face, sharing a common edge as in Fig. 1, don't fill the central dihedral angle of 360 degrees completely, but leave a small crack of 7.4 degrees. Thus in the metastable states above, a majority of bonds have five tetrahedra around them, with six and four tetrahedra around the remaining bonds. These six- and four-fold bonds can be identified as disclinations in the field of an order parameter measuring the degree of five-fold ordering. [5] The core energy required for these disclinations adds up, and for greater than 200 particles the state of minimum energy is a FCC crystallite. The FCC crystal has some octahedra and cannot therefore be described by the close packing of tetrahedra.

The physics of tetrahedral close packing is useful in characterizing the topology of dense random packing [6] or molecular dynamics simulations [7] of supercooled liquids. Via the Voronoi construction one can assign near neighbor bonds in these simulations. This construction will necessarily create only tetrahedral groups of particles. The distortions of the tetrahedra from equilateral are found to be small and the assignments of nearest neighbors meaningful. The simplest level of analysis involves just keeping the statistics of the four-fold, five-fold and six-fold configurations of tetrahedra. The number of five-fold bonds is $50 \pm 5\%$, and the number of six-fold bonds is $27 \pm 5\%$ in these simulations. The mean number of tetrahedra around a bond, $\bar{q}$, is rather more constant, lying in the range 5.11 ± 0.1. For simulations performed with particles of slightly different sizes $\bar{q}$ rises only to 5.13. The constancy of $\bar{q}$ follows rather nicely from a mean field argument due to Coxeter. [8] Coxeter argued in terms of a statistical model which allowed a fractional number of equilateral tetrahedra around every bond. This number is simply 2/ (dihedral angle of a tetrahedron) = 5.104, rather close to the numbers discussed above.

The power of Coxeter's argument is demonstrated even more strikingly by an analysis of certain crystalline intermetallics which form tetrahedrally close packed phases called the Frank-Kasper phases. [9] A complete catalog of all the known polytypes of such tetrahedrally close packed intermetallics has been given by Shoemaker and Shoemaker. [10] Intermetallics which form such phases are invariably good glass formers at other compositions. In the crystalline phases one can think of each unit

328

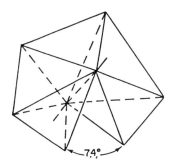

Fig. 1. Five perfect tetrahedra with common faces.

cell as trying to approach Coxeter's irrational value of $\bar{q}$ by a finite
rational fraction. In every one of these phases the value of q is within
0.1% of Coxeter's value.

Encouraged by the success of these ideas, it is natural to search
for a more quantitative understanding of the structural correlations in
super-cooled liquids. The approach we shall take is inspired by Landau's
theory of melting. [11] Landau's theory is good only near weak first
order transitions, where the predominant fluctuations in the liquid are
crystalline clusters of a size of the order of the correlation length.
Simple metal or metalloid alloys do not have weak first order transitions,
so Landau's theory is not very useful for these materials. I shall formu-
late the theory in a way in which the generalization to the case of tetrahedral close packing is straightforward.

Landau begins by writing down the free energy density of the liquid
in terms of the Fourier components of the density $\rho_q(r)$. These Fourier
components are determined by integrating over a coherence volume ΔV around
the point $\underset{\sim}{r}$.

$$\rho_q(\underset{\sim}{r}) = \int_{\Delta V} d\bar{r} \rho(\underset{\sim}{r} - \bar{\underset{\sim}{r}}) \exp(i\underset{\sim}{q} \cdot (\underset{\sim}{r} - \bar{\underset{\sim}{r}})) \tag{1}$$

Expanding the free energy density in powers of the ρ_q:

$$F\{\rho_q\} = \frac{1}{2} \sum_{\underset{\sim}{q}} \sum_{\{\underset{\sim}{G}\}} \left\{ \left[\frac{K_G}{2G^2} (q^2 - G^2)^2 + r_G \right] |\rho_q|^2 \right\}$$
$$+ \sum_{\underset{\sim}{q}_1 + \underset{\sim}{q}_2 + \underset{\sim}{q}_3 = 0} W_{q_1 q_2 q_3} \rho_{q_1} \rho_{q_2} \rho_{q_3} + \cdots \tag{2}$$

The sum over $\underset{\sim}{G}$ extends over reciprocal lattice vectors of possible crystal
structures. Near a transition to a BCC crystal, we would expect the dominant fluctuations to be BCC microcrystallites, and therefore the masses r_G
will be very large except for the special set of $\underset{\sim}{G}$ values which belong to
the reciprocal lattice of a BCC crystal. The physics determining the

329

terms in the free energy can be broken down to the following steps:

(i)   Identification of the ordered state (the BCC crystal).

(ii) Expansion of the density in a coherence volume around $\underset{\sim}{r}$ in terms of the representations of the translation group - $\exp(i\underset{\sim}{q}\cdot\underset{\sim}{r})$.

(iii) The complete point group symmetry of the ordered state determines which masses r_G will be large.

The theory can also be used to obtain the structure factor of liquid near the freezing transition. One obtains Lorentzian peaks in the structure factor at the reciprocal lattice vectors of the BCC lattice. Of course nothing like this happens in real life. One is not close to a weak first order transition and the peaks in the structure factor cannot be indexed at all by any crystalline reciprocal lattice. Nevertheless, as we shall show, the program of Landau can be extended to the case of tetrahedral close packing by a rather close adherence to the three basic steps above with an additional piece of physics: frustration.

The first step is the identification of the ordered state. In flat three-dimensional space there is no configuration with perfect five-fold tetrahedral ordering: the tetrahedra are always distorted. On the three-dimensional surface of a four-dimensional sphere (S3), however, there exists a configuration of 120 particles, in which every tetrahedron is perfect, every bond has five tetrahedra around it and every particle is surrounded by an icosahedral shell. This ideal solid is known as polytope $\{3,3,5\}$. The first step in the Landau program would be to expand the density in a coherence volume around the point $\underset{\sim}{r}$ in flat three-dimensional space, in terms of the irreducible representations of the symmetry group of the ideal space. Since the ideal space is curved, the best we can do is by expanding the projection of the coherence volume around $\underset{\sim}{r}$ onto a tangent sphere [12] (see Fig. 2). The relevant symmetry group is SO(4), the irreducible representations are the hyperspherical harmonics $Y_n(\hat{u})$ labeled by the representation n, and the order parameters are the $Q_n(\underset{\sim}{r})$.

$$Q_n(\underset{\sim}{r}) = \int_{\Delta V} d\hat{u} Y_n(\hat{u}) \ (\underset{\sim}{r},\hat{u}) \tag{3}$$

Polytope-120 has a point group symmetry which results in its density having non-zero components only for a special set of values of n = 12,20,24,30,32.... So we can immediately conclude that the masses r_n in the free energy for the Q_n, will be very large except for these special values of n.

The non-trivial part of continuing Landau's program is writing down the correct form of the gradient term. Sethna [13] proposed that the tangent spheres at neighboring points should be related by a rolling opera-

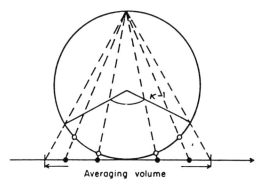

Fig. 2. Projection of the flat space density onto the tangent sphere over the coherence volume.

tion. If the sphere is rolled along a straight line, it will unroll upon flat space the configuration of tetrahedra along its equator. These tetrahedra form a Bernal spiral (Fig. 3). The rolling operation implies a free energy density of the form:

$$F = \frac{1}{2} \sum_n \int d\underline{r} \, (K_n | \partial_\mu - i\kappa L_{o\mu}) \underline{Q}_n |^2 + r_n | \underline{Q}_n |^2) + \ldots \tag{4}$$

Here $L_{o\mu}$ are the generators of the rolling operation for $SO(4)$, and κ is the inverse of the radius of the sphere.

Where is the frustration in this free-energy? When the sphere is rolled in one direction, the density along its equator can be unrolled upon the flat space with no distortions. Frustration, however, arises when we attempt to roll the sphere in a closed loop (Fig. 4). Due to the non-commutivity of the generators of rolling in different directions, the final configuration of the order parameter is related to the initial one by a rotation. This implies that there must be disclination lines threading the loop. The appearance of these defects in the ground state is a direct consequence of the frustration of tetrahedral close-packing.

A graphic illustration of this type of frustration can be seen by a close analysis of some Frank-Kasper phases. The Frank-Kasper phases with trigonal symmetry (the Laves phase and the μ phase) are made up of layers of Kagome nets (Fig. 5) alternating with layers of defect sites. Every

Fig. 3. Tetrahedra forming a Bernal spiral.

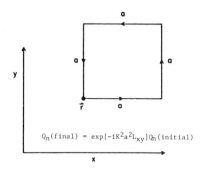

$$Q_n(\text{final}) = \exp[-iK^2a^2L_{xy}]Q_n(\text{initial})$$

Fig. 4. Rolling the sphere around a closed loop. The final state of the order parameter is related to the initial state by a rotation about an axis perpendicular to the plane of the loop.

hexagon in the Kagome net has a disclination line threading it. Rolling polytope-$\{3,3,5\}$ along the three sets of parallel lines which make up the Kagome net will reproduce the environment around these lines. Rolling the polytope in a loop will enclose the disclination lines threading the Kagome net at a density close to the value given by Fig. 4.

The free energy in Eq. 4 can now be used to describe structural correlations in the supercooled liquid. [14] As in the original Landau theory of weak first-order melting, we expect peaks in the structure factor at the 'reciprocal lattice vectors' of the ideal crystal, i.e., for each representation n which is non-vanishing upon polytope $\{3,3,5\}$. The energy spectrum of the density fluctuations is determined by the eigenvalues of the quadratic form in the free energy as a function of wavevector. The wavevector at which the minimum eigenvalue occurs is the position of the peak in the structure factor. The quadratic form will have such a minimum eigenvalue for each of the special n values 12,20,24,30,32 .... The positions of the peaks in the structure factor are absolutely determined, modulo the constant k. Using the value of k (known within 10%) one gets good agreement with experiment. However, the ratio of the peak positions is known absolutely, and Table 1 shows a comparison of the predictions of the theory with experiments on vapor-deposited amorphous metal films and with computer simulations. The agreement for the first two peaks is better than 1%, and the small discrepancy in the third peak can be well understood from the effects of higher order terms. [14] Treating K_n and r_n as adjustable constants, one obtains a fit to the entire structure factor. The results of such a fit are shown in Fig. 6.

So we see that our earlier statement that the peaks in the structure factor cannot be indexed to reciprocal lattice vectors of a simple lattice wasn't quite correct. The lattice is polytope $\{3,3,5\}$ which lives on a curved space and because of frustration, the reciprocal lattice vectors are non-abelian matrices $L_{o\mu}!$.

Table 1. Relative peak positions in the structure factor.

	Theory	Amorphous Cobalt[3]	Amorphous Iron[4]	Computer Simulations[5]
$\dfrac{q_{20}}{q_{12}}$	1.71	1.69	1.72	1.7
$\dfrac{q_{24}}{q_{12}}$	2.04	1.97	1.99	2.0

2. QUASI-CRYSTALS

Not all quasi-crystals can be described using the physics of tetrahedral close-packing. However, there is an important class of quasi-crystals which lends itself very naturally to description in such terms. These are the quasi-crystals with a structure similar to that of $Mg_{32}(AlZn)_{49}$. The most stable quasi-crystal known to date (made up of aluminum, lithium and copper) belongs to this class. The material $Mg_{32}(AlZn)_{49}$ also forms a BCC crystalline structure with a very large unit-cell (the Bergmann phase). Each unit cell in the Bergmann phase has 81 atoms, with the central 45 atoms forming an icosahedral cluster whose local coordination is exactly the same as polytope $\{3,3,5\}$. The remainder of the unit cell can be described rather nicely with a network of disclination lines. Henley and Elser [15] have recently shown that the quasi-crystalline phase also has a simple description in terms of disclination lines. One can therefore look upon this quasicrystalline phase as another way of relieving the frustration associated with tetrahedral close packing in flat space.

This relationship between quasi-crystals and tetrahedral close packing can also be seen within the framework of Landau theory. [16] The quasi-crystalline phase is most naturally described within the framework of density waves as in the original Landau theory of melting. So we have to transcribe the free energy Eq. 4 into density waves. Most of the work to achieve this has already been done--the spectrum of the quadratic form of the density wave expansion is simply the inverse of the structure factor in Fig. 6. So if the quasi-crystalline density wave is going to be a minimum of the free energy, the strongest density waves must have their wavevectors lying near the peaks in the structure factor of the metallic glass. We find that this is indeed the case. If the strongest peak in the diffraction pattern of the quasi-crystal occurs at $\mathbf{G}_0$, the three next

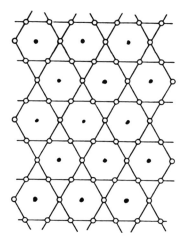

Fig. 5. A kagone set of atoms. The dark circles represent disclination lines perpendicular to the plane of the net.

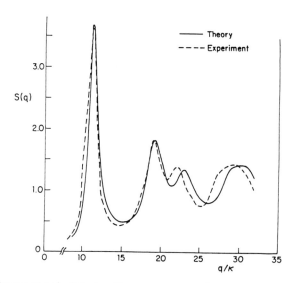

Fig. 6. Theoretical fit to the structure factor of a thin film of vapor-deposited cobalt.

strongest peaks are at $\underset{\sim}{G}_A = 1.051 \underset{\sim}{G}_o$, $\underset{\sim}{G}_B = 1.701 \underset{\sim}{G}_o$ and $\underset{\sim}{G}_C = 2 \underset{\sim}{G}_o$, all
falling very close to the peaks in the structure factor.

These considerations can be formalized within the framework of a
density-functional expansion. [16] As a result of such an analysis we
find that the expansion does indeed have a metastable quasicrystalline
state, but the BCC and FCC lattices are lower in energy. At first this
may seem surprising because the reciprocal lattice vectors of the BCC and
FCC lattices are not in registry with the peaks in the structure factor.
However, because of its incommensurability, the quasicrystalline phase has
density at very small wavevectors. These density waves extract a high
cost in free energy because the structure factor is very small at these
small wavevectors. This drives the energy of the quasi-crystal above that
of the conventional crystalline phases. Thus we see that the energy of
the quasi-crystalline phases is determined by a competition between the
favorability of short range icosahedral order (the wavevectors near the
peaks in the structure factor decrease the energy) and the frustration at
larger distances (the small wavevectors drive the energy up).

3. DYNAMICS OF METALLIC GLASSES

We have so far entirely side-stepped the question of how the
structural correlations lead to glass formation in the supercooled liquid.
We will attempt to develop a simple model of such glassy behavior [17] and
find some features which are experimentally relevant.

The philosophy of the approach is to assume that, as a result of
some as yet undetermined microscopic process, the icosahedral order-
parameter has a dramatic slowing down in its dynamics as one lowers the
temperature below the glass transition temperature. The relevant question
is: does this microscopic slowing down lead to a dramatic enhancement of
the macroscopic viscosity of the system and the appearance of a shear modulus? Let us consider the relaxation of the icosahedral order parameter,
which we shall now call ψ (it turns out that only one scalar component of
the icosahedral order parameter is important. In addition the identification of ψ as an icosahedral parameter is not particularly important for
the remaining analysis). Its convective coupling to shear momentum fluctuations is described by

$$\frac{\partial \psi}{\partial t} = - \Gamma \frac{\delta F}{\delta \psi} - g \underset{\sim}{J} \cdot \nabla \psi$$

$$\frac{\partial \underset{\sim}{J}}{\partial t} = \eta \nabla^2 \underset{\sim}{J} + g \nabla \psi \cdot \frac{\delta F}{\delta \psi} \tag{5}$$

Here Γ, the relaxation rate of the order parameter, becomes anomalously small near the glass transition. The shear viscosity is η, and g is a coupling constant. Does the effective viscosity at large distances become large, when the microscopic dynamics becomes sluggish? The answer depends sensitively upon the value of the width of the main peak in the structure factor. If the width is smaller than a constant c (which can be determined with no adjustable parameters), then the short distance slowing down propagates to large distances and the glass transition occurs. If the width is too large (i.e., the correlation length is small), the short distance freezing in is irrelevant and the system remains a liquid. Remarkably, the structure factors we have so far considered have a width which is just below this bound, leading us to conclude that the increase in the correlation length as the temperature is lowered may be one of the controlling factors in the glass transition.

To conclude: the statics of tetrahedral close packing has yielded valuable information about the position of the peaks in the structure factor of metallic glasses, and dynamic considerations have been used to put a constraint upon the width of the dominant peak.

4. ACKNOWLEDGMENT

The work described in Sections 1 and 2 was done in collaboration with David R. Nelson.

5. REFERENCES

1. W. Klement, R. H. Willens and P. Duwez, Nature 187, 809 (1960).
2. D. Turnbull, J. Chem. Phys. 20, 411 (1952).
3. F. C. Frank, Proc. R. Soc. London, Ser. A, 215, 43 (1952).
4. M. R. Hoare, Adv. Chem. Phys. 40, 49 (1979).
5. D. R. Nelson, Phys. Rev. B28, 5515 (1983).
6. C. H. Bennet, J. Appl. Phys. 43, 2727 (1972); J. F. Sadoc, T. Dixmier, and A. Guinier, J. Non-Cryst. Solids 12, 46 (1973).
7. K. Kimura and F. Yonezawa in "Topological Disorder in Condensed Matter," F. Yonezawa and T. Ninomiya (eds.) (Springer, 1983).
8. H. S. M. Coxeter, III, J. Math, 2, 746 (1958).
9. F. C. Frank and J. S. Kasper, Acta Cryst. 11, 184 (1958); 12, 483 (1959).
10. D. P. Shoemaker and C. B. Shoemaker, Acta Cryst. B42, 3 (1986).
11. See, e.g., G. Baym, H. A. Bethe, and C. Pethic, Nucl. Phys. A175, 1165 (1971).
12. D. R. Nelson and M. Widom, Nucl. Phys. B240 (FS12), 113 (1984).
13. J. P. Sethna, Phys. Rev. Lett. 51, 2198 (1983); Phys. Rev. B31, 6278 (1985).
14. S. Sachdev and D. R. Nelson, Phys. Rev. Lett. 53, 1947 (1984); Phys. Rev. B32, 1480 (1985).
15. C. L. Henley and V. Elser, Phil. Mag. B53, L59 (1986).
16. S. Sachdev and D. R. Nelson, Phys. Rev. B32, 4592 (1985).
17. S. Sachdev, Phys. Rev. B33, 6395 (1986).

LOCAL ATOMIC ENVIRONMENTS IN THE MANGANESE-ALUMINUM ICOSAHEDRAL PHASE

Gabrielle Gibbs Long and Masao Kuriyama

Institute for Materials Science and Engineering
National Bureau of Standards
Gaithersburg, MD 20899

Abstract   The atomic scale structure of icosahedral phase MnAl was
derived from experimental diffraction data. The resulting structure
is compared to theoretical work and to other experimental data. A
large array of this self-modulated structure, which is space-filling,
was used for site topology studies.

1.   INTRODUCTION

About a year an a half have passed since Schechtman et al. [1] reported
evidence of a new rapidly solidified phase of manganese-aluminum alloy which
surprisingly displays both icosahedral point group symmetry and sharp diffraction patterns. Even before the experimental reports of icosahedral
phases, Mackay [2] proposed a new crystallography for which sharp five-fold
patterns could exist. He suggested that two particular rhombohedra having
identical faces and complementary included angles (63.435° ... and 116.565°
...) be fitted together according to a set of non-periodic packing rules.
Kramer and Neri [3] demonstrated the association between the icosahedral
group A(5) and space-filling operations via the two rhombohedra designated
above.

After the discovery of an icosahedral phase in a binary metal alloy,
there have been numerous attempts to characterize the underlying structure
of such a material. As noted below, most of the work has been theoretical,
whereas the efforts of the present authors have been based on diffraction
experiments. In any event, all of these works have many elements in common.
This is particularly interesting in view of the very different starting
points of these studies. Using insights from icosahedral ordering in metallic glasses, Nelson and Sachdev [4] predicted a specific decoration of
rhombohedra in real space, and to each site they associated a probability of

337

occupancy. Coming from another direction, Audier and Guyot [5] and Henley
[6] proposed decorations of the two rhombohedra similar to reference 4 by
noting the resemblance between crystalline-AlMnSi and the icosahedral
phase. A phenomenological study by Bak [7] showed that the icosahedral
phase is essentially a multiple-q density wave structure. A microscopic
view of the icosahedral vertex mass density wave description was carried
out by Watson and Weinert [8]. They use their large sample of approximately 40,000 atoms to calculate radial distribution functions and the
local environments of atoms via Voronoi constructs. Since the three-dimensional materials are interpreted as the six-dimensional patterns produced
by a hypercrystal [9], one may hope to decorate the hyperunit cells. This
is presently not possible, but Jaric [10] suggests that the hyperlattice be
decorated by point-like atoms, then a slab can be cut parallel to physical
space, and finally the contents of the slab can be projected into physical
space. Finally, using a modification of the projection method [11], Elser
and Henley [12] described adjustments to α-AlMnSi which can be made to
relate it to icosahedral structures. Other descriptions of the icosahedral phase are not space-filling and therefore cannot be used for studies
of the atomic environments.

All of the above are theoretical approaches which have been used to
gain some understanding of the nature of icosahedral phase alloys. In
contrast, the analysis by the present authors [13,14,15] has been based on
experiments. The insights which are gained come from using analysis techniques appropriate for diffraction patterns taken from incommensurate modulated crystals. The derived structure, which is found to be neither dis-
placively modulated nor occupationally modulated in the usual sense, is a
special case of modulated crystal in which successively larger periodic
arrays are required. In this sense, the structure is actually self-
modulated. The atomic motifs, their relationship to the theoretical work
just mentioned and the topology of the icosahedral structure are the subjects which will be discussed in this paper. More complete references can
be found in reference 17.

2. THE ICOSAHEDRAL PHASE AND INCOMMENSURATE MODULATED STRUCTURES

The (kinematical) scattering amplitude for a system, such as an incommensurate modulated crystal that lacks lattice translational invariance is

$$S(k) = V^{-1} \int d^3r \phi(r) \exp[-i(H + \sum_i n_i q_i) \cdot r]$$
$$\times [(2\pi)^3 \delta(k - H - \sum_i n_i q_i)].$$

$$(1)$$

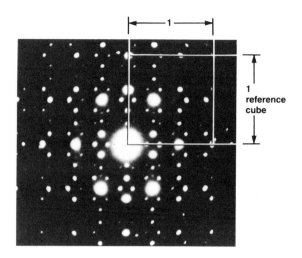

Fig. 1. An arbitrary reference frame is chosen using a two-fold electron
diffraction pattern taken from icosahedral MnAl.

H is the reciprocal lattice vector for a reference crystal lattice and the
q_i's are the deviations from the reference reciprocal lattice--the "modulations." The scattering amplitude is the product of two factors: the
"intrinsic" structure factor due to the lattice and the geometric structure
factor due to the decoration. For crystals which lack translational invariance (Eq. 1) these two generally cannot be separated.

A two-fold electron diffraction pattern taken from iMnAl (Fig. 1) can
be used to identify a reference reciprocal lattice and the special "modulations" that lead to the icosahedral symmetry. All of the spots in this
pattern, and indeed all of the spots in the three-fold and five-fold patterns as well, can be indexed using six rationally independent q_i vectors:

$$q_1 = q \sin \theta \ (\tau,1,0)$$
$$q_2 = q \sin \theta \ (1,0,\tau)$$
$$q_3 = q \sin \theta \ (0,\tau,1)$$
$$q_4 = q \sin \theta \ (-1,0,\tau) \qquad \qquad (2)$$
$$q_5 = q \sin \theta \ (0,\tau,-1)$$
$$q_6 = q \sin \theta \ (\tau,-1,0)$$

where $\tau = (1 + \sqrt{5})/2$. Only one length scale and two angles (63.435°...
and 116.565°...) are involved. It is important to notice [14] then
that the reference reciprocal lattice spots are created by the sums of
pairs of q_i's and another set of possible reference reciprocal lattice
spots is given by the differences of those same pairs. This means that
the underlying reference lattice cannot be identified--either all of the

339

spots are due to the "modulations" or all of the spots are due to the "reference reciprocal lattices," but there are no unambiguous mixtures of the two.

When the q_i in momentum space are converted into distances r_i in real space, there are again one length (0.456 nm) scale and the same two icosahedral angles involved. The position of each atom in the real space structure is given by a particular value of $\sum_{i=1}^{6} n_i r_i$ where the n_i are integers. The direction cosines for the r_i are:

$$
\begin{array}{ll}
r_1 & (\cos\theta, \sin\theta, 0) \\
r_2 & (\sin\theta, 0, \cos\theta) \\
r_3 & (0, \cos\theta, \sin\theta) \\
r_4 & (-\sin\theta, 0, \cos\theta) \\
r_5 & (0, \cos\theta, -\sin\theta) \\
r_6 & (\cos\theta, -\sin\theta, 0)
\end{array}
\qquad (3)
$$

where $\theta = 31.717°\ldots$. Since the set of allowed values of $\sum_{i=1}^{6} n_i r_i$ is quite complicated, it is helpful to select out first the simplest atomic motifs. This can be accomplished by separating the three r_i with the acute included angle and the three r_i with the obtuse included angle into two sets. The result is a derivation of the two Mackay rhombohedra mentioned above. Sites on the faces of the acute rhombohedron are identified when one adds r_4, r_5 and r_6 to r_1, r_2 and r_3 (see Fig. 2). Each of these face sites is $1/\tau$ the distance measured from the rhombohedron point to the opposite vertex. These important vertex and face sites correspond to the high occupancy sites in reference 4. We will return to a derivation of the lower occupancy sites below.

The acute and obtuse rhombohedra can be assembled such that the face sites are unambiguous if the faces of the acute rhombohedra are matched only to faces of the obtuse rhombohedra and vice versa. This forms a space filling structure within which four major directions 70.53° apart are defined. These correspond to four of the three-fold directions in the icosahedral point group. Along each of these directions, alternating acute and obtuse rhombohedra are aligned tip-to-tip. If each rhombohedron is decorated with atomic sites as defined in Fig. 2, then the remaining six three-fold symmetric directions of the icosahedral point group can be seen (Fig. 3). Thus it now becomes clear that it is artificial to denote various sites as "face" or "vertex" in the sense that they are all donated by $\sum_{i=1}^{6} n_i r_i$. This simple motif, in which each atomic site is described by this summation rule, can serve as a basis for the structure of iMnAl.

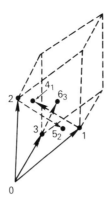

Fig. 2. The simplest atomic correlations near an origin 0 are shown. 1, 2, and 3 are atomic sites located distances r_1, r_2, and r_3 from 0. 4_1 is located a distance r_4 from site 1; 5_2 is located a distance r_5 from site 2; and 6_3 is located a distance r_6 from site 3.

Within the atomic motif described above, there are four groups of eight sites, each aligned along a different 70.53° axis. Each group of eight positions consists of two partial sublattices. When these sublattices are completed (using symmetry arguments given in reference 13), there are twice as many sites as originally. Within realistic atomic densities, only about half of these sites can be occupied by atoms. The additional sites derived from completing the sublattices are the same as the lower occupational probability sites derived in reference 4, where the authors of reference 4 already noted that all of the sites cannot be simultaneously occupied. It now remains to derive the occupancy rules for the icosahedral structure. The present authors have suggested that it would be reasonable to attempt an occupancy rule for which there is layering along the three-fold directions.

In the two-fold pattern (Fig. 4), four reciprocal lattice reference cubes can be seen. These scale as $1:\tau:\tau^2:\tau^3$, with the smallest being relevant to atomic ordering [16] and the largest being originally used [13] for indexing but perhaps more relevant to what we have called "internal modulations" [15]. There appear to be modulations on a scale smaller than atomic correlations rather than larger. Neither displacive nor simple occupational modulations can satisfy the requirements of the nested reference reciprocal lattices of Fig. 4. We have said instead that there is a staggering of layers with locally interpenetrating rhombohedral motifs as shown in Fig. 5. This means that the same atomic motif derived above starts, for example, at site 1, then starts again at A→1, and then later again elsewhere. The real space packing follows a sliding origin sequence

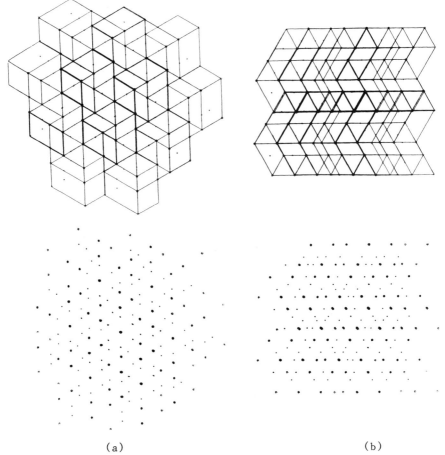

(a)

(b)

Fig. 3a. Patterns of atomic sites
in real space for a cut
perpendicular to a major
three-fold axis. The
solid lines show the
edges of rhombohedra
in the upper figure;
below, the edges are
suppressed.

Figure 3b. Patterns of atomic sites
in real space for a cut
perpendicular to a non-
major three-fold axis. It
can be verified that
three-fold symmetry is
preserved.

such that the unit is alternately stretched or compressed as the cell grows
larger. Since:

$$\tau^3 - \tau^{-3} = 4,$$

then $\tau^3 = 4.23...$ units are "compressed" into 4. On a larger scale, $\tau^4 +$
$\tau^{-4} = 11$, in which case there is "stretching" of the units. Thus, the

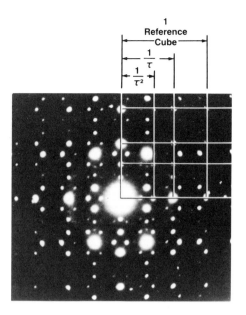

Fig. 4. Three possible reference frames are shown superimposed
on a two-fold patterns. The $1/\tau^3$ smallest frame, relevant to atomic correlations can be seen within the
$1/\tau^2$ frame.

nested reference frames require icosahedrally undistorted self-modulations
of the real space structure. The sliding origin description places certain
sites at positions which already existed and also creates new sites, from
the point of view of an absolute origin. Among the so-called new sites
which are seen in the atomic motif are sites on the edges of rhombohedra
identified in reference 6 from icosahedral modifications to the crystalline
α-phase of AlMnSi. The edge sites are actually sites which were already
present on the faces of rhombohedral motifs. When the motifs are
interpenetrating as in Fig. 5, it appears as if the sites are on the edges
of rhombohedra. It probably bears repeating that, while rhombohedra provide a useful means of visualization of sites, there is no intrinsic or
fundamental difference between different sites on edges, faces or vertices.

3. ATOMIC ENVIRONMENTS IN THE ICOSAHEDRAL PHASE

As mentioned above, the self-modulation of the icosahedral lattice
creates new sites from the point of view of an absolute origin. With
reference to a local origin, the number of site topologies may be expected
to be finite.

In periodic crystals, the number of Wigner-Seitz cells (or Voronoi
polyhedra) surrounding the atoms in the structure are finite. For icosahedral crystals, the number of local topologies is expected to be large.

One measure of the local environment of atoms in a structure is the coordination number, or the number of nearest neighbor atoms. Each nearest neighbor atom contributes one face to the atomic Wigner-Seitz polyhedron. Another measure of the local topology is a count of how many nearest neighbors are common to a particular near-neighbor pair. This is measured by the number of edges on the Wigner-Seitz cell face associated with the pair. Local atomic environments catalogued in this way have a variety of uses. They provide descriptions of the elementary building blocks of glasses. For example, the $(0,3,6,0)$ polyhedron, with no 3-sided faces, three 4-sided faces, six 5-sided faces and no 6-sided faces, is associated with the capped bipyramidal Bernal environment in metal-metalloid glasses. The Wigner-Seitz analysis is also relevant to the study of bonding and intrinsic geometry.

In the sliding origin description of iMnAl as described above, the site topologies for the simplest reference skeleton are $(0,6,6,1)$ for the "vertex" sites and $(0,4,6,2)$ for the "face" sites. For reasons of stoichiometry, it may be natural to assume that the (minority) manganese atoms occupy most of the vertex sites and aluminum (or silicon, in the case of iAlMnSi) atoms occupy the rest of the sites. However, as an atomic scale model is built up according to the self-modulated sliding origin rules, there are many new site topologies and sites visualized as "vertex" in the reference framework are sometimes occupied by aluminum atoms. This is rather important and represents an area in which the sliding origin model and references 5 and 6 diverge.

The Wigner-Seitz cells calculated for iMnAl are predominantly $(0,3,6,3)$, $(0,3,6,4)$, $(0,5,6,2)$ and $(0,4,6,2)$. The average coordination number is slightly greater than 12, suggesting close packing. This is consistent with results [17] calculated from the mass density wave description [8]. There are no $(0,0,12,v_6)$ cells, with $v_6 = 0,2,3$ or 4. Cells with these topologies are the building blocks of the Frank-Kasper phases. This means that the local structure of iMnAl or iMnAlSi is not Frank-Kasper-like. This may not be true for all icosahedral phase structure. For example, in iPdUSi the larger atom is the minority element and therefore the topology of that icosahedral material (which also exists over a very narrow compositional range) may be quite different.

The Mn partial radial distribution function for the sliding origin description has been compared out to 7 Å to experimental neutron diffraction results [19] (Figure 6). The small tick marks represent distances given by the vertex model such as in refs. 5, 6 and 16. The peaks between 0.5 and 0.6 nm are represented better by the sliding origin model, which is shown by the full lines. The long range positional correlations put the

344

icosahedral phase firmly into the crystal classification rather than glass despite the large number of site topologies that are present.

The Wigner-Seitz cells in iMnAl and in crystalline MnAl systems can be compared [18]. There is a broad distribution of crystalline cells, suggesting that manganese and aluminum atoms accommodate to one another in a

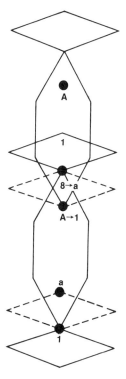

Fig. 5.   Possible stacking sequence along one of the 70.53° directions, illustrating interpenetration of rhombohedral motifs.

solid in a variety of ways. In the icosahedral phase, this is seen in the broad range of volumes per atom that is calculated. The polyhedra for the i-phase are shared by some of the crystal structures, and indeed are seen in other crystal structures of alloy systems which are glass formers.

4. DISCUSSION AND CONCLUSIONS

At the time of this writing, the sliding origin description [13,14,15] and the mass density wave description [8] are the only atomic scale models of iMnAl for which the atomic motifs fill space. For other descriptions, e.g., refs. 4, 5, and 6, there are remarkably similar arrangements of atomic sites on the icosahedral rhombohedra, but there are either holes and gaps or inconsistencies in the manner in which the structure is assembled.

Several remarkable new characteristics of the MnAl icosahedral structures are apparent. Unlike displacive or compositionally modulated structures, this structure appears to be self-modulated. A recurring feature in all of the crystalline-type models of this phase is that the atomic sites are not distorted from icosahedral point symmetries. In this, these models differ from those originating from a glassy description in which there can be a broader range of bond angles. Although it is natural to assume that iMnAl will have local ordering similar to that found in Frank-Kaspar phases, this work indicates that such ordering is probably not found in iMnAl at the local level, although it may still be relevant to other icosahedral phase alloys.

The radial distribution function calculated for the sliding origin descriptions of iMnAl shows a bimodal distribution of neighboring atoms

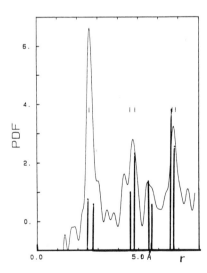

Fig. 6. Neutron diffraction results from reference 18. This is a partial radial distribution function of manganese atoms to other manganese, aluminum and silicon atoms in i-MnAlSi. The small tick marks indicate positions of manganese atoms according to predictions of vertex theories; the solid lines indicate positions of aluminum atoms and (minority) manganese atoms according to the sliding origin model.

between 0.25 and 0.30 nm in good agreement with EXAFS data [19]. On a larger scale, the calculated radial distribution functions are consistent with pulsed neutron diffraction data (18), picking up all the large peaks out to 0.7 nm, including some that are missing from the vertex model.

There is a broad distribution of Mn site Wigner-Seitz cell topologies, and an even broader one of Al site topologies. There is a wide range in site metrics and volumes. The volumes were found to span those appropriate to Mn and Al with 20% characteristic of Mn. Al atoms are found on both face and vertex sites in this model. This last has also been confirmed by the neutron results [18].

The many areas in which the sliding origin model of iMnAl is similar to other models are noteworthy. In this work, the key factor in assembling the common atomic motifs was to apply self-modulation of those motifs consistent with the observed diffraction patterns. That permitted the assembly of a large scale atomic model and the topological studies to explore the underlying nature of this fascinating new phase.

5. ACKNOWLEDGMENTS

The authors gratefully acknowledge the collaboration of Dr. L. Bennett of the U.S. National Bureau of Standards and Dr. Melamud of the Nuclear Research Center - Negev. Ber-Sheva, Israel in the Wigner-Seitz cell studies.

6. REFERENCES

1. D. Schectman and I. A. Blech, Met. Trans., 16A, 1005 (1985); D. Schectman I. A. Blech, D. Gratis, and J. W. Cahan, Phys. Rev. Lett. 53, 1951-1953 (1984).
2. A. L. Mackay, Physica 114A, 609-613 (1982).
3. P. Kramer and R. Neri, Acta Cryst. A40, 580-587 (1984).
4. D. R. Nelson and S. Sachdev, Phys. Rev. B32, 4592-4605 (1985).
5. M. Audier and P. Guyot, Phil. Mag. B53, L43-L51 (1986).
6. C. L. Henley, J. Non-cryst. Sol. 75, 91 (1985).
7. P. Bak, Phys. Rev. Lett. 54, 1517-1520 (1985).
8. R. E. Watson and M. Winert, Mater. Sci. and Eng. 79, 105 (1986).
9. A. Janner and T. Janssen, Acta Cryst. A36, 399-408 and 408-415 (1980).
10. M. V. Jaric, to be published in J. de Phys. (Serie colloques).
11. M. Dunneau and A. Katz, Phys. Rev. Lett. 54, 2688 (1985).
12. V. Elser and C. L. Henley, Phys. Rev. Lett. 55, 2883-2886 (1985).
13. M. Kuriyama, G. G. Long, and L. Bendersky, Phys. Rev. Lett. 55, 849-851 (1985).
14. G. G. Long and M. Kuriyama, Acta Cryst. A42, 156-164 (1986).
15. M. Kuriyama and G. G. Long, Acta Cryst. A42, 164-172 (1986).
16. V. Elser, Acta Cryst. A42, 36 (1986).
17. L. H. Bennett, M. Kuriyama, G. G. Long, M. Melamud, R. E. Watson and M. Weinert, Phys. Rev. B34, 8270-8272 (1986).
18. S. Nanos, W. Dmowski, T. Egami, J. W. Richardson, Jr., and J. D. Jorgensen, submitted to Phys. Rev. B.

19. P. A. Heiney, P. A. Bancel, A. I. Goldman and P. W. Stephans, Bull. Am. Phys. Soc. <u>31</u>, 221 (1986).
20. P. Bak, Phys. Rev. Lett. <u>56</u>, 861-864 (1986).

X-RAY SCATTERING FROM TWO-DIMENSIONAL LIQUIDS MODULATED BY A PERIODIC

HOST: THEORY, SIMULATION AND EXPERIMENT

George Reiter

Physics Department
University of Houston
Houston, Texas

1. INTRODUCTION

A fluid layer on a surface, such as Helium on graphite, or intercalated between the planes of a crystal, such as Rubidium in graphite, provide examples, upon solidification, of incommensurate solids. Despite extensive theoretical and experimental work [1], the structure of these solid phases is still controversial. This is due to the difficulty of extracting the structure from the available experimental evidence [2], the difficulty of doing an accurate theory in the regimes of interest, and the lack of knowledge of essential physical parameters, in particular, the substrate fluid potential and the interaction potential between the atoms or ions of the fluid. The latter quantity is not actually used in most theoretical predictions [3], which use instead effective parameters which would be difficult to relate to the physical parameters even if known, making possible a wide range of theoretical predictions. We present here theoretical, numerical and experimental results for Rb intercalated in graphite in the fluid phase and show that it is possible to extract from the x-ray scattering data the substrate-intercalant potential. We expect to be able to extract as well the intercalate-intercalate interaction potential, making possible molecular dynamics simulations that should reveal the structure of the solid phase as well as permit an interpretation of neutron scattering data on the dynamics of the intercalate. The essential features of the x-ray scattering patterns are exhibited and explained theoretically.

We will begin with a discussion of the physical systems used in the experiments, and describe the observations that need to be explained. The bulk of the paper will be concerned with the theory of these effects, and we will describe briefly the results of some molecular dynamics simulations that have been performed.

2. INTERCALATED GRAPHITE

When graphite is heated in the presence of alkali atom vapor, the alkali atoms are absorbed by the graphite and fill in the planes between adjacent graphite layers normal to the c-direction. The basic mechanism for this is the transfer of the outer alkali electron to the graphite, lowering the energy of the whole system. The alkali atoms may be between every graphite plane, in which case the intercalant is called stage 1, or every other plane (stage 2), or every third plane (stage 3), etc., depending upon the conditions of temperature and pressure. We will be concerned with stage 2 material. Intercalate structures can be formed in graphite crystals, or in highly oriented pyrolitic graphite, AOP, which consists of crystallites of characteristic size about 3000A, oriented so that their c-axes are identical but with all possible orientations of the a and b axes. Experiments measuring correlations along the c-direction show that these are negligible, so that we will neglect any interactions between planes of alkali metal. These are essential for treating the mechanism of staging, but are too weak to affect the x-ray scattering patterns. The physical system can then be thought of as a stack of independent alkali-graphite sandwiches.

Figure 1 shows the x-ray scattering pattern from a single crystal of stage 2 Rb at room temperature [4]. The solidification temperature is 140 K, so the alkali is a liquid. We have included also a sketch to highlight the important features of the pattern. The most striking feature is the halos that show up about the [0,1] Bragg peaks. At the temperature is lowered through the freezing point of the layers, the halos will go continuously to the satellite peaks of the incommensurate solid phase. It should not be surprising, then, that the structure of the halos is determined by the pair correlation function of the liquid. It should also be noted that the halo about the origin, which is just the diffuse scattering from the liquid from which the pair correlation function is measured, has an angular modulation, as do the halos. This angular modulation becomes more severe as the temperature is lowered, the high points becoming the Bragg peaks of the two-dimensional solid. A third effect, which cannot be seen in the photo, is that there is a contribution to the graphite Bragg peaks from the alkali. It is this last effect that enables one to measure the intercalate-substrate potential, as the intensity of this contribution depends upon the strength of alkali density fluctuations at the graphite reciprocal lattice points, which depends in turn on the strength of the

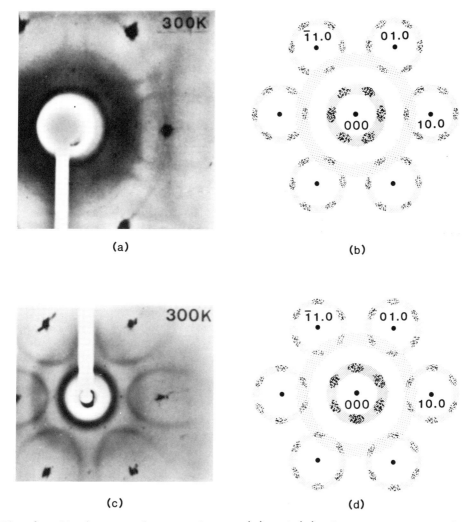

(a)

(b)

(c)

(d)

Fig. 1. Single crystal x-ray photos, (a) and (c), from Rousseaux et al.
(see Ref. 6) and schematic representations (b) and (d) of the
principal effects including angular variation of the diffuse
scattering and halos. The distortion in the data, due to
cylindrical film geometry is not reproduced in (b) or (d).
(a) and (b): $C_{24}K$ with peaking in [10.0]; (c) and (d): $C_{24}Rb$
with peaking in [11.0].

coupling. To summarize, the three effects of the coupling of the alkali
fluid to the graphite substrate are:

1. An alkali contribution to the graphite Bragg peaks.

2. Halos about some of the graphite Bragg peaks.

3. An angular modulation of the alkali diffuse scattering
    (liquid structure factor).

These effect may all be given a satisfactory theoretical explanation.

Theoretical Description of Intercalated Layer

We will consider a fluid layer contained between two graphite layers. A surface layer has a nearly identical treatment. The potential felt by a Rb atom due to its interaction with the graphite layer can be expanded rigorously as

$$V(\vec{r}) = \sum_{H,K} V_{HK}(z) e^{2\pi i (H\vec{b}_1 + K\vec{b}_2) \cdot \vec{r}} \tag{1}$$

where the $\vec{b}_i$ are the graphite lattice basis vectors. We will assume that only $V_0(z)$ has a significant z dependence so that

$$V(\vec{r}_{\parallel}, z) = V_0(z) + \sum_{H,K} V_{HK} e^{2\pi i (H\vec{b}_1 + K\vec{b}_2) \cdot \vec{r}_{\parallel}} \tag{2}$$

This is an essential simplification, since the z motion is now separable from the in-plane motion, and contributes experimentally only to a Debye Waller factor. The V_{HK} that are measured may be thought of as effective, averaged over z, parameters. Likewise, in the interaction between Rb atoms, we will neglect any z dependence so that the Hamiltonian becomes

$$\mathcal{H} = H(z) + H^{2D}_{Rb-Rb} + H_{Rb-G} \tag{3}$$

We will choose a Yukawa potential for the Rb-Rb interaction, $\frac{Ae^{-br}}{r}$. A and b will be chosen to give a good fit to the observed structure. This is consistent with physical intuition, and some calculations of Plischke [5].

As a result of the modulation potential in (2) there are static density fluctuations induced in the fluid. With $\rho_q = \sum_i e^{i\vec{q}\cdot\vec{r}_i}$, we will have

$$\langle \rho_q \rangle = \sum \langle e^{iq_\perp z_i} \rangle_0 \langle e^{i\vec{q}_{\parallel}\cdot\vec{r}_i} e^{-\beta H_{Rb-G}} \rangle_0 / \langle e^{-\beta H_{Rb-G}} \rangle_0 = N e^{-M_\perp} \langle \rho'_q \rangle \tag{4}$$

where M is the Debye Waller factor for out of plane motion and

$$\langle \rho'_q \rangle = \frac{\left\langle e^{i\vec{q}_{\parallel}\cdot\vec{r}_1} \prod_{i=1}^{N} e^{-\beta V_{Rb-G}(\vec{r}_i)} \right\rangle_0}{\left\langle \prod_{i=1}^{N} e^{-\beta V_{Rb-G}(\vec{r}_i)} \right\rangle_0} \tag{5}$$

where $\langle \ \rangle_0 = \text{Tr} \, e^{-\beta \mathcal{H} H^{2D}_{Rb-Rb}}$. If one neglects the correlations in the fluid entirely, then

$$\langle \rho'_q \rangle = \frac{F^1(q)}{F^1(0)} \tag{6}$$

where

$$F^1(q) = \frac{1}{V_c} \int e^{i\vec{q} \cdot \vec{r}_\parallel} e^{-\beta \sum\limits_{H,K} i\vec{q}_{H,K} \cdot \vec{r}_\parallel V_{HK}} d\vec{r}_\parallel \tag{7}$$

V_c is the volume of the graphite unit cell, and the integration is over one such cell. $F^1(q)$ is only non-zero at graphite reciprocal lattice vectors. As it turns out, there are small (10%) corrections to (7) at the temperature the experiments were done (room temperature) and these become more significant as the temperature is lowered to the freezing point for the layer (140 K). These may be included by means of a cluster expansion which neglects three particle and higher correlations but includes pair correlations. The result, derived in ref. 6 is that (6) is modified to

$$\langle \rho'_q \rangle = \frac{F^1(q)}{F^1(0)} e^{[F^2(q) - F^2(0)]} \tag{8}$$

where

$$F^2(q) = \sum_{q_{HK}} F'(q_{H,K}) F'(q-q_{HK}) [S(q_{HK}) - 1]/F^1(q)F^1(0) \tag{9}$$

$S(q)$ is the liquid structure factor for the layer in the absence of a substrate. This is not experimentally accessible, but simulations we have carried out show that at the temperatures of the experiment, and measured interaction strength, the differences are not significant. We show in Fig. 2 the measured $S(q)$, with Bragg peaks eliminated, and an obvious effect of the substrate, the contribution from the halos, removed. As the experiments are done on HOPG the result is a circularly averaged correlation function. Shown also are two sets of $\langle \rho'_q \rangle$ that would result from this structure factor if the spacing of the graphite were varied, for two different values of the interaction strength, V_{10} with all other coefficients assumed to be zero. The physical value is $Q_{10} = 2.94 \text{ Å}^{-1}$. The dominant feature of the figure, the peak at the position of the first maximum in the liquid structure factor, shows that the liquid would be strongly modified, even at these temperatures, if the interparticle spacing of the fluid coincided with that of the graphite. For sufficiently small V_{HK}, the response is $\rho'_{10} = \beta V_{10} S(q_{10})$, with all others zero, but for the larger value of

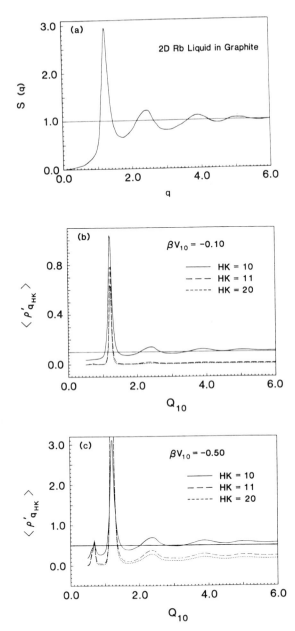

Fig. 2. (a) The measured liquid structure factor, with the effects of substrate-induced halos removed, for Stage 2 Rubidium in graphite, at room temperature. (b,c) The effect of varying the graphite lattice spacing, keeping the interparticle spacing in the liquid fixed, for two different values of the modulation potential. Only one Fourier component has been included in the potential.

V_{10}, one can see that there is a significant contribution to ρ'_{11}, ρ'_{20}, even though the potential has no Fourier component at these wavevectors.

Formula 8 may be inverted numerically, if a set of $\langle \rho'_q \rangle$ together with $S(q)$ are measured, to obtain the V_{HK} with which they are consistent, providing a means of measuring the V_{HK}. This has been done for two sets of data. One of these was obtained by standard x-ray crystallography on the intercalated materials by K. Ohshima et al.[7]. The other was obtained by C. Thompson [12], using the variation in scattering amplitude for the Rb atoms near an absorption edge to separate out the Rb and graphite contributions. Both works were done in collaboration with S. C. Moss. The measured values were fit to a set of V_{HK}, and the average of the two sets of fitted parameters are displayed in Table 1, along with the values of $\langle \rho'_q \rangle$ that result from these V_{HK}, and the experimental values. Note that there is only one significant Fourier coefficient, V_{10}. Using the measured value for this coefficient, molecular dynamics simulations have been done by J. D. Fan and O. Karim [9]. The system has 736 particles, with the density chosen to agree with the experimental values. The interparticle potential coefficients A and b were adjusted to give a good fit between the measured and calculated $S(q)$, and the values of $\langle \rho'_q \rangle$ were evaluated. The results are also shown in Table 1. The excellent agreement with the measured values shows that the three particle correlations neglected in deriving (8) are not significant at these temperatures. The potential that results from the coefficients of Table 1 is shown in Fig. 3. The barrier height to go from the center of the carbon hexagon over the ridge separating two carbon atoms is .092 eV. This is to be compared to the barrier heights of between .05 and .1 eV inferred by Zabel et al [10] from neutron scattering data.

3. THEORY FOR THE DIFFUSE SCATTERING

To lowest order in the strength of the substrate potential, the change in the diffuse scattering is

$$\delta[\langle \rho_q \rho_{-q} \rangle - |\langle \rho_q \rangle|^2] = \sum_{HK} (\beta V_{HK})^2 \langle \rho_q \rho_{-q} (\rho_{q_{HK}} \rho_{-q_{HK}} - \langle \rho_{q_{HK}} \rho_{-q_{HK}} \rangle) \rangle, \quad (10)$$

To understand this quantity, it is useful to observe that the change can be thought of as arising from an effective two particle potential

$$\delta H = \sum_{r_1, r_2} \delta V(\vec{r}_1 - \vec{r}_2) = \sum_{r_1, r_2} \sum_{H, K} e^{i \vec{q}_{HK} \cdot (\vec{r}_1 - \vec{r}_2)} \delta V'_{HK} \quad (11)$$

$$= \sum_{H, K} \delta V'_{HK} \rho_{q_{HK}} \rho_{-q_{HK}}$$

Table 1. The Measured Rubidium Contribution to
the Graphite Structure Factor Together
with the Average of the Values of βV_{HK}
Calculated via Eqs. (6)-(9). The last
column uses βV_{10} in an MD simulation to
compare with the input data.

HK	$< \rho'_{q_{HK}} >$		βV_{HK}	$< \rho'_{q_{HK}} >_{MD}$
	Ref. 7	Ref. 12		Ref. 9
10	0.48 + .03	0.48 + .03	-0.45	0.48
11	0.19 ± .03	0.22 ± .03	-0.06	0.18
20	0.14 ± .02	-	-0.01	0.13
21	0.04 ± .02	0.04 ± .02	0.03	0.04
30	0.04 + .02	0.03 ± .02	-0.01	0.02
22	0.02 ± .02	0.00 ± .04	-0.02	0

Fig. 3. The substrate modulation potential, $V(\vec{r})$, for a Stage 2 Rb liquid
in graphite. The contours are in steps of 0.0037eV and the change
from attractive ($V(r) < 0$) to repulsive is noted as a dashed
contour. $V(r=0) = - 0.0735eV$

with $\delta V_{HK}^1 = -\beta V_{HK}^2$. The first order change in the scattering produced by this effective, orientation dependent potential would be just that given by Eq. 10.

The pair correlation function is related to the two-particle density function by $p(\vec{r}_1,\vec{r}_2) = \rho^2 g(\vec{r}_1 - \vec{r}_2)$ and to $\langle \rho_q \rho_{-q} \rangle$ by

$$\langle \rho_q \rho_{-q} \rangle = N\left[1 + \rho \int e^{i\vec{q}\cdot\vec{r}} g(\vec{r})d\vec{r}\right] = N[1 + \rho g(\vec{q})].$$

and hence

$$\frac{\partial \langle \rho_q \rho_{-q} \rangle}{\partial V_{HK}} = -\beta \langle \rho_q \rho_{-q} (\rho_{q_{HK}} \rho_{-q_{HK}} - \langle \rho_{q_{HK}} \rho_{-q_{HK}} \rangle) \rangle$$

$$= N\rho \left.\frac{\partial g(\vec{q})}{\partial V'_{HK}}\right|_{V'_{HK}=0}$$

The left hand side, when multiplied by $-\beta V_{HK}^2$ and summed over H, K is then the expression (10). The right hand side can be calculated approximately analytically by using one of the self-consistent integral equations that give $g(r)$ in terms of the two-particle potential. We choose the Percus-Yevick [11] equation

$$g(\vec{r}) = e^{-\beta V(\vec{r})}\left[1 + \rho \int d\vec{r}' g(\vec{r}-\vec{r}')(1 - e^{\beta V(\vec{r}-\vec{r}')})\ [g(\vec{r}')-1]\right]$$

and by differentiation obtain

$$\frac{\partial g(r)}{\partial V'_{HK}} = -\beta e^{i\vec{q}_{HK}\cdot\vec{r}} g(\vec{r}) - \beta \rho e^{-\beta V(\vec{r})}\int d\vec{r}' g(\vec{r}-\vec{r}') e^{\beta V(\vec{r}-\vec{r}')}[g(\vec{r}')-1]e^{i\vec{q}_{HK}\cdot(\vec{r}-\vec{r}')}$$

$$+\rho \int d\vec{r}'\frac{\partial g(\vec{r}-\vec{r}')}{\partial V'_{HK}}(1 - e^{\beta V(r-r')})[g(\vec{r}')-1] + \rho \int d\vec{r}' g(\vec{r}-\vec{r}')(1 - e^{\beta V(r-r')})\frac{\partial g(\vec{r}')}{\partial V'_{HK}}.$$

This is an integral equation for $\dfrac{\partial g}{\partial V'_{HK}}$. The leading term, when Fourier transformed, leads to $-\beta g(q+q_{HK})$. This corresponds to a halo centered around $-q_{HK}$ with an intensity that is proportional to $(\beta V_{HK})^2$. Since only V_{10} is significant in our measurements, we have an immediate explanation of why there are no halos seen about other graphite Bragg points in Fig. 1. The remaining terms that are all proportional to the density are responsible for the angular modulation of the central ring and the halos. In fact, if we take the low density expression

$$g(\vec{r}) = \exp\left[-\beta\left[V_0(\vec{r}) + \sum_{H,K} \delta V'_{HK} e^{i\vec{q}_{HK}\cdot\vec{r}}\right]\right]$$

357

and expand it for small $\delta v'_{HK}$, we see that it produces only halos, but no modulation. An accurate treatment of the modulation requires solving the integral equation, which we have not done yet. An approximate calculation is described in ref. 5. We find the leading term is of the form

$$\delta < \rho_q \, \rho_{-q} > \, = N(\beta V_{10})^2 a_1(q) \cos 6\theta$$

where $a_1(q)$ is proportional to ρ^2 and is given in term s of integrals of Bessel functions against the products of pair correlation functions. The sign of $a(q)$ is highly sensitive to the relative size of intercalate and graphite. A positive sign leads to enhancement of the 10 direction, as observed in Fig. 1, a negative sign the 11 direction, as observed in Cesium intercalates.

4. CONCLUSION

The intensities of the density fluctuations that result in the intercalated fluid from its interactions with the graphite substrate, can be calculated accurately, using the measured pair correlation function, and the calculations used to obtain the interaction potential from the measured contribution of the intercalate to the Bragg peaks at the graphite positions. The other prominent features of the scattering from the liquid, the halos and the angular modulation of the diffuse scattering, can be at least qualitatively understood from perturbation calculations and standard self-consistent theories of the liquid phase.

5. ACKNOWLEDGEMENTS

This work was carried out in close collaboration with Simon Moss. I particularly wish to acknowledge Lee Robertson for doing the numerical work needed, and J.D. Fan for the data from his simulations. This work was supported by the National Science Foundation under Grant DMR-8603662.

6. REFERENCES

1. M.S. Dresselhaus, G. Dresselhaus. J.E. Fischer, and M.J. Moran (eds.). "Intercalated Graphite" (North-Holland, New York, 1983); T.F. Rosenbaum, S.E. Nagler, P.M. Horn, and R. Clarke, Phys. Rev. Lett. 50, 1971 (1983).
2. R. Clarke, N. Caswell, S.A. Solin and P.M. Horn, Phys. Rev. Lett. 43, 1552 (1979).
3. R.G. Caflisch, A. Nihat Berker, M. Kardar, Phys. Rev. 31, 4527 (1985).
4. G.S. Parry, Mater, Sci. Eng. 31, 99 (1977)
5. M. Plischke and W.D. Leckie, Can. J. Phys. 60, 1139 (1982).
6. G. Reiter and S.C. Moss, Phys. Rev. B33, 7209 (1986).
7. K. Ohshima, S.C. Moss and R. Clarke, Synth. Met. 12, 125 (1985).
8. C. Thompson, S.C. Moss, G.Reiter and M.E. Misenheimer, Synth.Met.12,57(1985).
9. J.D. Fan and O. Karim, work in progress.
10. H. Zabel, A. Magerl, A.J. Dianoux and J.J. Rush, Phys.Rev.Lett. 50,2094(1983).
11. S.A. Rice and P. Gray, The Statistical Mechanics of Simple Liquids. (Wiley-Interscience, New York, 1965), p100.
12. C. Thompson, Univ. of Houston, Dept. of Physics, Ph.D. thesis, 1987(unpub.).

ARTIFICIALLY STRUCTURED INCOMMENSURATE MATERIALS

Roy Clarke and R. Merlin

Department of Physics
The University of Michigan
Ann Arbor, MI 48109

Abstract. Modern ultrahigh vacuum deposition techniques such as
Molecular Beam Epitaxy offer interesting opportunities for the fabri-
cation of materials in which the incommensurate nature of the structure is expected to dominate the physical properties. We have
recently demonstrated the MBE growth of heterostructures in which
layers of GaAs and AlAs were deposited in a Fibonacci sequence. This
yields a quasi-periodic structure with the ratio of incommensurate
periods equal to the golden mean, τ. Here we present an overview of
the unique structural, electronic, and vibrational properties of this
new class of materials emphasizing the role of the incommensurate
structure normal to the layers. Inevitably, defects are introduced
by growth fluctuations but do not appear to disrupt significantly the
special characteristics which originate from the quasiperiodic ordering.

1. INTRODUCTION

The MBE technique lends itself ideally to the artificial structuring
of layered materials on nanometer length scales. Over the past fifteen
years much effort has gone into making heterostructures which have periodically repeating layers, i.e., superlattices. Recently, a new class of
heterostructure was demonstrated in which the constituent layers are deposited not in a periodic fashion but according to a predetermined, yet non-
repeating, mathematical sequence based on the Fibonacci series [1]. This
yields a so-called 'quasiperiodic' (or incommensurate) structure with very
interesting and unusual properties. Some special properties of incommensurate materials, and some of the limitations likely to be encountered in

their measurement, have been reviewed by Professor Sokoloff at this Workshop [2].

It has long been recognized that quasiperiodic ordering could offer interesting possibilities for experimental studies of novel physical phenomena [3]. The motivation for this was, in large part, theoretical work on quasiperiodic one-dimensional (1D) wave equations revealing spectra and eigenstates that are quite unlike those of periodic or random ID systems [4].

The major problem in fabricating quasiperiodic structures has been the fact that simple incommensurate modulations require increasingly larger layer thicknesses to approach the irrational limit. Layer deposition in sequences generated by special growth rules provide a solution to this problem [1]. Heterostructures grown according to these sequences show a degree of quasiperiodicity that is determined not by the width of individual layers (which is arbitrary), but by the thickness of the sample [1].

A number of studies [6-10] have focused on the class of structures derived from the Fibonacci sequence. The general properties of these materials and results of our specific experiments on Fibonacci GaAs-$Al_xGa_{1-x}As$ heterostructures are discussed in this paper.

2. STRUCTURAL PROPERTIES AND X-RAY SCATTERING

The Fibonacci sequence ABAABABAA... is defined by the production rules A $\rightarrow$ AB and B $\rightarrow$ A, with A as the single element of the first generation. To obtain a Fibonacci superlattice, one simply replaces A and B by two arbitrary blocks of layers [1,8]. This arrangement leads to a structure (see Fig. 1) which shows two basic reciprocal periods in a ratio given by the golden mean $\tau = (1 + \sqrt{5})/2$, perpendicular to the layers [1]. The associated structure factor consists of a dense set of components such that δ-function diffraction peaks are expected at wavevectors given by k = $2\pi d^{-1}$ (m+nτ), where m and n are integers and d = $\tau d_A + d_B$; d_A and d_B are the thicknesses of the building blocks A and B [1]. Using the projection method of Elser [11] and Zia and Dallas [12] it can easily be shown that the amplitude of the structure factor of the individual peaks is determined by the function $\sin\phi_{nm}/\phi_{nm}$, where

$$\phi_{nm} = \frac{\pi\tau d_B}{\tau d_A + d_B} (m\tau - n). \tag{1}$$

The argument ϕ_{nm} is related to the transverse (or "phason") momentum, $g_\perp$, such that X-ray peaks associated with $g_\perp \approx 0$ will be the most intense features of the diffraction pattern. Thus, even though the expected pattern is dense, it should be possible in practice to distinguish more-or-less

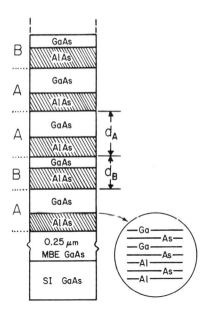

Fig. 1. Schematic arrangement of GaAs and AlAs layers deposited in a Fibonacci sequence ABAABABA... For the sample considered in section 2, the AlAs strata are nominally of identical thicknesses (~17Å); $d_A \approx 59$Å, $d_B \approx 37$ Å.

isolated peaks characteristic of the quasiperiodic ordering. Note that the observation of a true phason-like excitation in this system is highly unlikely since atomic displacements of many lattice spacings would be required and the atomic mobility is extremely small.

X-ray diffraction patterns have been obtained [5] from a Fibonacci heterostructure consisting of A $\equiv$ [17Å AlAs-42Å GaAs] and B $\equiv$ [17Å AlAs-20Å GaAs]; the sample was grown by molecular beam epitaxy on a (001) GaAs substrate. The patterns for $\vec{k}$ parallel to [001], in Fig. 2, demonstrate many of the unusual properties of quasiperiodic ordering. The synchrotron X-ray data indicate that, at least up to the instrumental resolution ($\simeq 0.0015$Å^{-1} FWHM), the diffraction peaks do indeed form a dense set. Moreover, the measurements agree remarkably well with the calculated profile for an ideal Fibonacci structure (solid curve in Fig. 2). [5]. This, and also numerical simulations [5], show that quasiperiodic ordering is largely insensitive to the unavoidable random fluctuations in the growth parameters such as the deposition fluxes and the timing of the source shutters. This aspect is demonstrated in an interesting way by the low-angle X-ray scattering (Fig. 3), which consists entirely of high $g_\perp$ (i.e., high order m,n) contributions. Even in the perfect Fibonacci superlattice these peaks are expected from (1) to be quite weak. That they are not suppressed entirely indicates that the underlying quasiperiodic order is not seriously disrupted by growth fluctuations. In other words, the defects introduced by the

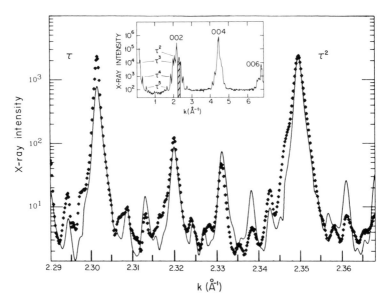

Fig. 2. High resolution (0.0015Å⁻¹ FWHM) synchrotron X-ray data (dots)
and calculated diffraction profile for the ideal Fibonacci
structure (solid line); the sample is described in Section 2.
The inset shows indexing of strong peaks in a low-resolution
scan (0.03Å⁻¹ FWHM). The shaded portion of the inset is
shown on an expanded scale in the main body of the figure.

method of growth are probably not phason-like in character [12]. A further interesting effect is observed in the extended low-resolution scan of
the inset; the dominant zinc-blende peaks and their satellites can be
expressed as $k=2\pi d^{-1}\tau p$, with integer p. This behavior results from d_A/d_B
being close to τ in our sample [5]. It is <u>not</u> a general feature of Fibonacci heterostructures.

3. ELECTRONIC AND OPTICAL PROPERTIES

Tight-binding models describing electrons in Fibonacci structures have
been studied by several authors [7-10]. The spectrum is a Cantor set characterized by clusters of eigenvalues that divide into three subclusters [7-
10]. The wavefunctions are critical (i.e., neither localized nor extended)
exhibiting either self-similar or chaotic behavior [9,10]. This applies
only to bulk states; finite samples can further show solutions localized at
the surfaces [9].

Investigations of the electronic structure of Fibonacci GaAs-Al_xGa_{1-x}
As heterostructures using standard optical probes reveal mainly excitonic
features. The example of a structure with A $\equiv$ [20Å $Al_{0.3}Ga_{0.7}As$-40Å GaAs]
and B $\equiv$ [20Å $Al_{0.3}Ga_{0.7}As$-20Å GaAs] is given in Fig. 4(a). Fig. 4(b) shows
results of effective-mass tight-binding calculations of the electron,

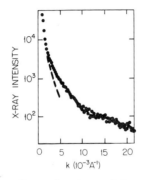

Fig. 3. Low-angle portion of synchrotron data showing contribution of
overlapping large-$g_\perp$ peaks. The dashed line represents the
instrumental resolution function.

heavy-hole and light-hole spectra of this sample, for motion normal to the
layers. These results were used to further determine the Im(χ) vs. photon-
energy plots shown in Fig. 4(a). The comparison with the experimental
data indicates a correlation between the position of the exciton peaks and
the largest calculated plateaux; the latter reflect major gaps in the 1D
spectrum of Fig. 4(b). Since surface states may also occur at these gaps,
it is not clear whether the excitons derive from bulk critical states or
from quasi-2D states localized at the surface.

4. RAMAN SCATTERING BY ACOUSTIC PHONONS AND VIBRATIONAL PROPERTIES

The spectrum of phonons in 1D Fibonacci lattices shows a self-similar
hierarchy of gaps which decrease in size with the phonon frequency
[7,9,10]. As expected, the eigenfunctions are extended in the continuum
limit. Their high-frequency behavior is not as yet well understood
although there is some evidence favoring localization [7]. If this turns

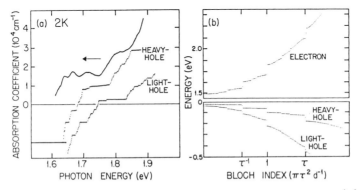

Fig. 4. (a) Optical absorption coefficient and calculated Im(χ) for
transitions involving heavy- and light-hole states, in
arbitrary units. (b) Energy vs. Bloch index. Heterostructure
parameters are indicated in Section 3.

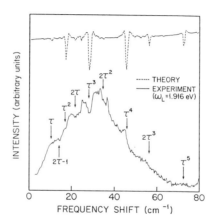

Fig. 5. Room temperature resonant Raman spectrum of the heterostructure
considered in Section 4 corrected for thermal factors, and calculated density of states of LA modes propagating along [001]
(dashed curve). Arrows denote expected midfrequencies of main
gaps in units of $\pi c d^{-1}$; c is the average sound velocity.

out to be correct, a transition between extended and localized states may
take place at intermediate frequencies.

Raman scattering has been used to study longitudinal acoustic (LA)
phonons propagating parallel to the growth axis in Fibonacci
GaAs-Al$_x$Ga$_{1-x}$As heterostructures [6]. The results reveal important differences between resonant and non-resonant spectra. For the latter, the scattering is largely determined by structural effects as in the case of periodic superlattices [14]. Off-resonance data from the sample described in
Section 2 show doublets centered at frequencies that follow a $_\tau$P-behavior
[6], consistent with the X-ray findings. The Raman spectrum of the same
sample obtained under resonant conditions is shown in Fig. 5. The scattering reflects now a weighted density of states of LA modes, providing an
experimental demonstration of the richness of the phonon spectrum. Reso-
nances with electronic states localized at the surface can possibly account
for this behavior [6].

The contributions of K. M. Mohanty and J. D. Axe to the synchrotron
results are gratefully acknowledged. We also wish to thank F.-Y Juang and
P. K. Bhattacharya for growing the excellent samples used in this study,
J. Todd for help with the X-ray analysis and K. Bajema for his contribution to the optical and Raman scattering work.

The research was supported in part by ARO Contracts DAAG-29-83-K-0131
and DAAG-29-85-K-0175.

5. REFERENCES

1. R. Merlin, K. Bajema, R. Clarke, F.-Y Juang and P. K. Bhattacharya,
    Phys. Rev. Lett. 55, 1768 (1985).

2. J. B. Sokoloff (see these proceedings).

3. J. B. Sokoloff, Phys. Rev. $B22$, 5823 (1980); S. Das Sarma, A. Kobayashi and R. E. Prange, Phys. Rev. Lett. 56, 1280 (1986).

4. See, e.g., S. Ostlund and R. Pandit, Phys. Rev. $B29$, 1394 (1984).

5. J. Todd, R. Merlin, R. Clarke, K. M. Mohanty and J. D. Axe, to be published.

6. K. Bajema and R. Merlin, to be published.

7. J. P. Lu, T. Odagaki and J. L. Birman, Phys. Rev. $B33$, 4809 (1986).

8. T. Odagaki and L. Friedman, Solid State Commun. 57, 915 (1986).

9. F. Nori and J. P. Rodriguez, to be published.

10. M. Kohmoto and J. R. Banavar, to be published; see also M. Kohmoto, Phys. Rev. Lett. 51, 1198 (1983).

11. V. Elser, Acta Cryst. $A42$, 36 (1986).

12. R. K. P. Zia and W. J. Dallas, J. Phys. $A18$, L341 (1985).

13. P. M. Horn, W. Malzfeldt, D. P. DiVincenzo, J. Toner and R. Gambino, Phys. Rev. Lett. 57, 1444 (1986).

14. See, e.g., C. Colvard, T. A. Gant, M. V. Klein, R. Merlin, R. Fischer, H. Morkoc and A. C. Gossard, Phys. Rev. $B31$, 2080 (1985).